Principles and Techniques of Food Analysis

Principles and Techniques of Food Analysis

Edited by Samuel Ford

SYRAWOOD
PUBLISHING HOUSE
New York

Published by Syrawood Publishing House,
750 Third Avenue, 9th Floor,
New York, NY 10017, USA
www.syrawoodpublishinghouse.com

Principles and Techniques of Food Analysis
Edited by Samuel Ford

International Standard Book Number: 978-1-64740-369-0 (Hardback)

Cataloging-in-publication Data

Principles and techniques of food analysis / edited by Samuel Ford.
p. cm.
Includes bibliographical references and index.
ISBN 978-1-64740-369-0
1. Food--Analysis. 2. Food--Composition. 3. Chemistry, Technical. I. Ford, Samuel.
TX541 .P75 2023
664.07--dc23

TABLE OF CONTENTS

Preface VII

Chapter 1 **Assessing Nutritional Traits and Phytochemical Composition of Artisan Jams Produced in Comoros Islands: Using Indigenous Fruits with High Health-Impact as an Example of Biodiversity Integration and Food Security in Rural Development** 1
Dario Donno, Maria Gabriella Mellano, Saandia Hassani, Marta De Biaggi, Isidoro Riondato, Giovanni Gamba, Cristina Giacoma and Gabriele Loris Beccaro

Chapter 2 **Development of an Accelerated Solvent Extraction-Ultra-Performance Liquid Chromatography-Fluorescence Detection Method for Quantitative Analysis of Thiamphenicol, Florfenicol and Florfenicol Amine in Poultry Eggs** 20
Bo Wang, Xing Xie, Xia Zhao, Kaizhou Xie, Zhixiang Diao, Genxi Zhang, Tao Zhang and Guojun Dai

Chapter 3 **Transfer of a Multiclass Method for over 60 Antibiotics in Food from High Resolution to Low Resolution Mass Spectrometry** 36
Danilo Giusepponi, Fabiola Paoletti, Carolina Barola, Simone Moretti, Giorgio Saluti, Federica Ianni, Roccaldo Sardella and Roberta Galarini

Chapter 4 **Ion Mobility Spectrometry in Food Analysis: Principles, Current Applications and Future Trends** 55
Maykel Hernández-Mesa, David Ropartz, Ana M. García-Campaña, Hélène Rogniaux, Gaud Dervilly-Pinel and Bruno Le Bizec

Chapter 5 **Identification of the *Pol* Gene as a Species-Specific Diagnostic Marker for Qualitative and Quantitative PCR Detection of *Tricholoma matsutake*** 83
Luying Shan, Dazhou Wang, Yinjiao Li, Shi Zheng, Wentao Xu and Ying Shang

Chapter 6 **Phytochemical and Biological Characteristics of Mexican Chia Seed Oil** 95
Yingbin Shen, Liyou Zheng, Jun Jin, Xiaojing Li, Junning Fu, Mingzhong Wang, Yifu Guan and Xun Song

Chapter 7 **Chemical Composition, Antioxidant and Antihyperglycemic Activities of the Wild *Lactarius deliciosus*** 111
Zhou Xu, Liang Fu, Shiling Feng, Ming Yuan, Yan Huang, Jinqiu Liao, Lijun Zhou, Hongyu Yang and Chunbang Ding

Chapter 8 **Extraction of Carotenoids and Fat-Soluble Vitamins from *Tetradesmus Obliquus* Microalgae: An Optimized Approach by using Supercritical CO_2** 125
Laura Chronopoulou, Chiara Dal Bosco, Fabrizio Di Caprio, Letizia Prosini, Alessandra Gentili, Francesca Pagnanelli and Cleofe Palocci

Chapter 9 **Alternative Ultrasound-Assisted Method for the Extraction of the Bioactive Compounds Present in Myrtle (*Myrtus communis* L.)** 139
Ana V. González de Peredo, Mercedes Vázquez-Espinosa, Estrella Espada-Bellido, Marta Ferreiro-González, Antonio Amores-Arrocha, Miguel Palma, Gerardo F. Barbero and Ana Jiménez-Cantizano

Chapter 10 **^{1}H-NMR Profiling and Carbon Isotope Discrimination as Tools for the Comparative Assessment of Walnut (*Juglans regia* L.) Cultivars with Various Geographical and Genetic Origins** 160
Raluca Popescu, Roxana Elena Ionete, Oana Romina Botoran, Diana Costinel, Felicia Bucura, Elisabeta Irina Geana, Yazan Falah Jadee 'Alabedallat and Mihai Botu

Chapter 11 **Development of an Impedimetric Aptasensor for Label Free Detection of Patulin in Apple Juice** 173
Reem Khan, Sondes Ben Aissa, Tauqir A. Sherazi, Gaelle Catanante, Akhtar Hayat and Jean Louis Marty

Chapter 12 **Fatty Acid, Lipid Classes and Phospholipid Molecular Species Composition of the Marine Clam Meretrix *lyrata* (Sowerby 1851) from Cua Lo Beach, Nghe an Province** 185
Quoc Toan Tran, Thi Thanh Tra Le, Minh Quan Pham, Tien Lam Do, Manh Hung Vu, Duy Chinh Nguyen, Long Giang Bach, Le Minh Bui and Quoc Long Pham

Chapter 13 **Novel Electrochemical Sensors Based on Cuprous Oxide-Electrochemically Reduced Graphene Oxide Nanocomposites Modified Electrode toward Sensitive Detection of Sunset Yellow** 201
Quanguo He, Jun Liu, Xiaopeng Liu, Yonghui Xia, Guangli Li, Peihong Deng and Dongchu Chen

Chapter 14 **Use of ^{1}H NMR to Detect the Percentage of Pure Fruit Juices in Blends** 216
Lucia Marchetti, Federica Pellati, Stefania Benvenuti and Davide Bertelli

Chapter 15 **Essential Oils from *Humulus Lupulus* $scCO_2$ Extract by Hydrodistillation and Microwave-Assisted Hydrodistillation** 227
Katarzyna Tyśkiewicz, Roman Gieysztor, Marcin Konkol, Jan Szałas and Edward Rój

Permissions

List of Contributors

Index

PREFACE

Food analysis is a discipline that is concerned with the development, application and study of analytical procedures for characterizing the properties and constituents of food items. It aims to provide information about different properties of food such as composition, structure, physicochemical properties, and sensory attributes. Several techniques are used to carry out food analysis. Mass spectrometry (MS) is one of the most prominent food analysis techniques used extensively for the study of different food qualities such as nutraceutical active compounds, nutritional value and micronutrients. Enzyme-linked immunosorbent assay (ELISA) is another analytical procedure commonly used to detect hidden allergenic proteins in food products such as hormones, antibodies and peptides. Lateral flow device (LFD) testing is an immunological assay technique used to detect gluten in products, with an antigen-antibody reaction used to identify the protein. This book provides comprehensive insights into the basic concepts and techniques of food analysis. With its detailed analyses and data, it will prove immensely beneficial to professionals and students involved in food analysis at various levels.

The information shared in this book is based on empirical researches made by veterans in this field of study. The elaborative information provided in this book will help the readers further their scope of knowledge leading to advancements in this field.

Finally, I would like to thank my fellow researchers who gave constructive feedback and my family members who supported me at every step of my research.

Editor

1

Assessing Nutritional Traits and Phytochemical Composition of Artisan Jams Produced in Comoros Islands: Using Indigenous Fruits with High Health-Impact as an Example of Biodiversity Integration and Food Security in Rural Development

Dario Donno [1,*], Maria Gabriella Mellano [1], Saandia Hassani [2], Marta De Biaggi [1], Isidoro Riondato [1], Giovanni Gamba [1], Cristina Giacoma [3] and Gabriele Loris Beccaro [1]

1 Dipartimento di Scienze Agrarie, Forestali e Alimentari, Università degli Studi di Torino, 10095 Grugliasco, Italy; gabriella.mellano@unito.it (M.G.M.); marta.debiaggi@unito.it (M.D.B.); isidoro.riondato@unito.it (I.R.); giovanni.gamba@unito.it (G.G.); gabriele.beccaro@unito.it (G.L.B.)
2 École National de Cuisine et d'Application-Codcom, 167 Moroni, Comoros; saandiacodcom@gmail.com
3 Dipartimento di Scienze della Vita e Biologia dei Sistemi, Università degli Studi di Torino, 10123 Torino, Italy; cristina.giacoma@unito.it
* Correspondence: dario.donno@unito.it

Academic Editors: Alessandra Gentili and Chiara Fanali

Abstract: In the Comoros Islands, as in other developing countries, malnutrition and food insecurity affect a very large percentage of the population. Developing fruit-based products in order to make profit, reduce poverty and improve indigenous people diet could be very important for local population of countries as Comoros Islands. The aim of the present work was to study the chemical composition of jams and jellies produced from seven fruit species harvested in Grand Comore Island. The following parameters were studied sugars and organic acids, total phenolics, total anthocyanins and high-performance liquid chromatography (HPLC) fingerprint of the main phytochemicals. Antioxidant activity was also measured. A multivariate approach (Principal Component Analysis) was performed in order to better characterize the products and to set a potential analytical tool for jam characterisation. Results showed that the analysed products are a good source of polyphenolic constituents, as caffeic and gallic acids, catechin and quercetin and volatile compounds, as limonene and γ-terpinene: these molecules may be considered as suitable markers for these fruit-derived products as characterizing the chromatographic patterns. The characterisation of these products and their nutritional and nutraceutical traits is important as valorisation of local food production for poverty reduction and rural development. Further benefits of this approach include the maintenance of local agro-biodiversity as raw material for fruit-based products and the strengthening of food security practices.

Keywords: fruit jams; food security; phenolic acids; quercetin; agro-biodiversity; HPLC fingerprint

1. Introduction

The Comoros Islands are situated off the coast of East Africa (290 km), at the northern entrance of the channel of Mozambique between Madagascar and the southern-east African mainland. The archipelago is composed of four main islands: Grand Comore (Ngazidja), Anjouan (Ndzuani), Mohéli (Mwali) and Mayotte (Maore). Comoros Islands are characterized by high plant biodiversity but less than one-sixth of the land remains covered by forest due to a rapid deforestation mainly caused by domestic firewood consumption [1].

Comoros, one of the world's poorest countries, has an economy based on subsistence agriculture and fishing. In the Comoros Union, as in all the developing countries, malnutrition and food insecurity are the main challenges: in particular energy, malnutrition in children and micronutrient deficiencies (e.g., vitamin deficiency and nutritional anaemias), are important public health issues influencing productivity, maternal/infant health and intellectual development. The improvement of productivity and post-harvest techniques is pivotal to increase Comorian smallholder farmers' income in order to help fighting poverty and ensure a medium-high nutrition [2].

On the other hand, an abundance of tropical fruit species, often underexploited, grown in semi-natural conditions occurring in Comoros. Tropical fruits attract special attention as they usually have stronger antioxidant properties than common fruits, thanks to bioactive compounds as polyphenols (anthocyanins, flavonoids, phenolic acids and tannins), carotenoids, organic acids and vitamins (B2, B6, C, E, P, PP), as reported in several studies [3,4]. Health-promoting components occurring in these fruits, in particular polyphenols, also show anti-microbial, anti-carcinogenic and anti-viral effects [5]. Therefore, tropical fruits show a high considerable horticultural and nutritional importance for the diet of the rural population [6] in specific critical periods of the year providing sustenance to millions of people. Unfortunately, a significant amount (30–40%) of total fruit production in developing countries is wasted due to poor postharvest handling and inadequate marketing and/or storage facility [7]. Moreover, most of population in these countries lives in rural areas and depends on subsistence agriculture for their livelihoods. In these settlements, agriculture is dependent on rainfall. Inadequacy of food during the dry season and in early summer before the harvest period exposes people to inadequate intake of both macro- and micro-nutrients [6].

Fruit processing industries making preserves, jams, sauces and jellies play an important role in reducing these losses of fruit production, thriving largely as a domestic-product market [8]. In particular, some of the most popular postharvest-stable products made from fruit are jams and jellies, both at the household and commercial levels. Jams and jellies are defined as a mixture, brought to a suitable gelled consistency, of sugars, pulp of one or more fruits and water [9]. To manufacture these products, fruits and sugars are combined in a similar ratio, followed by cooking, to produce a tasty product of sufficiently high sugar content with satisfactory keeping qualities [10]. Jam processing is an important strategy to preserve perishable fruits and improve food security in developing countries [11].

Considering new potential agro-industrial and commercial activities related to jam and jelly production, including value-added product labelling, it is of essential importance to guarantee both high quality and compliance with the product specification. For these purposes, increasingly sophisticated analytical methodologies, based on chemical markers, have been developed. Chemical fingerprint methods include the analysis of organic acids, pigments (as carotenoids and anthocyanins), sugars, phenolics and other bioactive compounds [12]. In particular, secondary plant metabolites are very suitable as chemotaxonomic markers [13]. Quantitative differences may occur depending on fruit genotype (e.g., species and cultivar), maturity stages and environmental growth [14], storage conditions [15] and on presence of the skin in fruit-based products [16]. In previous research, they have been successfully used for the determination of the adulteration of some fruit jams and jellies as reported by Dragovic-Uzelac et al. [17].

The aim of the present work was to study chemical compositions of jams and jellies from seven tropical fruit species, harvested near Moroni in Comoros Islands. The following parameters were studied: sugar and organic acid contents, total phenolics (TPC), total anthocyanins (TAC) and fingerprint of the main phytochemicals with demonstrated health-promoting activity by high-performance liquid chromatography (HPLC). Furthermore, antioxidant activity was measured in these products. A multivariate approach (Principal Component Analysis—PCA) was performed in order to better characterize fruit-based products and set a potential analytical tool for analysis and characterisation of local jams and jellies. This study could contribute to the commercial valorisation of these fruit-derived products in rural communities in Comoros Islands thereby reducing post-harvest losses, promoting food security, enhancing small farmers' income and contributing to a sustainable rural development.

2. Results and Discussion

2.1. Nutraceutical Properties

Food processing plays an important role in the bioactive compound degradation, because several transformations of phenolics occur to produce yellowish or brownish pigments [18]: the final product outward appearance is a crucial in determination of consumers' choices and anthocyanins are the main food colorants responsible for intense colour (associated with raw material freshness and good quality) [19]. Moreover, physical and biological factors as temperature increase and enzymatic activity may be very important in degradation of polyphenolic compounds [10].

In this research, the used methods allowed a rapid measurement of TAC and TPC: the Folin-Ciocalteu method suffers many interferences but it can be a complementary technique applied to confirm and support chromatography results, as reported in this study. The TPC ranged from 10.98 ± 2.18 $mg_{GAE}/100g_{Pr}$ (mango jam, CM2) to 625.34 ± 67.86 $mg_{GAE}/100g_{Pr}$ (red guava jam, CM8), while TAC ranged from 0.59 ± 0.25 $mg_{C3G}/100g_{Pr}$ (mango jam, CM2) to 9.56 ± 0.46 $mg_{C3G}/100g_{Pr}$ (orange jam, CM7): values obtained from the analysed extracts (Table 1) were higher than values reported by Poiana et al. [20] and Rababah et al. [21]; the differences in phenolic and anthocyanin content could be due to the effects of several internal and external factors on plant material (genetic variability, climatic conditions and environmental factors) [22]. In particular, Comoros pedoclimatic conditions such as volcanic soil, high temperatures, well distributed rains, influence polyphenolic content in fresh fruits and related products.

Bioactive compound redox properties allow them to act as reducing agents, hydrogen donators and singlet oxygen quenchers [23]. In this research, the Ferric Reducing Antioxidant Power (FRAP) assay was used to evaluate antioxidant capacity of fruit jams and jellies, studying the ability of antioxidants to reduce Fe^{3+} ions to Fe^{2+} ions. Jams and jelly FRAP value ranged from 4.71 ± 2.07 mmol Fe^{2+} $kg_{Pr}{}^{-1}$ (mango jam, CM1) to 25.50 ± 0.28 mmol Fe^{2+} $kg_{Pr}{}^{-1}$ (red guava jam, CM8) as shown in Table 1, in accordance with other studies [24,25].

Table 1. Nutraceutical traits of the analysed fruit-derived products.

		Total Polyphenolic Content ($mg_{GAE}/100$ g_{Pr})			Antioxidant Activity (mmol Fe^{2+}/kg_{Pr})			Total Anthocyanin Content ($mg_{C3G}/100$ g_{Pr})		
Sample	**ID**	**Mean Value**	**SD**	**Tukey Test**	**Mean Value**	**SD**	**Tukey Test**	**Mean Value**	**SD**	**Tukey Test**
mango jam	CM1	11.11	1.34	a	4.71	2.07	a	1.48	0.26	ab
mango jam	CM2	10.98	2.18	a	7.86	0.36	b	0.59	0.25	a
tamarind jam	CM3	484.95	81.89	d	23.97	0.28	d	0.87	0.43	ab
banana flower	CM4	13.06	1.85	a	19.23	1.29	c	4.49	1.62	c
guava jelly	CM5	437.13	2.44	d	25.40	0.35	d	8.35	1.13	d
lychee jam	CM6	266.96	7.95	c	23.01	0.07	d	3.15	0.66	bc
orange jam	CM7	124.50	27.49	b	16.62	0.89	c	9.56	0.46	d
red guava jam	CM8	625.34	67.86	e	25.50	0.28	d	4.99	0.67	c

Mean value and standard deviation (SD) of each sample is given ($N = 3$). Different letters for each class indicate the significant differences at $p < 0.05$. GAE = gallic acid equivalent; C3G = cyanidin 3-*O*-glucoside. Pr = product.

Antioxidant activity of fruit-derived products was determined by different bioactive molecules (e.g., polyphenols, as anthocyanins and vitamin C) [26]; for example, antioxidant activity of cyanidin is about 4 times higher compared to ascorbic acid [27]. In this study, the TPC/TAC/antioxidant activity correlation was positive: results showed a significant Pearson correlation coefficients (R = 0.6636 for TPC/antioxidant activity and R = 0.4347 for TAC/antioxidant activity).

2.2. Phytochemical Composition

Antioxidant (polyphenols and vitamin C) and anti-inflammatory (terpenes) compounds are the main biologically active substances in fresh fruits and derived products: synergistic or additive health-promoting effects of different phytochemicals (phytocomplex) contribute to biological activity better than a single molecule or a group of few compounds [28]. In this study 22 biologically active compounds together with 9 nutritional substances were selected as markers for HPLC fingerprinting due to their importance in humans [29]. In Supplementary material (Suppl. Figure S1) HPLC chromatograms of orange jams are reported as an example of analysed fruit-based products.

An important question in HPLC analysis is whether the peak comprises one or more components. In quality control and research analysis, impurities hidden behind the peak of interest can falsify results and an undetected component might lead to a loss of essential information. In this research, a peak purity check was assessed in order to control if peaks were pure or contained impurities comparing spectra recorded during the elution of each peak. No-coeluting peaks were detected. Moreover, HPLC-DAD does not allow a definitive identification of phytochemicals. Indeed, liquid chromatography (LC) coupled to mass (MS) or mass/mass spectrometry (MS^2) is one of the most effective technique for analysis on complex plant extract/fresh fruit/derived product providing a rapid and accurate identification of phytochemicals, as phenolics. For this reason, future developments are necessary but in this preliminary study HPLC-DAD was a simply, rapid and effective approach to describe considered samples in relation to the research aim. Additional markers with demonstrated biological activity could be also taken into consideration for a better identification of the chromatographic pattern of fruit-derived products, together with a mass spectrometry detection of unknown peaks.

The chemical fingerprint of analysed jams and jellies is reported in Tables 2–4 (phenolics, other health-promoting components and nutritional substances, respectively). The total bioactive compound content (TBCC) was calculated as the sum of the main molecules (polyphenols, monoterpenes and vitamin C) selected for their biological proved effects on humans and detected in the extracts: TBCC value ranged from 94.25 ± 4.13 mg/100 g_{Pr} (banana flower, CM4) to 357.50 ± 2.10 mg/100 g_{Pr} (orange jam, CM7). This is only a preliminary study on tropical fruit-based jams: in further fingerprint studies, other markers with health-promoting capacity or positive nutritional value should be added for a complete chromatographic pattern characterisation coupling a mass spectrometry (MS) detection of unknown peaks to a UV-visible determination.

Table 2. Phytochemical fingerprint of the polyphenolic compounds in analysed jams and jellies.

		Mango Jam, CM1		Mango Jam, CM2		Tamarind Jam, CM3		Banana Flower, CM4		Guava Jelly, CM5		Lychee Jam, CM6		Orange Jam, CM7		Red Guava Jam, CM8	
		Mean Value	SD	Mean Value	SD	Mean Value	SD	Mean Value	SD	Mean Value	SD	Mean Value	SD	Mean Value	SD	Mean Value	SD
Cinnamic acids	caffeic acid	0.678	0.170	0.761	0.101	0.498	0.134	0.749	0.154	0.687	0.034	0.764	0.116	0.571	0.110	0.744	0.204
	chlorogenic acid	n.d.	/	n.d.	/	11.454	0.731	n.d.	/	13.766	0.546	n.d.	/	12.971	0.185	n.d.	/
	coumaric acid	n.d.	/	n.d.	/	n.d.	/	0.556	0.509	n.d.	/	n.d.	/	n.d.	/	5.583	0.797
	ferulic acid	n.d.	/	n.d.	/	1.377	0.217	n.d.	/	n.d.	/	n.d.	/	1.001	0.110	n.d.	/
Flavonols	hyperoside	n.d.	/	n.d.	/	n.d.	/	0.356	0.033	n.d.	/	1.027	0.116	n.d.	/	n.d.	/
	isoquercitrin	n.d.	/	n.d.	/	0.491	0.006	n.d.	/	0.525	0.034	n.d.	/	1.140	0.110	0.541	0.155
	quercetin	n.d.	/	n.d.	/	11.441	0.963	n.d.	/	7.562	0.034	n.d.	/	15.694	1.107	7.621	0.204
	quercitrin	0.161	0.030	0.007	0.003	1.640	0.257	0.173	0.019	1.001	0.034	1.334	0.116	1.152	0.110	1.477	0.256
	rutin	0.936	0.036	0.858	0.034	n.d.	/	n.d.	/	n.d.	/	n.d.	/	n.d.	/	n.d.	/
Benzoic acids	ellagic acid	1.167	0.264	1.228	0.154	23.302	1.962	1.604	0.365	4.444	0.316	3.799	0.116	57.105	1.637	4.561	0.607
	gallic acid	0.668	0.105	1.535	0.232	8.571	0.314	0.611	0.070	42.023	0.582	30.650	0.886	87.724	1.812	40.955	1.727
Catechins	catechin	0.872	0.023	1.749	0.175	40.644	1.558	1.216	0.162	0.327	0.034	n.d.	/	0.436	0.110	4.287	0.204
	epicatechin	3.683	0.328	1.747	0.400	n.d.	/	1.348	0.184	2.523	0.034	n.d.	/	2.015	0.110	4.031	0.157
Tannins	castalagin	n.d.	/	n.d.	/	n.d.	/	2.595	0.425	4.384	0.034	n.d.	/	15.210	2.414	9.773	0.295
	vescalagin	3.639	0.174	0.359	0.008	n.d.	/	1.554	0.371	4.339	0.205	n.d.	/	3.031	0.110	10.012	0.204

Mean value and standard deviation (SD) of each sample is given ($N = 3$). Results are expressed as mg/100 g_{Pr}. Pr = product. n.d. = not detected.

Table 3. Phytochemical fingerprint of monoterpenes and vitamin C in analysed jams and jellies.

		Mango Jam, CM1		Mango Jam, CM2		Tamarind Jam, CM3		Banana Flower, CM4		Guava Jelly, CM5		Lychee Jam, CM6		Orange Jam, CM7		Red Guava Jam, CM8	
		Mean Value	SD	Mean Value	SD	Mean Value	SD	Mean Value	SD	Mean Value	SD	Mean Value	SD	Mean Value	SD	Mean Value	SD
Monoterpenes	limonene	103.529	3.116	31.201	0.538	n.d.	/	n.d.	/	102.750	0.582	33.602	0.687	18.818	0.472	51.319	3.204
	phellandrene	27.843	2.380	7.809	0.540	7.497	0.219	n.d.	/	4.671	0.034	4.668	0.116	n.d.	/	4.946	0.204
	sabinene	8.140	0.045	12.187	1.146	11.273	1.180	n.d.	/	12.847	1.020	12.811	0.687	n.d.	/	12.332	0.797
	γ-terpinene	24.552	2.464	48.067	2.301	n.d.	/	17.949	2.097	39.002	0.034	19.069	0.492	37.084	0.505	43.005	0.155
	terpinolene	n.d.	/	n.d.	/	26.944	2.718	8.228	0.213	7.731	0.034	7.473	0.886	7.926	0.110	7.715	0.564
Vitamin C	ascorbic acid	15.058	0.362	10.178	0.172	20.397	0.117	49.699	0.356	18.074	0.138	18.978	0.057	54.062	0.074	19.104	0.487
	dehydroascorbic acid	0.813	0.130	1.815	0.163	3.767	0.283	3.123	0.213	1.138	0.192	1.196	0.225	32.003	0.401	1.279	0.234

Mean value and standard deviation (SD) of each sample is given ($N = 3$). Results are expressed as mg/100g_{Pr}. Pr = product. n.d. = not detected.

Table 4. Phytochemical fingerprint of nutritional substances in analysed jams and jellies.

		Mango Jam, CM1		Mango Jam, CM2		Tamarind Jam, CM3		Banana Flower, CM4		Guava Jelly, CM5		Lychee Jam, CM6		Orange Jam, CM7		Red Guava Jam, CM8	
		Mean Value	**SD**	**Mean Value**	**SD**	**Mean Value**	**SD**	**Mean Value**	**SD**	**Mean Value**	**SD**	**Mean Value**	**SD**	**Mean Value**	**SD**	**Mean Value**	**SD**
Organic acids	citric acid	16.217	4.058	29.696	0.469	522.583	2.566	694.580	3.627	326.950	3.030	268.422	2.629	287.911	4.614	200.408	2.204
	malic acid	357.878	3.679	302.528	2.531	n.d.	/	21.926	1.696	n.d.	/	n.d.	/	n.d.	/	n.d.	/
	oxalic acid	4.741	0.198	4.573	0.416	21.444	0.530	8.441	0.102	8.194	0.117	n.d.	/	6.629	0.110	23.717	0.797
	quinic acid	n.d.	/	n.d.	/	n.d.	/	n.d.	/	n.d.	/	138.233	1.263	391.128	2.192	n.d.	/
	succinic acid	n.d.	/	n.d.	/	n.d.	/	n.d.	/	n.d.	/	n.d.	/	406.648	4.401	n.d.	/
	tartaric acid	n.d.	/	n.d.	/	n.d.	/	n.d.	/	76.724	1.513	105.358	4.157	193.497	3.661	134.637	1.729
Sugars	fructose	11.701	0.038	11.316	0.100	15.353	0.140	12.275	0.287	21.341	0.034	14.814	0.116	13.620	0.110	14.428	0.204
	glucose	14.505	0.321	13.620	0.401	27.637	2.326	12.342	0.111	12.615	0.034	11.044	0.116	13.090	0.110	13.413	0.155
	sucrose	13.444	0.339	14.831	0.680	13.509	0.231	11.131	1.084	11.801	0.867	14.573	0.492	14.935	0.110	14.904	0.157

Mean value and standard deviation (SD) of each sample is given ($N = 3$). Results are expressed as mg/100g_{Pr}. Pr = product. n.d. = not detected.

In Figure 1 identified health-promoting substances were grouped into bioactive classes for the evaluation of the single contribution of each class to total phytocomplex/TBCC (mean values were considered). The most important class in tamarind (CM3) and orange (CM6) jams was polyphenols (58.93% and 58.07%, respectively), expressed as the sum of anthocyanins, phenolic acids, flavanols, catechins and tannins, while monoterpenes were the first class in mango jam (83.79%), guava jelly (60.48%) and lychee jam (56.04%). Banana flower showed a high percentage of vitamin C (56.05%), while red guava jam presented similar content of polyphenols and monoterpenes (40.37% and 50.93%, respectively) with a positive percentage of vitamin C (8.70%).

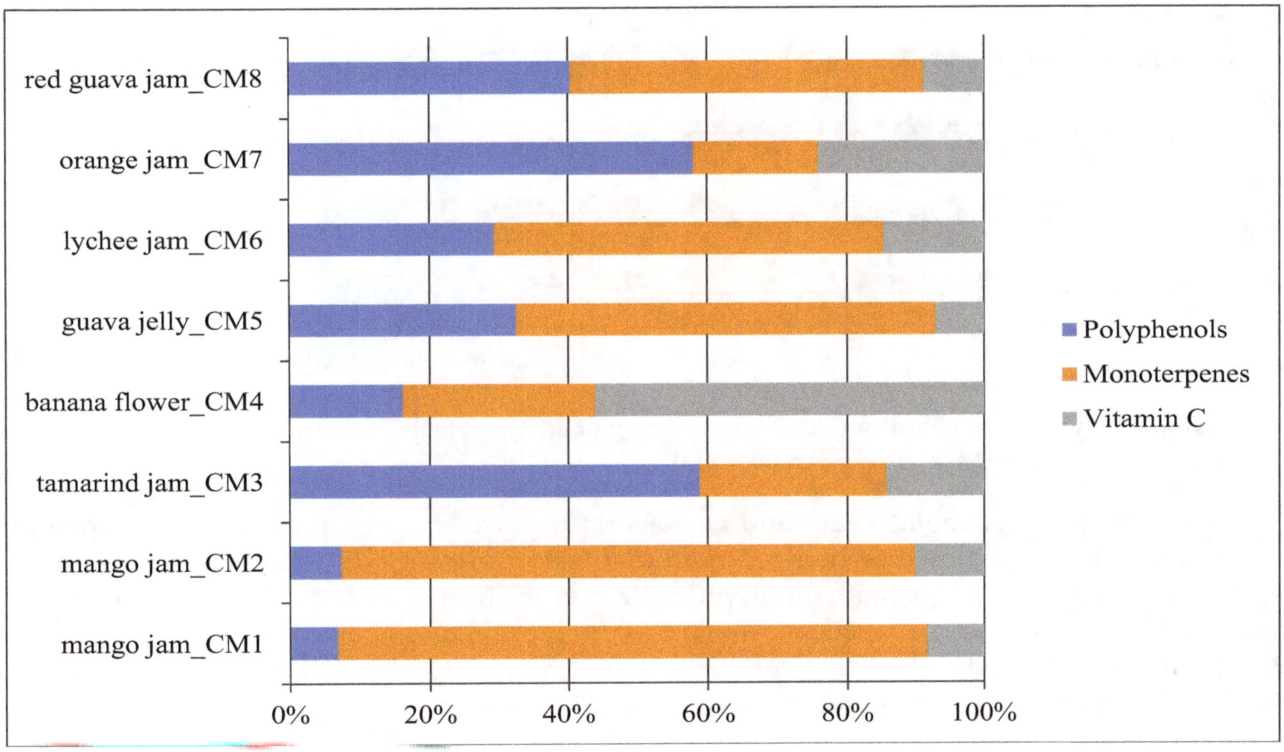

Figure 1. Phytocomplex representation of analysed fruit-derived products. The mean value of each analysed sample is given ($N = 3$).

Results showed that analysed fruit jams and jellies are a very good source of polyphenolic compounds. Figure 2 reports the single contribution of each polyphenolic class to total polyphenols detected by HPLC. Catechins were the most important class in mango and tamarind jams (36.93% and 40.53%, respectively), while phenolic acids (cinnamic acids plus benzoic acids) were the most important classes in guava jelly (67.75%) and in lychee (86.46%), orange (76.77%) and red guava (54.82%) jams. Banana flower showed a high percentage of tannins and anthocyanins (27.21% and 29.42%, respectively), followed by catechins (16.81%) and benzoic acids (14.53%).

Analysed jams and jellies presented interesting quali-quantitative polyphenolic profiles if compared to commercial products derived from common temperate fruits; in particular, they showed higher TPC values and relative antioxidant activity than strawberry (101.40 $mg_{GAE}/100\ g_{Pr}$) [25], apricot (51.49 $mg_{GAE}/100\ g_{Pr}$) [21], berry fruits (336.67 $mg_{GAE}/100\ g_{Pr}$) [24], peach (18.85 $mg_{GAE}/100\ g_{Pr}$) and apple (20.07 $mg_{GAE}/100\ g_{Pr}$) [16]. Moreover, jams from tropical fruits showed higher values of specific phenolic markers as coumaric, caffeic and ferulic acids (lychee, banana and guava), catechin (tamarind) and rutin (mango) than commercial products derived from berry fruits (the content values in berry fruits were 0.39 mg/100 g_{Pr} for coumaric acid, 1.38 mg/100 g_{Pr} for caffeic acid, 0.13 mg/100 g_{Pr} for ferulic acid, 3.93 mg/100 g_{Pr} for catechin and 0.26 mg/100 g_{Pr} for rutin) [19]. These results may contribute to better valorise the products derived from local biodiversity compared to imported commercial ones and improve food industry in the Comoros Islands with the potential exportation of these productions.

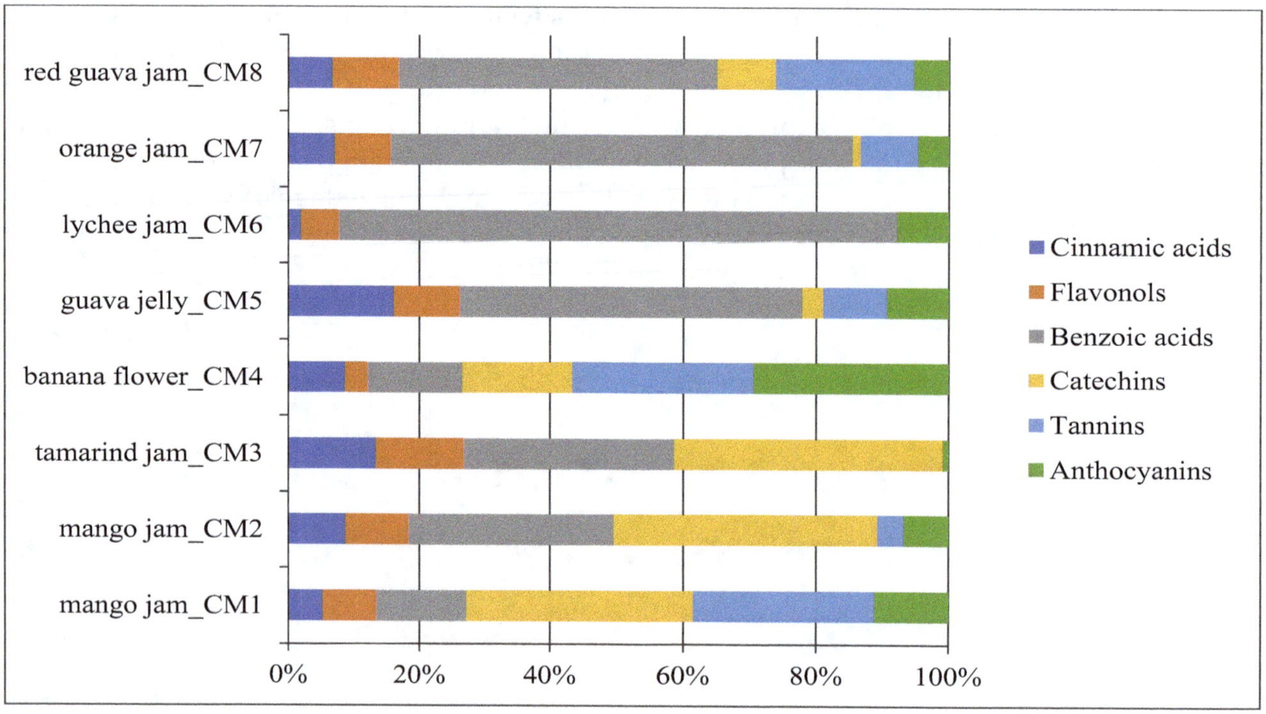

Figure 2. Polyphenolic phytocomplex representation of analysed fruit-derived products. The mean value of each analysed sample is given ($N = 3$).

In some studies, phenolic compound characterisation was mainly used for chemotaxonomic purposes; moreover, some researchers reported that fruit processing during jam/jelly production did not change much the qualitative polyphenolic profile [10,30]. In this research, each product showed a specific phenolic composition characterised by the presence of one or more specific markers. Chlorogenic acid proved to be characteristic of tamarind jam (11.45 ± 0.73 mg/100 g_{Pr}), guava jelly (13.77 ± 0.55 mg/100 g_{Pr}) and orange jam (12.97 ± 0.19 mg/100 g_{Pr}) and it could be used as a marker to prove the addition of these fruits to other jams. Moreover, chlorogenic acid is considered a preferential substrate for the catecholase activity of polyphenol oxidase and it may be important during fruit processing [31]. Caffeic acid was detected in all the fruit-derived products in a close range between 0.49 ± 0.13 mg/100 g_{Pr} (tamarind jam, CM3) and 0.76 ± 0.12 mg/100 g_{Pr} (lychee jam, CM6) similar to other studies [32,33]. Coumaric acid was the most important cinnamic acid in red guava jam-CM8 (5.58 ± 0.80 mg/100 g_{Pr}), while ferulic acid was present in tamarind jam-CM3 (1.38 ± 0.22 mg/100 g_{Pr}) and orange jam-CM7 (1.00 ± 0.11 mg/100 g_{Pr}) as reported by Jimohand Onabanjo [34] and Marquina et al. [35]. Caffeic, ferulic and coumaric acids could be involved in the oxidation processes and colour development during technological processing [36]. Flavonoids (flavanols and catechins) were also demonstrated to be important as markers for orange and tamarind jam quality control [37], because flavonoids are not affected by the manufacturing process. Flavonols quench active oxygen species [38] and inhibit in vitro oxidation of low-density lipoproteins reducing thrombotic tendency. In this research, the most effective flavonol selected as marker was quercetin in guava jelly (7.56 ± 0.03 mg/100 g_{Pr}) and jams of tamarind (11.44 ± 0.96 mg/100 g_{Pr}), orange (15.69 ± 1.11 mg/100 g_{Pr}) and red guava (7.62 ± 0.20 mg/100 g_{Pr}), while rutin proved to be characteristic of mango jams (0.94 ± 0.04 mg/100 g_{Pr} for CM1 and 0.86 ± 0.03 mg/100 g_{Pr} for CM2) according to other studies [39,40]. The identification of catechin (maximum value of 40.64 ± 1.56 mg/100 g_{Pr} in tamarind jam) and epicatechin (maximum value of 4.03 ± 0.16 mg/100 g_{Pr} in red guava jam) could be useful: indeed, they are involved in the lipid peroxidation inhibition and human cancer cell line proliferation as other similar compounds [41]. The presence of tannins in adequate amounts in orange and red guava jams (18.24 ± 2.52 mg/100 g_{Pr} and 19.79 ± 0.50 mg/100 g_{Pr}, respectively) are positive as they are free radical quenchers [42]. High levels of ellagic acid in tamarind jam (23.30 ± 1.96 mg/100 g_{Pr}) and orange jam (57.11 ± 1.64 mg/100 g_{Pr}) as well as high content of gallic acid in guava jelly (42.02 ± 0.58 mg/100 g_{Pr}) and orange jam (87.72 ± 1.81 mg/100 g_{Pr}) were also detected: these molecules are endowed with numerous biological properties, as anticancer, anti-inflammatory and anti-HIV replication activities [43]. These preliminary results on phenolic composition demonstrate the need of identifying more bioactive substances for control of the authenticity of fruit-based products.

Anthocyanins have frequently been also considered for the cited purposes, because their specific patterns may allow the classification of fruit species and relative derived products and the characterization of their nutraceutical and nutritional traits (e.g., detection of admixtures of fruits with a more stable colour during jam processing). As previously discussed, anthocyanins are of prominent importance in guava jelly (8.35 ± 1.13 mg_{C3G}/100g_{Pr}), orange jam (9.56 ± 0.46 mg_{C3G}/100g_{Pr}) and red guava jam (4.99 ± 0.67 mg_{C3G}/100g_{Pr}) because i) they are important for quality traits, due to their levels directly related to the product colour and ii) they have been proved to show several health-promoting activities and a high potential phytochemical value [44]. As opposed to other polyphenolic compounds, composition in anthocyanins may be subject to modification during processing and storage steps as reported by Garzon and Wrolstad [45]: in particular, enzymation during jam processing may change anthocyanin patterns [46]. During jam and jelly storage new pyranoanthocyanins may also be formed by direct reaction of anthocyanins with cinnamic acids, as shown by Schwarz et al. [47]. For this reason, anthocyanins may be only used as quantitative markers in the quality control of jams and similar products.

Monoterpenes represent an important fraction of the TBCC in analysed fruit-based products, in particular limonene in mango jam (103.53 ± 3.12 mg/100 g_{Pr}) and guava jelly (102.75 ± 0.58 mg/100 g_{Pr}) and γ-terpinene in jams of mango (48.07 ± 2.30 mg/100 g_{Pr}) and red guava (43.01 ± 0.16 mg/100 g_{Pr}): the plant terpenoids are a large class of phytochemicals used for their aromatic qualities and antioxidant and anti-inflammatory activity [48]. Monoterpenes are non-nutritive dietary substances with antibacterial and antitumor activity found in the essential oils of several plants [49] and several studies reported their chemopreventive activity against several cancers [50].

Vitamin C value was obtained as the sum of ascorbic and dehydroascorbic acids due to their biological activity in humans [51]. In this research banana flower and orange jams showed a high vitamin C content (52.82 ± 0.57 mg/100 g_{Pr} and 86.07 ± 0.48 mg/100 g_{Pr}, respectively). Other analysed products showed good values of vitamin C (about 10–25 mg/100 g_{Pr}) according to previous similar studies [52,53].

Organic acids in fruits are little influenced by changes during processing and storage and show a good stability if compared to pigments and flavour compounds. Accordingly, their identification may be suitable for the estimation of fruit amount and for the fruit quality control [54]. However, since organic acids (e.g., citric acid) are indispensable technological components of most derived-products, they are not applicable as quality control markers in fruit jams and jellies. Furthermore, organic acid composition is influenced by genetic factors (e.g., cultivar) and degree of ripeness, limiting their applicability as a quantitative marker in fruit-derived products [55]. In any case, they are important antioxidants with multi-purpose uses in pharmacology as reported by Eyduran et al. [56]. They were also utilised as food acidifiers by food companies [57]. In this research orange jam showed high variability in organic acid composition: in particular, succinic acid (406.65 ± 4.40 mg/100 g_{Pr}) and quinic acid (391.13 ± 2.19 mg/100 g_{Pr}) were found to be suitable markers for jam characterization because they were not detected in other analysed fruit-based products and represent specific molecules of orange jam chromatographic pattern as shown by Cejudo-Bastante et al. [58] and Flores et al. [59]. Similarly, malic acid could be a specific marker for mango jams and tartaric acid for guava jelly and jams of lychee and red guava. Figure 3 shows total organic acid content in analysed fruit-based products.

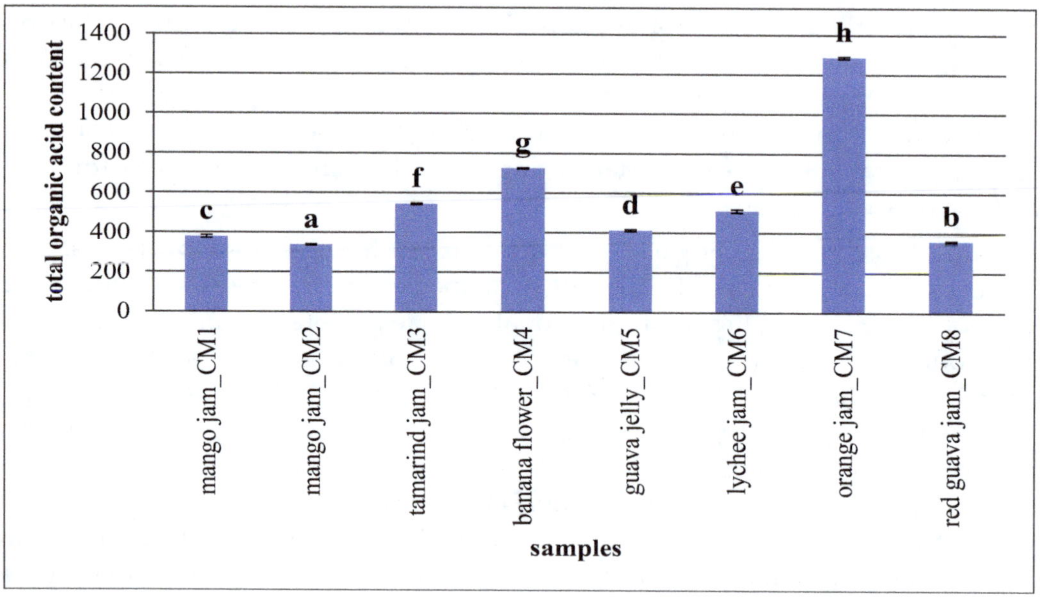

Figure 3. Total organic acid content of analysed fruit-derived products. The mean value of each analysed sample is given ($N = 3$). Different letters for each class indicate the significant differences at $p < 0.05$. Results are expressed as mg/100 g of product.

Apart from organic acids, the sugar pattern was also utilised for fruit species differentiation, while their use as quantifiers of fruit content is very limited. Identical sugar profiles, used often for the detection of illegal adulterations as the admixture of sugar solutions or fruit juices, were observed for fruits from different countries as well as for different genotype [55]. In this research sugar pattern (Figure 4) was studied in order to evaluate the nutritional potential of analysed fruit-derived products. Tamarind jam showed the highest sugar content (56.50 ± 2.70 g/100 g_{Pr}), expressed as sum of fructose (15.35 ± 0.14 g/100 g_{Pr}), glucose (27.64 ± 2.33 g/100 g_{Pr}) and sucrose (13.51 ± 0.23 g/100 g_{Pr}). Other analysed products presented a total sugar content of about 40 g/100 g_{Pr} as reported in other studies [60,61].

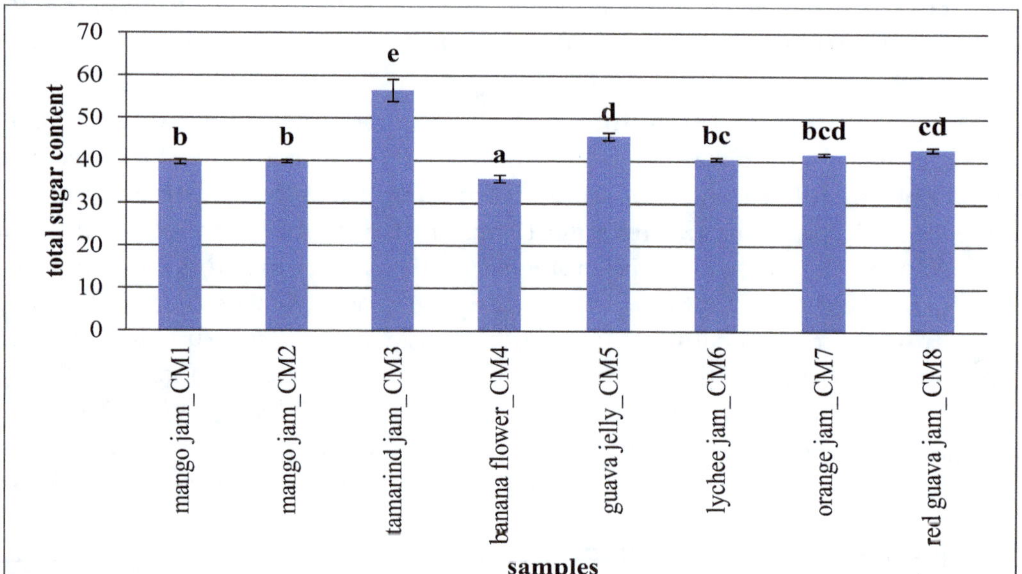

Figure 4. Total sugar content of analysed fruit-derived products. The mean value of each analysed sample is given ($N = 3$). Different letters for each class indicate the significant differences at $p < 0.05$. Results are expressed as g/100 g of product.

2.3. *Multivariate Analysis*

Single markers of the same phytochemical group were combined in bioactive classes for multivariate data handling. For the visualisation of potential differences in the products and easily characterize analysed jams and jellies, PCA was performed on all the data and it reduced the initial variables (TPC, antioxidant activity, TAC, content of 9 chemical classes) to three principal components (86.74% of total variance), placing the eight fruit-based products in the PCA score plot (Figure 5) in relation to phytochemical composition, nutritional properties and nutraceutical traits. PC1 and PC2 well represent the system information (70.20% of total variance); the PCA results showed five groups, highlighted in Figure 5 with circles, without statistical meaning, according to the phytochemical results; the groups were named α (guava jelly—CM5, red guava jam—CM8), β (tamarind jam—CM3), χ (lychee jam—CM6), δ (orange jam—CM7) and ε (mango jam, genotype 1—CM1, mango jam, genotype 2—CM2, banana flower—CM4).

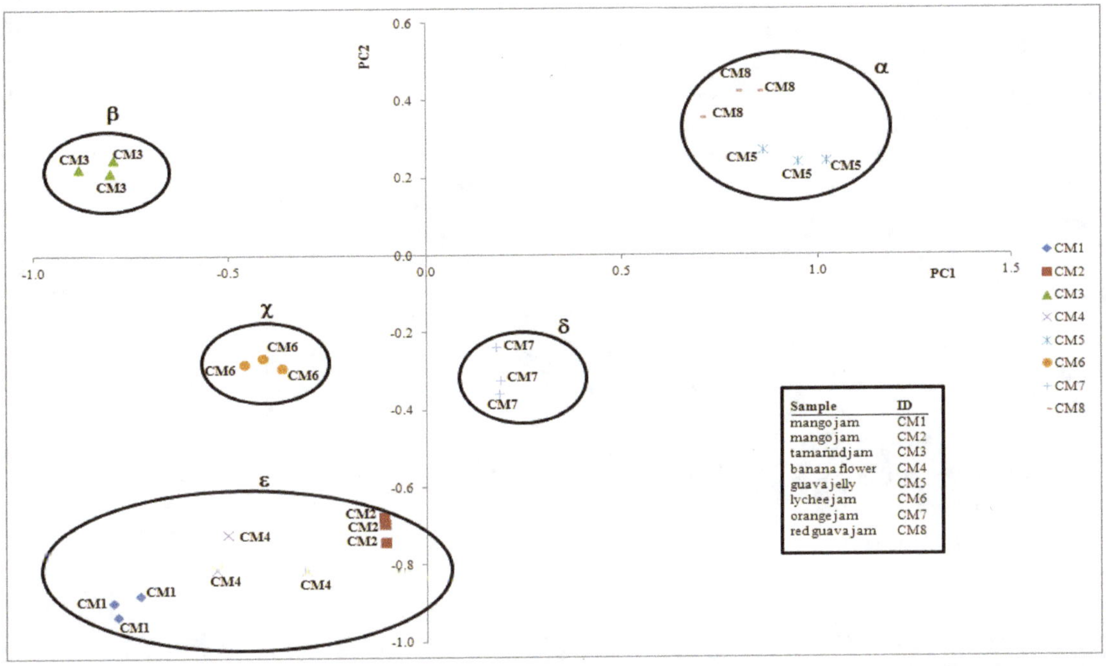

Figure 5. PCA score plot of fruit-derived products (eight samples and three replications per sample). The ellipses around each object group only indicate the position of a category in the plot without statistical meaning, based on the phytochemical results.

PCA loading plot showed a correlation between most of the polyphenolic classes (phenolic acids, tannins, anthocyanins and flavanols), vitamin C, organic acids and PC1 (43.15% of total variance) and a correlation between TPC, catechins, antioxidant activity, sugars and PC2 (27.05% of total variance). Monoterpenes showed an intermediate position between PC1 and PC2 (Figure 6). Polyphenolic compounds (in particular, benzoic acids, tannins and anthocyanins) and vitamin C presented a high discriminating power among different samples, as reported in other studies [37,55], as well as catechins.

These results showed that PCA classification obtained from main bioactive groups, nutritional traits and nutraceutical properties characterized the analysed products according to the different chemical pattern and provided information on the phytochemical markers with the most influence on the phytocomplex. A chemometric method was applied coupled to an HPLC fingerprint technique for a better recognition of analysed products as reported by Tzouros and Arvanitoyannis [62]. Different marker compounds were found to be the most discriminating variables, which could be applied to accurate composition control of fruit jams and jellies; in particular, the phytocomplex graphical view showed that genotypes included in the α and δ PCA group (guava jelly/red guava jam and orange jam, respectively) present the highest amount of antioxidant compound classes (polyphenols and vitamin C) together with the highest amount of volatile molecules (organic acids and monoterpenes), the most responsible compounds for the product aroma. The combination of chromatographic fingerprinting and chemometrics could be an effective potential tool for quality control of fruit-based products avoiding potential voluntary or involuntary adulterations and contaminations [63]. The analysis of other samples from different origin and nature are required to ensure that the proposed methodology is applicable to the authentication of jellies and jams from Comoros Islands or to the detection of their adulteration. In the present research, this methodology only showed that jams from Comoros Islands may be differentiated among them in relation to their fruit phytochemical composition. Therefore, further additional experiments are required for demonstrating that this methodology can be used for adulteration detection and/or authentication. These hyphenated techniques could also be positively used for the evaluation and differentiation of several products in local markets setting a potential tool to obtain label certifications for the valorisation of local productions.

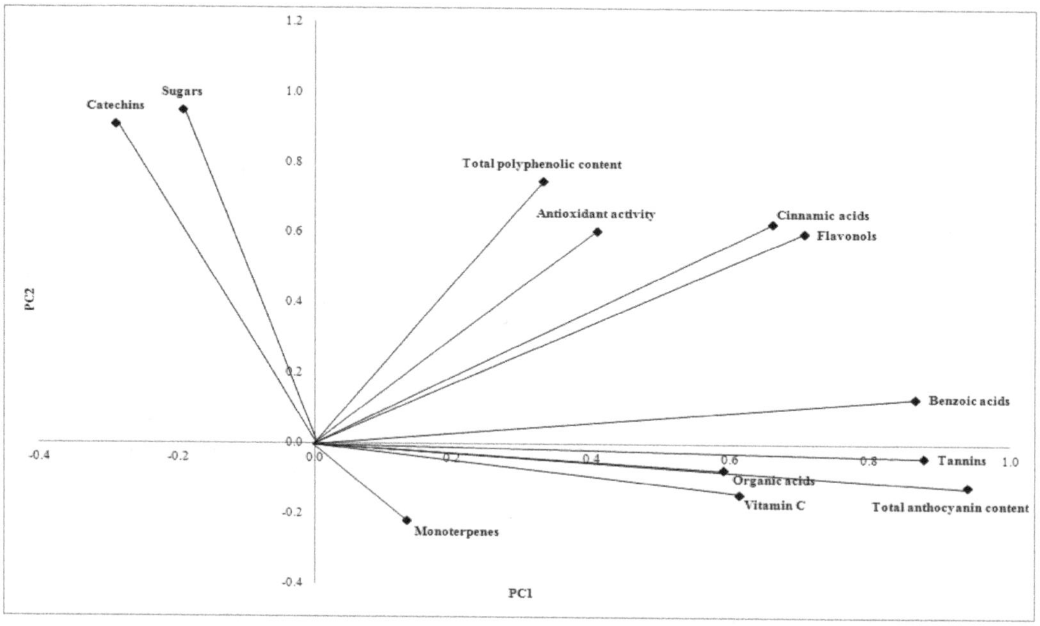

Figure 6. PCA loading plot of fruit-based products showing correlation among nutritional traits, nutraceutical properties and bioactive compound classes and PCs.

3. Materials and Methods

A detailed description of the used extraction protocols and analytical methods is reported in the Supplementary material.

3.1. Plant. Material

The investigated material consisted of fruit-derived products prepared from 7 species (*Mangifera indica* L.—two genotypes, *Tamarindus indica* L., *Musa × paradisiaca* L., *Psidium cattleyanum* Sabine, *Litchi chinensis* Sonn., *Citrus sinensis* (L.) Osbeck and *Psidium guajava* L.). Fruits were picked at commercial maturity stage during the 2017 from commercial orchards managed by the *Coopérative et Mutuelle des Comores pour le Développement* (Codcom) in Moroni, Comoros Islands. CODCOM is a non-governmental organization involved in poverty reduction in Comoros and in the countries of the Indian Ocean. Physiologically mature fruits were randomly selected from three plants for each biological replication ($n = 3$). Fruits were manually harvested from the plants based on selected qualitative parameters (e.g., colour, firmness and total soluble solids), considering also literature and experience of the local researchers and technicians. Fruits were then sorted, washed and stored in a 4 °C cold room for less than 5 days until jam and jelly preparation. Inedible parts were rejected.

3.2. *Preparation of Fruit-Derived Products*

A standard commercial procedure to manufacture jams and jellies was followed.

Fruit (1 kg) was reduced to fine particles with a commercial food processor for approximately 30 s. Jam formulation was 50% fruit, 48% sugar and 2% pectin mix (composed of dextrose, pectin and fumaric acid). pH was checked and adjusted if needed by addition of 50% citric acid solution (10–15 mL) to a target pH of 3.0. An essence of vanilla or cinnamon was added to improve the sensory properties of the final products. Sugar was added and the mixture was boiled for 30–40 min to a final concentration of 65°Brix (approximately 105 °C final boiling point). The jam was packed at 90 °C in 350 mL glass jars, immediately sealed with plastisol lined metal lids and inverted for 5 min to sterilize the lids. The jars were then returned to normal position for air-cooling.

For the preparation of 1 kg fruit jelly, 500 g filtered fruit juice, 550 g sugar, 5 g citric acid was used. Five g of pectin were added. The remaining sugar was mixed with fruit juice and heated until the total soluble solid (TSS) become near to 55°Brix. Then sugar mixed pectin was added and continued the heating until TSS becomes near to 58°Brix. The citric acid was added and continued the heating. When TSS of the jelly becomes 67°Brix, jelly was poured in a sterilized glass bottle and parafinning the cap.

3.3. *Spectrophotometric Analysis*

Total polyphenolic content (TPC) was determined according to the Folin-Ciocalteu colorimetric method [64,65]; results were expressed as mg of gallic acid equivalents (GAE) per 100 g of product (Pr).

Total anthocyanin content (TAC) was determined using the pH-differential method [66,67] and expressed as milligrams of cyanidin-3-*O*-glucoside (C3G) per 100 g of product ($mg_{C3G}/100\ g_{Pr}$).

Antioxidant activity was evaluated by Ferric Reducing Antioxidant Power (FRAP) assay [68], with the results expressed as millimoles of ferrous iron (Fe^{2+}) equivalents per kilogram (solid food) of Pr.

3.4. *Chromatographic Analysis*

Chromatographic analysis was carried out using an Agilent 1200 high-performance liquid chromatograph coupled to an Agilent UV-Vis diode array detector (Agilent Technologies, Santa Clara, CA, USA), based on HPLC methods previously validated for fresh fruits, herbal medicines and other food products [2,69].Composition of solvents, used gradient elution conditions and UV-Vis wavelengths were listed and described in Table 5, while calibration parameters for all the used analytical standards were reported in Table 6 [70].

Table 5. Chromatographic conditions of the used methods.

Method	Compounds of Interest	Stationary Phase	Mobile Phase	Flow (mL min^{-1})	Wavelength (nm)
A	cinnamic acids, flavanols	KINETEX-C18 column (4.6 × 150 mm, 5 μm)	A: 10 mM KH_2PO_4/H_3PO_4, pH = 2.8 B: CH_3CN	1.5	330
B	benzoic acids, catechins, tannins	KINETEX-C18 column (4.6 × 150 mm, 5 μm)	A: $H_2O/CH_3OH/HCOOH$ (5:95:0.1 *v/v/v*), pH = 2.5 B: $CH_3OH/HCOOH$ (100:0.1 *v/v*)	0.6	280
C	monoterpenes	KINETEX-C18 column (4.6 × 150 mm, 5 μm)	A: H_2O B: CH_3CN	1.0	210, 220, 235, 250
D	organic acids	KINETEX-C18 column (4.6 × 150 mm, 5 μm)	A: 10 mM KH_2PO_4/H_3PO_4, pH = 2.8 B: CH_3CN	0.6	214
E	vitamins	KINETEX-C18 column (4.6 × 150 mm, 5 μm)	A: 5 mM $C_{16}H_{33}N(CH_3)_3Br$/50 mM KH_2PO_4, pH = 2.5 B: CH_3OH	0.9	261, 348
F	sugars	SphereClone-NH_2 column (4.6 × 250 mm, 5 μm)	A: H_2O B: CH_3CN	0.5	200, 267, 286

Elution conditions. Method A: gradient analysis: 5%B to 21%B in 17 min + 21%B in 3 min (2 min conditioning time); Method B: gradient analysis: 3%B to 85%B in 22 min + 85%B in 1 min (2 min conditioning time); Method C: gradient analysis: 30%B to 56%B in 15 min + 56%B in 2 min (3 min conditioning); Method D: gradient analysis: 5%B to 14%B in 10 min + 14%B in 3 min (2 min conditioning time); Method E: isocratic analysis: ratio of phase A and B: 95:5 in 10 min (5 min conditioning time); Method F: isocratic analysis: ratio of phase A and B: 5:85 in 12 min (3 min conditioning time). Chromatographic separation was performed on a Kinetex-C18 column (Phenomenex, Torrance, CA, USA) and a SphereClone-NH2 column (Phenomenex, Torrance, CA, USA).

Table 6. Calibration parameters for all the used analytical standards.

Class	Standard	Calibration Curve Equation	R^2	Calibration Curve Range (mg L^{-1})
Cinnamic acids	caffeic acid	y = 59.046x + 200.6	0.996	111–500
	chlorogenic acid	y = 13.583x + 760.05	0.984	111–500
	coumaric acid	y = 8.9342x + 217.4	0.997	111–500
	ferulic acid	y = 3.3963x − 4.9524	1.000	111–500
Flavonols	hyperoside	y = 7.1322x − 4.583	0.999	111–500
	isoquercitrin	y = 8.3078x + 26.621	0.999	111–500
	quercetin	y = 3.4095x − 98.307	0.998	111–500
	quercitrin	y = 2.7413x + 5.6367	0.998	111–500
	rutin	y = 6.5808x + 30.831	0.999	111–500
Benzoic acids	ellagic acid	y = 29.954x + 184.52	0.998	62.5–250
	gallic acid	y = 44.996x + 261.86	0.999	62.5–250
Catechins	catechin	y = 8.9197x + 66.952	1.000	62.5–250
	epicatechin	y = 12.88x − 43.816	0.999	62.5–250
Tannins	castalagin	y = 4.236x − 8.535	1.000	62.5–250
	vescalagin	y = 4.939x − 1.232	1.000	62.5–250
Monoterpenes	limonene	y = 0.1894x − 5.420	0.999	125–1000
	phellandrene	y = 8.783x − 145.3	0.998	125–1000
	sabinene	y = 18.14x − 1004	0.998	125–1000
	γ-terpinene	y = 0.4886x − 23.02	0.999	125–1000
	terpinolene	y = 26.52x + 876.8	0.999	125–1000
Organic acids	citric acid	y = 1.0603x − 22.092	1.000	167–1000
	malic acid	y = 1.415x − 80.254	0.996	167–1000
	oxalic acid	y = 6.4502x + 6.1503	0.998	167–1000
	quinic acid	y = 0.8087x − 38.021	0.998	167–1000
	succinic acid	y = 0.9236x − 8.0823	0.995	167–1000
	tartaric acid	y = 1.8427x + 15.796	1.000	167–1000
Vitamins	ascorbic acid	y = 42.71x + 27.969	0.999	100–1000
	dehydroascorbic acid	y = 4.1628x + 140.01	0.999	30–300
Sugar	fructose	y = 1.8548x + 1.2324	0.999	125–1000
	glucose	y = 0.1269x − 0.1107	0.998	125–1000
	sucrose	y = 0.296x − 3.2202	1.000	125–1000

Single bioactive molecules were identified and quantified using selected biomarkers with a positive role in human health ("multi-marker approach") according to Mok and Chau [71]. All the results were expressed as mg/100 g of product (Pr), except sugars (expressed as g/100 g of Pr). Total bioactive compound content (TBCC) were determined as sum of selected compounds and expressed as mg/100 g of product.

3.5. *Statistical Analysis*

Uni- and multivariate analysis (MVA) were carried out on all of the samples. Data were treated by one-factor analysis of variance (ANOVA) and the averages were compared with Tukey's HSD post-hoc comparison test at significance level $p < 0.05$ ($N = 3$). Correlation between antioxidant activity and TPC/TAC was evaluated with Pearson's coefficient (R) at $p < 0.05$ ($N = 3$). For discrimination of the investigated samples, principal component analysis (PCA) was performed on the column-centred data. All calculations were performed with statistical software package IBM SPSS Statistics 22.0 (IBM, Armonk, NY, USA).

4. Conclusions

In this study, spectrophotometric and chromatographic methods coupled to chemometrics were used for phytochemical analysis in order to detect and quantify bioactive substances and characterize nutraceutical properties and nutritional traits in fruit-based products from Comoros Islands. HPLC profile of phenolic compounds may be used as "fingerprint" in the detection ofquali-quantitative differences in jellies and jams.

Results showed that analysed fruit jams and jellies could be a good source of polyphenolic constituents, as benzoic and cinnamic acids, catechins and flavanols and volatile compounds, as organic acids and monoterpenes: these molecules were found to be suitable markers for product characterization because they were specific compounds of the obtained chromatographic patterns. For this reason, even if each analytical approach has its limitations which restrict its applicability, the identification of these chemical markers could be a simple, rapid and generally available potential tool for quality control and labelling of local jams and jellies.

Moreover, the characterisation of these products and their nutritional and nutraceutical traits could be important to valorise local food production and to raise incomes for local population in the Comoros Islands, particularly women that work in agri-food industry. The obtained incomes could be useful to reduce poverty; indeed, the advances in local food production can be important in poverty reduction and deserves greater attention in rural development: it is necessary to link the evidence of poverty impact to simple policy recommendations in order to potentially integrate the promotion of local fruit-based products into national-level planning.

Finally, further benefits of this approach include the potential for a better nutrition, maintenance of biodiversity and environmentally sustainable food systems.

Author Contributions: D.D., S.H., C.G. and G.L.B. conceived and designed the experiments; S.H. proceeded to collect plant material and prepare fruit-based products; D.D. performed the chemical experiments and instrumental HPLC analysis; D.D. and M.G.M. statistically analysed the data; M.D.B., I.R. and G.G. contributed to manuscript elaboration; D.D. wrote the paper. All authors have read and approved the final manuscript.

References

1. Sewall, B.J.; Granek, E.F.; Trewhella, W.J. The endemic Comoros Islands fruit bat *Rousettus obliviosus*: Ecology, conservation, and red list status. *Oryx* **2003**, *37*, 344–352. [CrossRef]
2. Soifoini, T.; Donno, D.; Jeannoda, V.; Rakotoniaina, E.; Hamidou, S.; Achmet, S.M.; Solo, N.R.; Afraitane, K.; Giacoma, C.; Beccaro, G.L. Bioactive compounds, nutritional traits, and antioxidant properties of *Artocarpus altilis* (Parkinson) fruits: Exploiting a potential functional food for food security on the Comoros Islands. *J. Food Qual.* **2018**, in press. [CrossRef]
3. Clevidence, B. Tropical and subtropical fruits: Phytonutrients and anticipated health benefits. In Proceedings of the III International Symposium on Tropical and Subtropical Fruits, Fortaleza, Ceara, Brazil, 12–17 September 2004; pp. 485–498.
4. Lim, Y.Y.; Lim, T.T.; Tee, J.J. Antioxidant properties of several tropical fruits: A comparative study. *Food Chem.* **2007**, *103*, 1003–1008. [CrossRef]
5. Sajise, P.; Ahmad, I. Conservation and sustainable use of tropical and sub-tropical fruits: Current status and prospects. In Proceedings of the International Workshop on Tropical and Subtropical Fruits, Chiang Mai, Thailand, April 2007; pp. 97–112.
6. Chivandi, E.; Mukonowenzou, N.; Nyakudya, T.; Erlwanger, K.H. Potential of indigenous fruit-bearing trees to curb malnutrition, improve household food security, income and community health in sub-saharan africa: A review. *Food Res. Int.* **2015**, *76*, 980–985. [CrossRef]
7. Reardon, T.; Barrett, C.B.; Berdegué, J.A.; Swinnen, J.F.M. Agrifood industry transformation and small farmers in developing countries. *World Dev.* **2009**, *37*, 1717–1727. [CrossRef]
8. Singh, S.; Jain, S.; Singh, S.; Singh, D. Quality changes in fruit jams from combinations of different fruit pulps. *J. Food Process. Preserv.* **2009**, *33*, 41–57. [CrossRef]
9. Codex_Alimentarius. Codex Standard for Jams, Jellies and Marmalades (Codex Stan 296-2009). Available online: http://www.codexalimentarius.org (accessed on 27 March 2018).
10. Kim, D.O.; Padilla-Zakour, O. Jam processing effect on phenolics and antioxidant capacity in anthocyanin-Rich fruits: Cherry, plum, and raspberry. *J. Food Sci.* **2004**, *69*, S395–S400. [CrossRef]
11. Rosa, A.; Atzeri, A.; Deiana, M.; Scano, P.; Incani, A.; Piras, C.; Cesare Marincola, F. Comparative antioxidant activity and ^{1}H NMR profiling of Mediterranean fruit products. *Food Res. Int.* **2015**, *69*, 322–330. [CrossRef]
12. Puiggròs, F.; Solà, R.; Bladé, C.; Salvadó, M.-J.; Arola, L. Nutritional biomarkers and foodomic methodologies for qualitative and quantitative analysis of bioactive ingredients in dietary intervention studies. *J. Chromatogr. A* **2011**, *1218*, 7399–7414. [CrossRef] [PubMed]
13. Donno, D.; Cavanna, M.; Beccaro, G.L.; Mellano, M.G.; Torello-Marinoni, D.; Cerutti, A.K.; Bounous, G. Currants and strawberries as bioactive compound sources: Determination of antioxidant profiles with HPLC-DAD/MS. *J. App. Bot. Food Qual.* **2013**, *86*, 1–10.
14. Gonzalez-Molina, E.; Moreno, D.A.; Garcia-Viguera, C. Genotype and harvest time influence the phytochemical

quality of fino lemon juice (*Citrus limon* (L.) Burm. F.) for industrial use. *J. Agric. Food Chem.* **2008**, *56*, 1669–1675. [CrossRef] [PubMed]

15. Tibaldi, G.; Fontana, E.; Nicola, S. Growing conditions and postharvest management can affect the essential oil of *Origanum vulgare* L. spp. *hirtum* (Link) ietswaart. *Ind. Crop. Prod.* **2011**, *34*, 1516–1522. [CrossRef]
16. Bengoechea, M.L.; Sancho, A.I.; Bartolomé, B.; Estrella, I.; Gómez-Cordovés, C.; Hernández, M.T. Phenolic composition of industrially manufactured purees and concentrates from peach and apple fruits. *J. Agric. Food Chem.* **1997**, *45*, 4071–4075. [CrossRef]
17. Dragovic-Uzelac, V.; Pospišil, J.; Levaj, B.; Delonga, K. The study of phenolic profiles of raw apricots and apples and their purees by HPLC for the evaluation of apricot nectars and jams authenticity. *Food Chem.* **2005**, *91*, 373–383. [CrossRef]
18. Clifford, M.N. Anthocyanins—Nature, occurrence and dietary burden. *J. Agric. Food Chem.* **2000**, *80*, 1063–1072. [CrossRef]
19. Banaś, A.; Korus, A.; Tabaszewska, M. Quality assessment of low-sugar jams enriched with plant raw materials exhibiting health-promoting properties. *J. Food Sci. Technol.* **2018**, *55*, 408–417. [CrossRef] [PubMed]
20. Poiana, M.-A.; Moigradean, D.; Dogaru, D.; Mateescu, C.; Raba, D.; Gergen, I. Processing and storage impact on the antioxidant properties and color quality of some low sugar fruit jams. *Rom. Biotechnol. Lett.* **2011**, *16*, 6504–6512.
21. Rababah, T.M.; Al-Mahasneh, M.A.; Kilani, I.; Yang, W.; Alhamad, M.N.; Ereifej, K.; Al-u'datt, M. Effect of jam processing and storage on total phenolics, antioxidant activity, and anthocyanins of different fruits. *J. Sci. Food Agric.* **2011**, *91*, 1096–1102. [CrossRef] [PubMed]
22. Heimler, D.; Romani, A.; Ieri, F. Plant polyphenol content, soil fertilization and agricultural management: A review. *Eur. Food Res. Technol.* **2017**, *243*, 1107–1115. [CrossRef]
23. Donno, D.; Randriamampionona, D.; Andriamaniraka, H.; Torti, V.; Mellano, M.G.; Giacoma, C.; Beccaro, G.L. Biodiversity and traditional medicinal plants from Madagascar: Phytochemical evaluation of *Brachylaena ramiflora* (DC.) Humbert decoctions and infusions. *J. Appl. Bot. Food Qual.* **2017**, *90*, 205–2013.
24. Šavikin, K.; Zdunić, G.; Janković, T.; Tasić, S.; Menković, N.; Stević, T.; Đorđević, B. Phenolic content and radical scavenging capacity of berries and related jams from certificated area in Serbia. *Plant. Food Hum. Nutr.* **2009**, *64*, 212–217. [CrossRef] [PubMed]
25. Da Silva Pinto, M.; Lajolo, F.M.; Genovese, M.I. Bioactive compounds and antioxidant capacity of strawberry jams. *Plant. Food Hum. Nutr.* **2007**, *62*, 127–131. [CrossRef] [PubMed]
26. Wang, H.; Cao, G.; Prior, R.L. Oxygen radical absorbing capacity of anthocyanins. *J. Agric. Food Chem.* **1997**, *45*, 304–309. [CrossRef]
27. Veberic, R.; Slatnar, A.; Bizjak, J.; Stampar, F.; Mikulic-Petkovsek, M. Anthocyanin composition of different wild and cultivated berry species. *LWT-Food Sci. Technol.* **2015**, *60*, 509–517. [CrossRef]
28. Jia, N.; Xiong, Y.L.; Kong, B.; Liu, Q.; Xia, X. Radical scavenging activity of black currant (*Ribes nigrum* L.) extract and its inhibitory effect on gastric cancer cell proliferation via induction of apoptosis. *J. Funct. Foods* **2012**, *4*, 382–390. [CrossRef]
29. Donno, D.; Beccaro, G.L.; Mellano, M.G.; Cerutti, A.K.; Bounous, G. Goji berry fruit (*Lycium* spp.): Antioxidant compound fingerprint and bioactivity evaluation. *J. Funct. Foods* **2015**, *18*, 1070–1085. [CrossRef]
30. Silva, B.M.; Andrade, P.B.; Gonçalves, A.C.; Seabra, R.M.; Oliveira, M.B.; Ferreira, M.A. Influence of jam processing upon the contents of phenolics, organic acids and free amino acids in quince fruit (*Cydonia oblonga* Miller). *Eur. Food Res. Technol.* **2004**, *218*, 385–389. [CrossRef]
31. Wojdyło, A.; Oszmiański, J.; Bielicki, P. Polyphenolic composition, antioxidant activity, and polyphenol oxidase (PPO) activity of quince (*Cydonia oblonga* Miller) varieties. *J. Agric. Food Chem.* **2013**, *61*, 2762–2772. [CrossRef] [PubMed]
32. Cabral, T.A.; de Morais Cardoso, L.; Pinheiro-Sant'Ana, H.M. Chemical composition, vitamins and minerals of a new cultivar of lychee (*Litchi chinensis* cv. Tailandes) grown in Brazil. *Fruits* **2014**, *69*, 425–434. [CrossRef]
33. Singh, D.; Wangchu, L.; Moond, S.K. Processed Products of Tamarind. *Nat. Prod. Radiance* **2007**, *6*, 315–321.
34. Jimoh, S.; Onabanjo, O. Potentials of *Tamarindus indica* (L.) in jam production. *J. Agric. Soc. Res.* **2012**, *12*, 29–43.
35. Marquina, V.; Araujo, L.; Ruíz, J.; Rodríguez-Malaver, A.; Vit, P. Composition and antioxidant capacity of the guava (*Psidium guajava* L.) fruit, pulp and jam. *Arch. Latinoam. Nutr.* **2008**, *58*, 98–102. [PubMed]

36. Sánchez-Salcedo, E.M.; Mena, P.; García-Viguera, C.; Martínez, J.J.; Hernández, F. Phytochemical evaluation of white (*Morus alba* L.) and black (*Morus nigra* L.) mulberry fruits, a starting point for the assessment of their beneficial properties. *J. Funct. Foods* **2015**, *12*, 399–408. [CrossRef]
37. Garcia-Viguera, C.; Tomás-Barberán, F.A.; Ferreres, F.; Artés, F.; Tomás-Lorente, F. Determination of *Citrus* jams genuineness by flavonoid analysis. *Eur. Food Res. Technol.* **1993**, *197*, 255–259.
38. Del Rio, D.; Rodriguez-Mateos, A.; Spencer, J.P.; Tognolini, M.; Borges, G.; Crozier, A. Dietary (poly) phenolics in human health: Structures, bioavailability, and evidence of protective effects against chronic diseases. *Antioxid. Redox Signal.* **2013**, *18*, 1818–1892. [CrossRef] [PubMed]
39. Kansci, G.; Koubala, B.B.; Mbome, I.L. Biochemical and physicochemical properties of four mango varieties and some quality characteristics of their jams. *J. Food Process. Preserv.* **2008**, *32*, 644–655. [CrossRef]
40. López, R.; Ramírez, A.; de Fariñas Graziani, L. Physicochemical and microbiological evaluation of 3 commercial guava jams (*Psidium guajava* L.). *Arch. Latinoam. Nutr.* **2000**, *50*, 291–295. [PubMed]
41. Seeram, N.P. Berry fruits: Compositional elements, biochemical activities, and the impact of their intake on human health, performance, and disease. *J. Agric. Food Chem.* **2008**, *56*, 627–629. [CrossRef] [PubMed]
42. Ammar, I.; Ennouri, M.; Bouaziz, M.; Ben Amira, A.; Attia, H. Phenolic profiles, phytchemicals and mineral content of decoction and infusion of *Opuntia ficus-indica* flowers. *Plant. Food Hum. Nutr.* **2015**, *70*, 388–394. [CrossRef] [PubMed]
43. Landete, J.M. Ellagitannins, ellagic acid and their derived metabolites: A review about source, metabolism, functions and health. *Food Res. Int.* **2011**, *44*, 1150–1160. [CrossRef]
44. Ştefănuţ, M.N.; Căta, A.; Pop, R.; Moşoarcă, C.; Zamfir, A.D. Anthocyanins HPLC-DAD and MS characterization, total phenolics, and antioxidant activity of some berries extracts. *Anal. Lett.* **2011**, *44*, 2843–2855. [CrossRef]
45. Garzon, G.; Wrolstad, R. Comparison of the stability of pelargonidin-based anthocyanins in strawberry juice and concentrate. *J. Food Sci.* **2002**, *67*, 1288–1299. [CrossRef]
46. Kammerer, D.; Carle, R.; Schieber, A. Detection of peonidin and pelargonidin glycosides in black carrots (*Daucus carota* ssp. *Sativus* var. *Atrorubens alef.*) by high-performance liquid chromatography/electrospray ionization mass spectrometry. *Rapid Commun. Mass Spectrom.* **2003**, *17*, 2407–2412. [CrossRef] [PubMed]
47. Schwarz, M.; Wray, V.; Winterhalter, P. Isolation and identification of novel pyranoanthocyanins from black carrot (*Daucus carota* L.) juice. *J. Agric. Food Chem.* **2004**, *52*, 5095–5101. [CrossRef] [PubMed]
48. Papaefthimiou, D.; Papanikolaou, A.; Falara, V.; Givanoudi, S.; Kostas, S.; Kanellis, A.K. Genus *Cistus*: A model for exploring labdane-type diterpenes' biosynthesis and a natural source of high value products with biological, aromatic, and pharmacological properties. *Front. Chem.* **2014**, *2*, 35. [CrossRef] [PubMed]
49. Trombetta, D.; Castelli, F.; Sarpietro, M.G.; Venuti, V.; Cristani, M.; Daniele, C.; Saija, A.; Mazzanti, G.; Bisignano, G. Mechanisms of antibacterial action of three monoterpenes. *Antimicrob. Agents Chemother.* **2005**, *49*, 2474–2478. [CrossRef] [PubMed]
50. Crowell, P.L. Prevention and therapy of cancer by dietary monoterpenes. *J. Nutr.* **1999**, *129*, S775–S778. [CrossRef] [PubMed]
51. Cazares-Franco, M.C.; Ramirez-Chimal, C.; Herrera-Hernandez, M.G.; Nunez-Colin, C.A.; Hernandez-Martinez, M.A.; Guzman-Maldonado, S.H. Physicochemical, nutritional and health-related component characterization of the underutilized mexican serviceberry fruit *Malacomeles denticulata* (Kunth) G. N. Jones. *Fruits* **2014**, *69*, 47–60. [CrossRef]
52. Mattila, P.H.; Hellström, J.; McDougall, G.; Dobson, G.; Pihlava, J.-M.; Tiirikka, T.; Stewart, D.; Karjalainen, R. Polyphenol and vitamin c contents in European commercial blackcurrant juice products. *Food Chem.* **2011**, *127*, 1216–1223. [CrossRef] [PubMed]
53. Georgé, S.; Brat, P.; Alter, P.; Amiot, M.J. Rapid determination of polyphenols and vitamin c in plant-derived products. *J. Agric. Food Chem.* **2005**, *53*, 1370–1373. [CrossRef] [PubMed]
54. Donno, D.; Cerutti, A.K.; Prgomet, I.; Mellano, M.G.; Beccaro, G.L. Foodomics for mulberry fruit (*Morus* spp.): Analytical fingerprint as antioxidants' and health properties' determination tool. *Food Res. Inter.* **2015**, *69*, 179–188. [CrossRef]
55. Fügel, R.; Carle, R.; Schieber, A. Quality and authenticity control of fruit purées, fruit preparations and jams—A review. *Trends Food Sci. Technol.* **2005**, *16*, 433–441. [CrossRef]
56. Eyduran, S.P.; Ercisli, S.; Akin, M.; Beyhan, O.; Gecer, M.K.; Eyduran, E.; Erturk, Y.E. Organic acids, sugars, vitamin c, antioxidant capacity and phenolic compounds in fruits of white (*Morus alba* L.) and black

(*Morus nigra* L.) mulberry genotypes. *J. Appl. Bot. Food Qual.* **2015**, *88*, 134–138.

57. Soyer, Y.; Koca, N.; Karadeniz, F. Organic acid profile of turkish white grapes and grape juices. *J. Food Compos. Anal.* **2003**, *16*, 629–636. [CrossRef]
58. Cejudo-Bastante, C.; Castro-Mejías, R.; Natera-Marín, R.; García-Barroso, C.; Durán-Guerrero, E. Chemical and sensory characteristics of orange based vinegar. *J. Food Sci. Technol.* **2016**, *53*, 3147–3156. [CrossRef] [PubMed]
59. Flores, P.; Hellín, P.; Fenoll, J. Determination of organic acids in fruits and vegetables by liquid chromatography with tandem-mass spectrometry. *Food Chem.* **2012**, *132*, 1049–1054. [CrossRef]
60. Chinnici, F.; Spinabelli, U.; Riponi, C.; Amati, A. Optimization of the determination of organic acids and sugars in fruit juices by ion-exclusion liquid chromatography. *J. Food Compos. Anal.* **2005**, *18*, 121–130. [CrossRef]
61. Touati, N.; Tarazona-Díaz, M.P.; Aguayo, E.; Louaileche, H. Effect of storage time and temperature on the physicochemical and sensory characteristics of commercial apricot jam. *Food Chem.* **2014**, *145*, 23–27. [CrossRef] [PubMed]
62. Tzouros, N.; Arvanitoyannis, I. Agricultural produces: Synopsis of employed quality control methods for the authentication of foods and application of chemometrics for the classification of foods according to their variety or geographical origin. *Crit. Rev. Food Sci. Nutr.* **2001**, *41*, 287–319. [CrossRef] [PubMed]
63. Reid, L.M.; O'Donnell, C.P.; Downey, G. Potential of SPME-GC and chemometrics to detect adulteration of soft fruit purees. *J. Agric. Food Chem.* **2004**, *52*, 421–427. [CrossRef] [PubMed]
64. Slinkard, K.; Singleton, V.L. Total phenol analysis: Automation and comparison with manual methods. *Am. J. Enol. Vitic.* **1977**, *28*, 49–55.
65. Sánchez-Rangel, J.C.; Benavides, J.; Heredia, J.B.; Cisneros-Zevallos, L.; Jacobo-Velázquez, D.A. The Folin–Ciocalteu assay revisited: Improvement of its specificity for total phenolic content determination. *Anal. Methods* **2013**, *5*, 5990–5999. [CrossRef]
66. Lee, J.; Durst, R.W.; Wrolstad, R.E. Determination of total monomeric anthocyanin pigment content of fruit juices, beverages, natural colorants, and wines by the pH differential method: Collaborative study. *J. AOAC Int.* **2005**, *88*, 1269–1278. [PubMed]
67. Giusti, M.M.; Wrolstad, R.E. Characterization and measurement of anthocyanins by UV-visible spectroscopy. *Curr. Protoc. Food Anal. Chem.* **2001**, *00*, F1.2.1–F1.2.13. [CrossRef]
68. Benzie, I.F.; Strain, J.J. Ferric reducing/antioxidant power assay: Direct measure of total antioxidant activity of biological fluids and modified version for simultaneous measurement of total antioxidant power and ascorbic acid concentration. *Meth. Enzymol.* **1999**, *299*, 15–27. [PubMed]
69. Donno, D.; Mellano, M.G.; Prgomet, Z.; Beccaro, G.L. Advances in *Ribes* × *nidigrolaria* Rud. Bauer and A. Bauer fruits as potential source of natural molecules: A preliminary study on physico-chemical traits of an underutilized berry. *Sci. Hortic.* **2018**, *237*, 20–27. [CrossRef]
70. Donno, D.; Mellano, M.G.; De Biaggi, M.; Riondato, I.; Rakotoniaina, E.N.; Beccaro, G.L. New findings in *Prunus padus* L. fruits as a source of natural compounds: Characterization of metabolite profiles and preliminary evaluation of antioxidant activity. *Molecules* **2018**, *23*, 725. [CrossRef] [PubMed]
71. Mok, D.K.W.; Chau, F.T. Chemical information of chinese medicines: A challenge to chemist. *Chemom. Intell. Lab. Syst.* **2006**, *82*, 210–217. [CrossRef]

2

Development of an Accelerated Solvent Extraction-Ultra-Performance Liquid Chromatography-Fluorescence Detection Method for Quantitative Analysis of Thiamphenicol, Florfenicol and Florfenicol Amine in Poultry Eggs

Bo Wang [1,2,†], Xing Xie [3,†], Xia Zhao [2,4], Kaizhou Xie [2,4,*], Zhixiang Diao [2,4], Genxi Zhang [2,4], Tao Zhang [2,4] and Guojun Dai [2,4]

1 College of Veterinary Medicine, Yangzhou University, Yangzhou 225009, China; yzwbo168@163.com
2 Joint International Research Laboratory of Agriculture & Agri-Product Safety, Yangzhou University, Yangzhou 225009, China; 18252711481@139.com (X.Z.); 18352764521@163.com (Z.D.); zgx1588@126.com (G.Z.); zhangt@yzu.edu.cn (T.Z.); gjdai@163.com (G.D.)
3 Key Laboratory of Veterinary Biological Engineering and Technology, Ministry of Agriculture, Institute of Veterinary Medicine, Jiangsu Academy of Agricultural Sciences, Nanjing 210014, China; yzxx1989@163.com
4 College of Animal Science and Technology, Yangzhou University, Yangzhou 225009, China
* Correspondence: yzxkz168@163.com
† These authors contributed equally to this work.

Academic Editors: Alessandra Gentili and Chiara Fanali

Abstract: A simple, rapid and novel method for the detection of residues of thiamphenicol (TAP), florfenicol (FF) and its metabolite, florfenicol amine (FFA), in poultry eggs by ultra-performance liquid chromatography-fluorescence detection (UPLC-FLD) was developed. The samples were extracted with acetonitrile-ammonia (98:2, *v/v*) using accelerated solvent extraction (ASE) and purified by manual degreasing with acetonitrile-saturated n-hexane. The target compounds were separated on an ACQUITY UPLC® BEH C_{18} (2.1 mm × 100 mm, 1.7 μm) chromatographic column using a mobile phase composed of 0.005 mol/L NaH_2PO_4, 0.003 mol/L sodium lauryl sulfate and 0.05% trimethylamine, adjusted to pH 5.3 ± 0.1 by phosphoric acid and acetonitrile (64:36, *v/v*). The limits of detection (LODs) and limits of quantification (LOQs) of the three target compounds in poultry eggs were 1.8–4.9 μg/kg and 4.3–11.7 μg/kg, respectively. The recoveries of the three target compounds in poultry eggs were above 80.1% when the spiked concentrations of three phenicols were the LOQ, 0.5 maximum residue limit (MRL), 1.0 MRL and 2.0 MRL. The intraday relative standard deviations (RSDs) were less than 5.5%, and the interday RSDs were less than 6.6%. Finally, this new detection method was successfully applied to the quantitative analysis of TAP, FF and FFA in 150 commercial poultry eggs.

Keywords: poultry eggs; thiamphenicol; florfenicol; florfenicol amine; ASE; UPLC-FLD

1. Introduction

Poultry eggs contain high levels of essential amino acids and vitamins as well as various major and trace elements required by the human body, and as a result they have become an increasingly popular consumer product [1]. Consumer demand for poultry products has promoted the growth of the poultry industry, and intensive farming has also increased the morbidity and mortality of poultry. To control the occurrence of diseases and reduce mortality in poultry, antibiotics are widely used to prevent poultry diseases, increase feed conversion rates, and promote animal growth [2].

Thiamphenicol (TAP) and florfenicol (FF) are synthetic chloramphenicols (CAPs) used as broad-spectrum antibiotics, and they have chemical structures and efficacies similar to that of CAP as well as good therapeutic effects on various bacterial strains common in poultry. These compounds are widely used in actual production [3]. The only difference between TAP and CAP is the structure of the substituents on their benzene rings. CAP has a nitro group on the phenyl ring, and TAP has a methyl sulfone group. However, TAP is much less toxic than CAP, its blood toxicity effects are reversible, and it does not cause aplastic anemia. On the other hand, TAP can inhibit the formation of red blood cells, white blood cells and platelets; it has a strong immunosuppressive effect; and it has weaker antibacterial effects than CAP, which limit its practical use [4–6]. Many countries use these compounds as veterinary drugs, but they are banned from use in food animals. FF is a fluorinated analogue of TAP with a molecular structure similar to that of CAP, but it lacks the nitro group on the aromatic ring. Studies have shown that this substituent is the key molecular characteristic of CAP causing dose-independent irreversible aplastic anemia in the human body [7]. However, FF can theoretically cause serious adverse reactions similar to those caused by CAP. Therefore, FF can only be used for the treatment of animal diseases [8]. FF shows the most potent antibacterial activity among CAP drugs, and it has many advantages (wide spectrum of antibacterial activity, good oral absorption, wide distribution in the body, high bioavailability, good safety profile, etc.), making it a broad-spectrum antibiotic with great potential for practical applications [9]. At present, the drug is on the market in many countries, and it is widely used in animal husbandry and aquaculture for disease prevention. However, reproductive toxicity tests have shown that FF has certain embryotoxic effects, and as a result the detection of its residue in animal foods such as livestock, poultry, and aquatic products has attracted increasing attention [10].

Because both TAP and FF have toxic side effects, the EU [11], US [12] and China's Ministry of Agriculture [13] have established maximum residue limits (MRLs) for these compounds in poultry tissues (TAP: 50 μg/kg, FF: 100 μg/kg in muscle), and stipulated that the limit of FF residue in poultry tissues is based on the total amount of both the prototype drug (FF) and its metabolite, florfenicol amine (FFA), and that these drugs should not be detected in poultry eggs. Therefore, developing different detection methods to determine whether veterinary drug residues meet the legal requirements before the animal foods are marketed is of great importance. Establishing and improving the detection methods for TAP, FF and FFA residues in poultry eggs is also necessary.

At present, there are many reported methods for detecting TAP, FF and FFA residues, including methods based on high-performance liquid chromatography (HPLC) [14–16], gas chromatography (GC) [17,18], liquid chromatography-tandem mass spectrometry (LC-MS/MS) [19–22], and gas chromatography-tandem mass spectrometry (GC-MS) [23,24]. When using GC to analyze CAP drug residues, the target needs to be derivatized, making the analysis process cumbersome. The most widely used analytical method involves LC-MS, and although this method offers qualitative and quantitative (mass spectrometry) accuracy and high sensitivity, the instrument is expensive, and the detection cost is high. Meanwhile, fluorescence detection is commonly used for veterinary drug residues and environmental analysis because of its advantages of speed, ease of operation, and low cost of detection [14,25]. Therefore, developing a simple, fast and low-cost analytical method that meets the detection requirements is of great importance. Moreover, in the reported detection methods, the sample matrices used for analysis were generally animal tissues [15,26] or aquatic products [21,27], and there are few reports on detection methods for poultry eggs [14]. Xie et al. [14] established an HPLC-FLD method for the determination of TAP, FF and FFA residues in eggs and sample pretreatment using a liquid-liquid extraction method, ethyl acetate:acetonitrile:ammonium hydroxide (49:49:2, *v/v/v*) as an extractant, delipided in n-hexane. Based on previous research, a comprehensive method using ultra-performance liquid chromatography-fluorescence detection (UPLC-FLD) for the determination of TAP, FF and FFA residues in poultry eggs (hen eggs, duck eggs, goose eggs, pigeon eggs and quail eggs) is reported here. Compared with the previously studied HPLC-FLD method [14], the UPLC-FLD method has the advantages of fast analysis speed (detection time < 5 min), strong separation ability (recoveries were 80.1%–98.6%), high sensitivity, and low consumption of reagents. This study intends to use accelerated solvent extraction (ASE) as the sample pretreatment procedure to extract the target analytes from the samples, aiming to establish an ASE-UPLC-FLD method for the determination of TAP, FF and FFA residues in poultry eggs. This technique will provide a new, simple, inexpensive, highly efficient and rapid method for the detection of these analytes. In addition, the effects of ultrasonic extraction, vortex oscillation extraction, vortex oscillation + ultrasonic extraction and ASE extraction are compared in this study. Compared to other extraction methods, ASE was investigated as a novel alternative technology, which has the advantages of automation (saving time and human effort), consuming less reagents, higher recovery rate, and suitability for batch processing of samples.

2. Results and Discussion

2.1. Selection of the Chromatographic Column and Mobile Phase

The composition of the mobile phase and the type of chromatographic column have a substantial influence on the separation and peak shape of the analytes. Among the reported methods, the most commonly used columns for the detection of CAP drugs are C_{18} columns [28,29], produced by various manufacturers. Meanwhile, the ACQUITY UPLC® BEH C_{18} (2.1 mm × 100 mm, 1.7 μm) column offers outstanding chemical stability, a wide range of pH conditions (pH 1-12) and a wide range of mobile phases, which provides a versatile and reliable separation technique for method development. Therefore, this study used an ACQUITY UPLC® BEH C_{18} (2.1 mm × 100 mm, 1.7 μm) column as the analytical column. For the mobile phase, acetonitrile-water [16,30] mixtures are often used to determine CAP residues in LC-MS methods. FFA is a weakly basic substance that does not generally remain on the C_{18} column, and it elutes with the dead volume. A common solution to this is to add ammonium formate or ammonium acetate [27,31] to the mobile phase system to enhance the retention of FFA on the C_{18} column. When detecting CAPs with a fluorescence detector, to enhance the retention of FFA on a C_{18} column, an ion-pair reagent is usually added to the aqueous phase to react with the FFA to form a weakly bound ion pair, and the pH is adjusted with a buffer to keep the whole system weakly acidic and prevent dissociation of the ion pair [14,32]. Yang et al. [33] reported a liquid chromatography-fluorescence detection method for the determination of TAP, FF and FFA residues in aquatic products. Sodium heptane sulfonate was added to the mobile phase, and the target compounds were well separated. Commonly used ion-pairing reagents are sodium heptane sulfonate and sodium lauryl sulfate, and these ion-pairing reagents provide a good separation for the target compounds. Because it is inexpensive, sodium lauryl sulfate is used as an ion pairing reagent in this test. Sodium lauryl sulfate and FFA form weakly polar pairs, which are distributed on the surface of the hydrophobic stationary phase and then eluted by the mobile phase. The buffer system uses phosphate-phosphoric acid and triethylamine to improve peak shape and reduce peak tailing. In this study, the amounts of ion-pairing reagent (sodium lauryl sulfate), buffer system (phosphate-phosphoric acid), and triethylamine were optimized. The effects of 1, 3, 5, and 10 mM sodium lauryl sulfate were investigated. As the concentration increased, the retention time of FFA increased, and a concentration of 3 mM could ensure that FFA eluted first. The effects of 0, 3, 5, 10, and 20 mM NaH_2PO_4 are compared in Figure 1a. The retention time of FFA was slightly shorter with increasing NaH_2PO_4 concentration, but the use of salt impacted the instrument, and the chromatographic column was easily blocked; thus, considering the FFA retention time and the effect on the instrument, a concentration of 5 mM NaH_2PO_4 was selected. The pH impacted the response and retention time of FFA, as shown in Figure 1b. Under neutral conditions, the retention time of FFA was shorter, and decreasing the pH gradually increased the response of FFA (Figure 2) but increased the retention time. When the pH was 5.4, the response of FFA reached its highest value, and further reducing the pH had little effect on the response. However, the retention time was too high, increasing the overall detection time, so a pH of 5.3 ± 0.1 was selected. The amount of triethylamine was also investigated. It was found that 0.01% triethylamine could reduce peak trailing. However, as the amount of triethylamine was increased (0.03%, 0.05%, and 0.1%), the retention time of FFA increased. Based on all these factors, the concentration of triethylamine was set as 0.05%. To separate the targets from impurities, the ratio of the solvents (63:37, 64:36, 65:35, 66:34, *v/v*) in the mobile phase was optimized. When the mobile phase ratio was 64:36 (*v/v*), the targets and impurities were separated, and the peak shapes were good. In summary, the final mobile phase conditions were water (containing 5 mM NaH_2PO_4, 3 mM lauryl sodium sulfate, 0.05% triethylamine, adjusted to pH 5.3 ± 0.1) and acetonitrile in a 64:36 (*v/v*) ratio. According to the chemical nature of the ACQUITY UPLC® BEH C18 (2.1 mm × 100 mm, 1.7 µm) column and final mobile phase composition, the elution order and resolution of the target compound were analyzed; TAP was preferentially eluted, followed by FF and then FFA, and examining the fluorescence intensity of the target compound showed that the resolution was greatly improved.

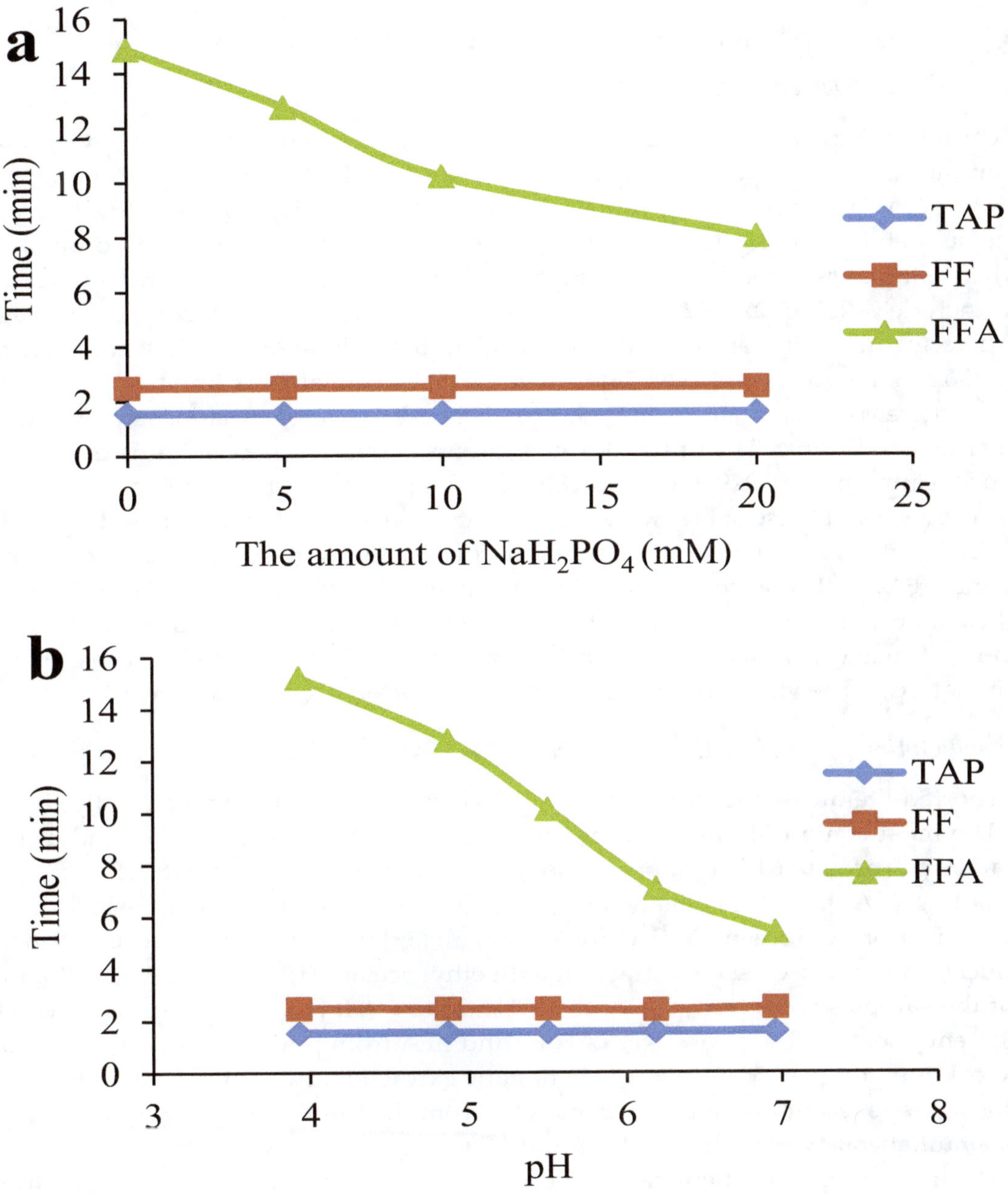

Figure 1. Effects of the amount of NaH_2PO_4 (**a**) and the mobile phase pH (**b**) on the retention times of the targets.

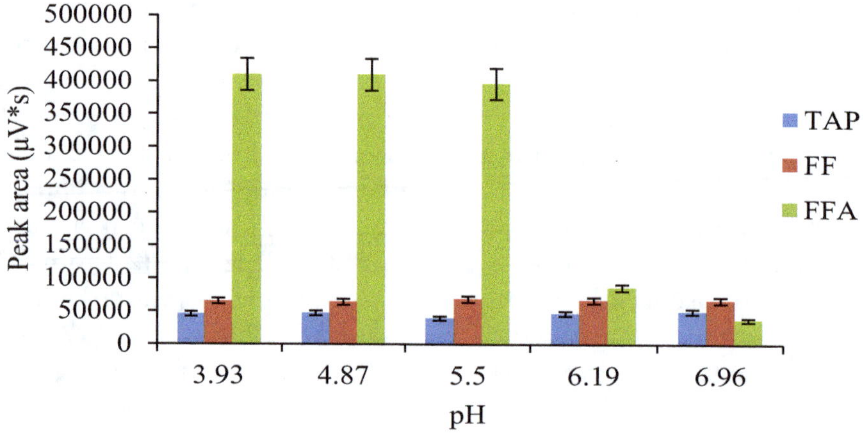

Figure 2. Effects of pH on the target responses.

2.2. *Determination of the Detection Wavelength*

Liquid chromatography-fluorescence detection methods are commonly used in veterinary drug residues in animal-derived foods, pesticide residues in agricultural products, and environmental analysis. LC-FLD can detect compounds containing fluorophores, which facilitates the development of simple and rapid methods for the determination of target compounds containing fluorophores. In the literature, when using ultraviolet detectors to identify CAP drugs, the most commonly used detection wavelengths include 220, 223, 224, 225, and 228 nm [15,16,33,34]; when detecting CAP drugs with a fluorescence detector, the excitation and emission wavelengths include 224 and 290 nm, 224 and 295 nm, and 225 and 290 nm, respectively [33,35,36]. However, the optimal detection wavelength for these targets may be different under different detection conditions because of variations in detection devices and experimental conditions. The optimal excitation and emission wavelengths of TAP obtained by fluorescence scanning were 229.8 nm and 285.3 nm, respectively; those of FF were 229.8 nm and 283.9 nm, respectively; and those of FFA were 227.9 nm and 283.9 nm. Therefore, based on the literature and the optimum excitation and emission wavelengths for each target, the optimal detection wavelength for TAP, FF and FFA under the conditions used in this study was simultaneously determined by scanning with a fluorescence detector. An excitation wavelength of 233 nm and an emission wavelength of 284 nm were ultimately selected to simultaneously measure TAP, FF and FFA. These wavelengths ensured both high response values of the targets and no interference from impurities.

2.3. *Selection and Optimization of the Extraction Solvent and Extraction Method*

In the reported literature, the extraction of CAP drugs is typically carried out by liquid-liquid extraction. The most commonly used extractants are acetonitrile, ethyl acetate and a mixture of acetonitrile and ethyl acetate (different proportions of ammonia-containing extractants will provide extracts containing FFA) [33,37,38]. There are few reports on the extraction of CAPs from matrices by automatic extraction equipment-ASE. Only Yang et al. [39] used ASE to extract CAP and FF from aquatic products, and they used static extraction with ethyl acetate at 100 °C for 5 min; their obtained recoveries of the samples were from 90.2% to 109%. Xiao et al. [40] used subcritical water as the extraction solvent to extract traces of CAP, TAP, FF and FFA from poultry tissues using pressurized liquid extractors operating at 150 °C and 100 bar (static extraction, two extraction cycles, 3 min each cycle), and the average recoveries of the four analytes from the samples were 86.8–101.5%. However, a method for simultaneously extracting TAP, FF and FFA from poultry eggs using a modern automatic extraction instrument-ASE instrument has not been reported. In this study, the effects of mixtures of acetonitrile and ammonia (98:2, *v/v*), ethyl acetate and ammonia (98:2, *v/v*) and acetonitrile and ammonia with ethyl acetate (49:49:2, *v/v/v*) were compared. The results showed that the above extractants can extract TAP, FF and FFA from the matrix and that the recoveries meet the detection requirements. However, acetonitrile offers better deproteinization, and the substance in eggs that causes the most interference is protein. To improve the detection results, acetonitrile containing ammonia (98:2, *v/v*) was selected as the extractant. Because FFA is a weakly alkaline compound with an amino group, it is more advantageous to extract FFA under alkaline conditions because greater similarity leads to better solubility. Under optimized ASE conditions, this study compared the effects of different ratios of extractants (acetonitrile: ammonia = 99:1, 98:2, 97:3, 96:4, and 95:5, *v/v*) on poultry egg recovery. Table 1 shows that as the proportion of ammonia increased, the recoveries of FFA also gradually increased; however, when the content of ammonia exceeded 2% of the total volume, the recoveries of FFA decreased, and excessive ammonia also reduced the recoveries of TAP and FF. In summary, the extraction outcome with acetonitrile:ammonia (98:2, *v/v*) was best, and the recoveries of all targets were above 91.0%.

Table 1. Effects of different ratios of extraction reagents (acetonitrile:ammonia) for ASE on the recoveries of 25 μg/kg TAP and 50 μg/kg FF and FFA from poultry eggs (%) (n = 6).

Matrix	Analyte	Extraction Reagents (acetonitrile:ammonia, *v/v*)				
		99:1	98:2	97:3	96:4	95:5
Hen eggs	TAP	93.4 ± 1.8	92.8 ± 1.1	89.4 ± 2.3	81.6 ± 2.2	73.9 ± 2.0
	FF	92.6 ± 2.7	93.2 ± 3.0	84.2 ± 2.2	74.3 ± 1.9	69.6 ± 2.0
	FFA	86.9 ± 2.1	92.4 ± 2.2	88.4 ± 3.1	78.8 ± 2.9	69.3 ± 1.9
Duck eggs	TAP	90.2 ± 2.3	91.3 ± 2.3	85.3 ± 2.4	80.8 ± 2.4	70.3 ± 2.2
	FF	93.0 ± 1.8	91.2 ± 2.5	84.2 ± 2.0	74.6 ± 1.9	68.7 ± 2.3
	FFA	85.0 ± 2.5	93.2 ± 1.8	84.3 ± 2.3	73.6 ± 1.9	61.5 ± 1.9
Goose eggs	TAP	90.7 ± 2.2	93.4 ± 2.6	86.7 ± 2.4	75.8 ± 2.7	67.9 ± 2.0
	FF	93.1 ± 1.8	91.3 ± 1.9	87.3 ± 2.3	71.4 ± 2.2	60.4 ± 2.4
	FFA	85.0 ± 1.8	92.3 ± 2.0	84.1 ± 1.8	73.0 ± 1.9	63.8 ± 2.2
Pigeon eggs	TAP	92.4 ± 2.6	92.6 ± 2.1	82.3 ± 3.1	75.5 ± 2.4	63.2 ± 2.3
	FF	92.3 ± 2.4	93.1 ± 2.0	83.8 ± 1.9	76.1 ± 3.0	61.0 ± 2.2
	FFA	88.2 ± 2.2	96.3 ± 2.1	86.5 ± 2.7	73.1 ± 2.1	64.8 ± 2.3
Quail eggs	TAP	91.3 ± 2.3	92.8 ± 2.3	87.3 ± 2.0	70.8 ± 1.9	65.7 ± 2.1
	FF	92.7 ± 2.2	94.2 ± 2.5	86.2 ± 2.5	74.5 ± 2.4	68.5 ± 2.5
	FFA	86.5 ± 1.8	93.0 ± 1.9	84.7 ± 2.5	76.2 ± 1.9	70.4 ± 1.8

In this study, under the conditions of acetonitrile:ammonia (98:2, *v/v*) as an extractant, the effects of ultrasonic extraction, vortex oscillation extraction, vortex oscillation + ultrasonic extraction and ASE extraction were compared. Table 2 shows that compared to those with other sample preparation processes, the recoveries with ASE extraction were best; in addition, the ASE method saves time and is suitable for batch processing of samples. The time required to prepare a sample with ASE (15 min) was half that required for vortexing + ultrasonic extraction (30 min). Moreover, using ASE to process samples avoided contact between the experimenter and the reagent, which is more consistent with the detection requirements of health and environmental protection. Therefore, this study ultimately selected ASE for sample extraction.

2.4. Optimization of the ASE Method

The sample extraction step is often considered to be the bottleneck of the entire analytical process. To simplify the pretreatment of the sample and improve the efficiency of sample preparation, a variety of pretreatment methods have been developed. Since the introduction of ASE in 1995, it has rapidly become an acceptable alternative to traditional extraction methods. ASE uses high-temperature and high-pressure conditions, which result in greatly improved extraction efficiency. This study explored the effects of various operating parameters (temperature, time, volume of solvent used and so on) on ASE performance. The effects of different temperatures (40, 60, 80, 100 and 120 °C) on the recoveries of the targets were studied. The recoveries of the targets gradually increased with increasing extraction temperature. When the temperature exceeded 80 °C, the recoveries decreased (Figure 3a). Studies have shown that increasing the time of static extraction provides the target sufficient time to diffuse into the extraction solvent, improving the efficiency of the extraction. However, this study compared different static extraction times (2, 3, 5, and 8 min) and found that for TAP and FF, static extraction for 5 min provides high recoveries (>90%), while the recovery of FFA decreases with increasing static extraction time (Figure 3b). As shown in Figure 3b, 3 min and 5 min had little effect on TAP and FF extraction, and a high recovery rate for FFA was observed when the static extraction time was 3 min. Therefore, to ensure that the recoveries of all the targets met the requirements of detection under these ASE conditions, 3 min was selected as the static extraction time. This study also explored the effect of the number of static extraction cycles and the volume of the extractant on the extraction outcome. Extraction of the target twice provided a better recovery than one extraction cycle. When the volume of the extractant was 40% by volume, the target could be efficiently extracted, so there was no need to increase the amount of extractant. In summary, the final ASE conditions were 80 °C, 1500 psi, 40% pool volume, static extraction for 3 min, and two static extraction cycles.

Table 2. Effects of different extraction methods on the recoveries of 25 μg/kg TAP and 50 μg/kg FF and FFA from poultry eggs (%) (n = 6).

Matrix	Analyte	Extraction Method			
		Ultrasonic	Vortex Oscillation	Vortex Oscillation + Ultrasonic	ASE
Hen egg	TAP	34.4 ± 2.5	78.6 ± 2.6	88.3 ± 2.2	92.8 ± 2.4
	FF	37.1 ± 2.3	81.5 ± 3.1	92.0 ± 2.7	96.0 ± 2.3
	FFA	54.4 ± 2.2	84.0 ± 2.9	91.9 ± 2.8	93.3 ± 2.0
Duck eggs	TAP	30.9 ± 2.0	70.0 ± 2.7	84.9 ± 2.2	90.2 ± 1.9
	FF	33.4 ± 2.5	79.5 ± 2.7	89.7 ± 2.4	92.5 ± 2.1
	FFA	49.0 ± 2.4	74.4 ± 3.2	86.0 ± 2.9	91.7 ± 2.2
Goose eggs	TAP	31.3 ± 2.9	71.9 ± 3.0	82.7 ± 2.6	93.3 ± 1.9
	FF	35.6 ± 2.7	72.0 ± 2.6	88.4 ± 3.0	92.4 ± 2.5
	FFA	40.5 ± 3.4	77.0 ± 2.8	85.0 ± 2.4	90.4 ± 2.1
Pigeon eggs	TAP	34.4 ± 3.0	69.9 ± 2.7	85.3 ± 2.9	86.9 ± 2.2
	FF	34.7 ± 2.6	70.0 ± 3.2	87.0 ± 2.6	89.0 ± 2.4
	FFA	40.7 ± 2.8	72.6 ± 2.9	80.3 ± 2.3	95.1 ± 1.6
Quail eggs	TAP	36.1 ± 2.9	72.1 ± 2.8	83.5 ± 2.7	90.1 ± 2.0
	FF	30.1 ± 3.0	73.4 ± 2.3	88.5 ± 3.3	92.6 ± 2.4
	FFA	45.5 ± 2.8	69.7 ± 3.1	83.7 ± 2.5	92.7 ± 2.7

2.5. Bioanalytical Method Validation

In the blank poultry eggs, TAP was spiked at a concentration from limit of quantification (LOQ)-250 μg/kg, and FF and FFA were added at LOQ-400 μg/kg. The peak area was correlated with the spiked concentration of the analyte, and the linearity was good. The linear equations, linear ranges and coefficients of determination of TAP, FF and FFA in poultry eggs are shown in Table 3.

The recoveries and precisions of TAP, FF and FFA in different blank poultry egg samples are shown in Tables 4 and 5, respectively. As shown in Tables 4 and 5, the recoveries of TAP, FF and FFA in poultry eggs were 80.1%–98.6%, the relative standard deviations (RSDs) were 1.2%–4.3%, the intraday RSDs were 1.2%–5.5%, and the interday RSDs were 1.8%–6.6%. According to the EU 2002/675/EC resolution and the FDA [41,42], the acceptable range of recoveries for multidrug residue testing procedures is 70–120%. The average recoveries of TAP, FF and FFA from different blank poultry egg samples were all above 80.0%, which are consistent with the EU's requirements for the recoveries of analytes. The limits of detection (LODs) and LOQs of TAP, FF and FFA in different blank poultry egg samples using the optimized pretreatment method and instrument analysis method are shown in Table 3. The LODs of TAP, FF and FFA were 3.3–3.4 μg/kg, 4.7–4.9 μg/kg and 1.8–1.9 μg/kg, respectively, and the LOQs were 9.7–9.9 μg/kg, 10.5–11.7 μg/kg and 4.3–4.8 μg/kg, respectively.

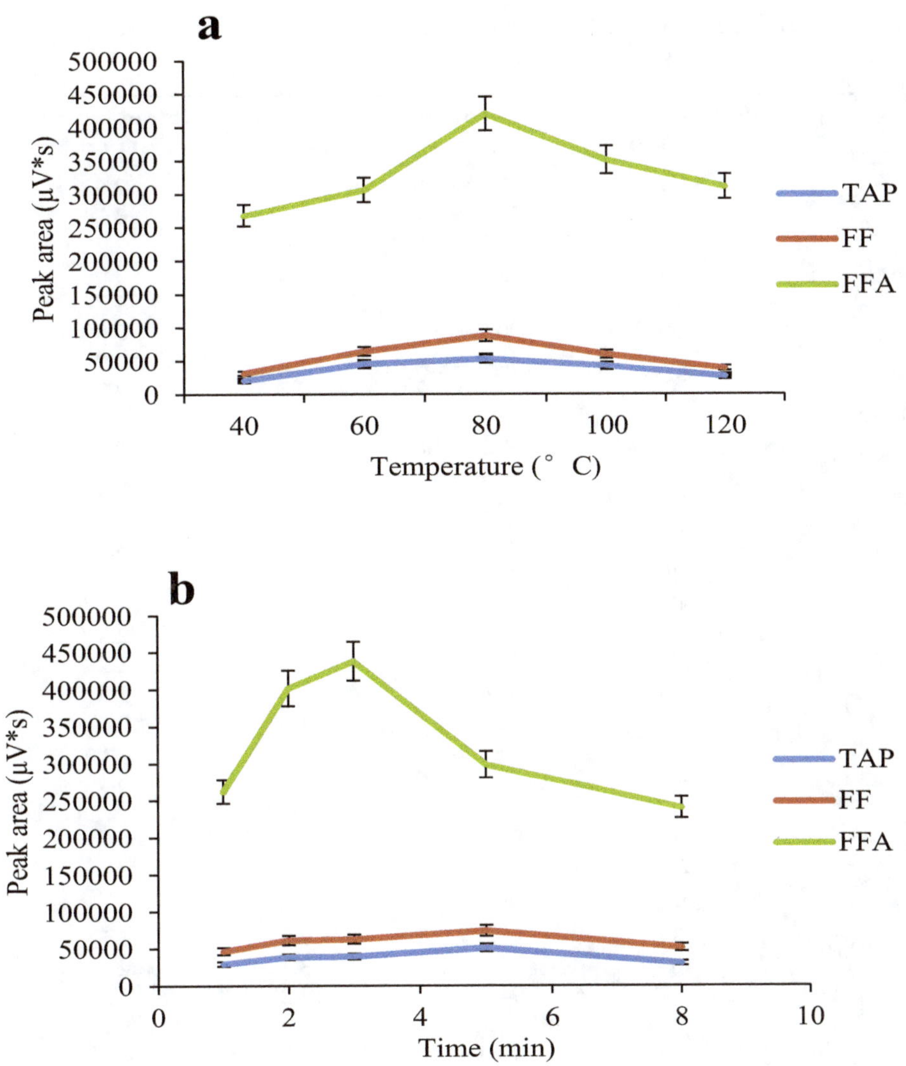

Figure 3. Effects of temperature (**a**) and time (**b**) on ASE extraction.

Table 3. The linear ranges, linear regression equations, determination coefficients, LODs and LOQs of TAP, FF and FFA from poultry eggs.

Matrix	Analyte	Linear Range (μg/kg)	Linear Regression Equation	Determination Coefficient (R^2)	LOD (μg/kg)	LOQ (μg/kg)
Hen eggs	TAP	9.7–250.0	y = 1030.2x + 295.52	0.9996	3.3	9.7
	FF	10.5–400.0	y = 737.43x + 714.06	0.9998	4.7	10.5
	FFA	4.3–400.0	y = 3020.3x + 164.01	0.9998	1.8	4.3
Duck eggs	TAP	9.9–250.0	y = 562.84x + 634.05	0.9997	3.4	9.9
	FF	11.7–400.0	y = 844.1x + 616.34	0.9997	4.9	11.7
	FFA	4.7–400.0	y = 4495.8x + 520.7	0.9996	1.9	4.7
Goose eggs	TAP	9.8–250.0	y = 618.73x + 139.95	0.9996	3.4	9.8
	FF	11.2–400.0	y = 713.68x + 738.78	0.9996	4.8	11.2
	FFA	4.7–400.0	y = 3081.1x – 258.32	0.9997	1.9	4.7
Pigeon eggs	TAP	9.9–250.0	y = 683.44x + 543.48	0.9998	3.4	9.9
	FF	11.2–400.0	y = 762.9x + 761.11	0.9993	4.8	11.2
	FFA	4.8–400.0	y = 4019.1x + 490.45	0.9998	1.9	4.8
Quail eggs	TAP	9.7–250.0	y = 667.44x + 483.91	0.9993	3.3	9.7
	FF	10.6–400.0	y = 753.98x – 193.97	0.9999	4.7	10.6
	FFA	4.6–400.0	y = 4824.6x – 229.06	0.9994	1.8	4.6

Table 4. Recoveries and precisions of TAP, FF and FFA from spiked blank hen eggs, duck eggs and goose eggs.

Matrix	Analyte	Spiking Level (μg/kg)	Recovery (%) (n = 6)	RSD (%) (n = 6)	Intraday RSD (%) (n = 6)	Interday RSD (%) (n = 18)
Hen eggs	TAP	9.7	85.6 ± 1.8	2.1	2.6	3.6
		25	90.5 ± 2.7	3.0	3.7	5.0
		50 $^{\alpha}$	92.7 ± 1.5	1.6	2.5	3.1
		100	91.5 ± 2.0	2.2	2.5	2.9
	FF	10.5	84.9 ± 3.4	4.0	4.7	5.1
		50	90.2 ± 2.5	2.8	3.8	4.8
		100 $^{\alpha}$	93.5 ± 2.6	2.8	4.0	4.3
		200	94.9 ± 2.7	2.8	2.3	3.4
	FFA	4.3	86.7 ± 3.6	4.2	3.9	5.3
		50	91.5 ± 1.8	2.0	3.1	3.6
		100 $^{\alpha}$	96.7 ± 3.5	3.6	3.2	2.8
		200	98.0 ± 1.8	1.8	2.4	2.7
Duck eggs	TAP	9.9	84.8 ± 2.0	2.4	3.9	4.2
		25	93.8 ± 1.8	1.9	3.6	4.6
		50 $^{\alpha}$	94.5 ± 2.4	2.5	1.9	2.6
		100	92.8 ± 1.1	1.2	2.4	2.4
	FF	11.7	85.1 ± 1.6	1.9	1.2	2.4
		50	89.4 ± 2.0	2.2	1.8	1.8
		100 $^{\alpha}$	94.8 ± 1.5	1.6	2.5	3.6
		200	94.5 ± 3.3	3.5	2.6	4.1
	FFA	4.7	87.5 ± 1.9	2.2	3.1	3.7
		50	96.5 ± 2.2	2.3	3.1	3.4
		100 $^{\alpha}$	96.9 ± 1.7	1.8	2.7	3.6
		200	96.1 ± 1.7	1.8	3.1	3.3
Goose eggs	TAP	9.8	85.5 ± 1.7	2.0	2.5	2.9
		25	93.3 ± 2.1	2.3	3.5	4.0
		50 $^{\alpha}$	93.1 ± 2.9	3.1	4.3	5.4
		100	94.5 ± 4.0	4.2	3.9	4.0
	FF	11.2	80.7 ± 3.5	4.3	5.2	6.2
		50	93.9 ± 2.2	2.3	2.7	3.1
		100 $^{\alpha}$	95.2 ± 1.5	1.6	2.2	3.0
		200	95.2 ± 2.5	2.6	3.1	3.4
	FFA	4.7	83.9 ± 2.8	3.3	3.5	4.1
		50	93.7 ± 2.5	2.7	3.2	3.8
		100 $^{\alpha}$	94.5 ± 2.0	2.1	3.2	3.5
		200	96.1 ± 2.6	2.7	3.3	4.4

Note: α. Maximum Residue Limits.

Under the optimized UPLC-FLD conditions, the retention times of TAP, FF and FFA from different poultry eggs were 1.50, 2.10 and 3.80 min, respectively. The peak shapes were good, and the blank samples had no interference peaks around these retention times. Taking hen egg samples as an example, the chromatograms of the standards, blank hen egg samples and blank hen egg samples spiked with standards are shown in Figures 4–6.

Table 5. Recoveries and precisions of TAP, FF and FFA from spiked blank pigeon eggs and quail eggs.

Matrix	Analyte	Spiking Level (μg/kg)	Recovery (%) (n = 6)	RSD (%) (n = 6)	Intraday RSD (%) (n = 6)	Interday RSD (%) (n = 18)
Pigeon eggs	TAP	9.9	80.1 ± 2.5	3.1	2.5	3.6
		25	93.6 ± 3.3	3.5	4.7	5.4
		50 $^{\alpha}$	92.3 ± 2.4	2.6	2.7	3.2
		100	94.4 ± 3.3	3.5	2.2	4.3
	FF	11.2	84.4 ± 2.2	2.6	3.2	3.3
		50	95.9 ± 2.5	2.6	4.4	4.2
		100 $^{\alpha}$	97.9 ± 4.1	4.2	3.3	5.0
		200	98.6 ± 2.8	2.8	3.1	3.1
	FFA	4.8	85.7 ± 2.0	2.3	3.8	4.1
		50	94.0 ± 2.3	2.4	2.9	3.2
		100 $^{\alpha}$	95.7 ± 3.2	3.3	4.2	5.4
		200	97.7 ± 1.8	1.8	2.3	3.7
Quail eggs	TAP	9.7	84.1 ± 3.0	3.6	3.9	4.0
		25	95.4 ± 2.0	2.1	3.3	3.9
		50 $^{\alpha}$	93.4 ± 2.8	3.0	3.9	4.4
		100	96.3 ± 2.2	2.3	3.7	5.4
	FF	10.6	86.5 ± 3.5	4.0	5.5	5.5
		50	94.9 ± 2.1	2.2	2.7	3.8
		100 $^{\alpha}$	95.9 ± 3.1	3.2	4.0	4.6
		200	96.7 ± 2.3	2.4	3.2	3.4
	FFA	4.6	87.5 ± 2.7	3.1	3.7	4.1
		50	96.0 ± 3.0	3.1	4.7	5.6
		100 $^{\alpha}$	95.5 ± 3.4	3.6	3.5	5.5
		200	96.2 ± 3.6	3.7	5.4	6.6

Note: α. Maximum Residue Limits.

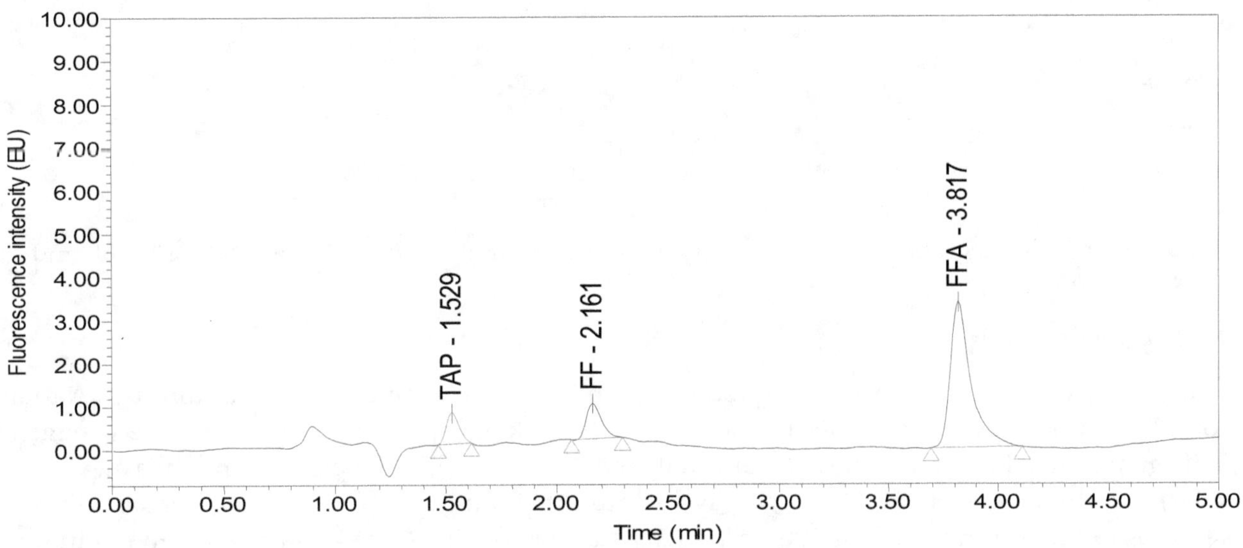

Figure 4. Chromatogram of the standards (25 μg/kg TAP and 50 μg/kg FF and FFA standards).

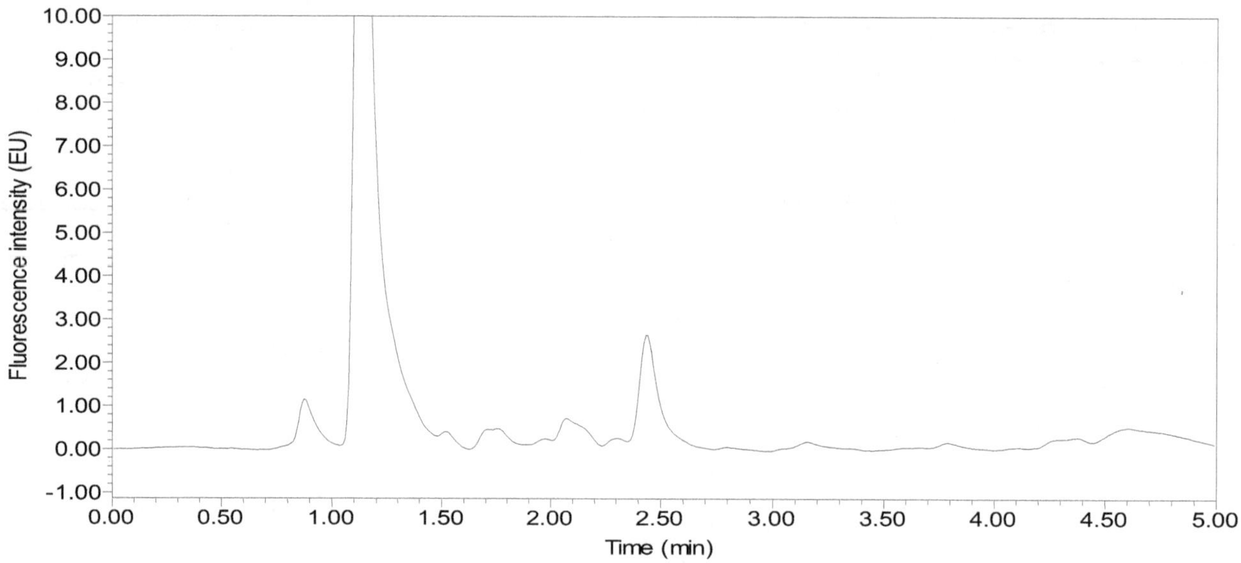

Figure 5. Chromatogram of the blank hen egg.

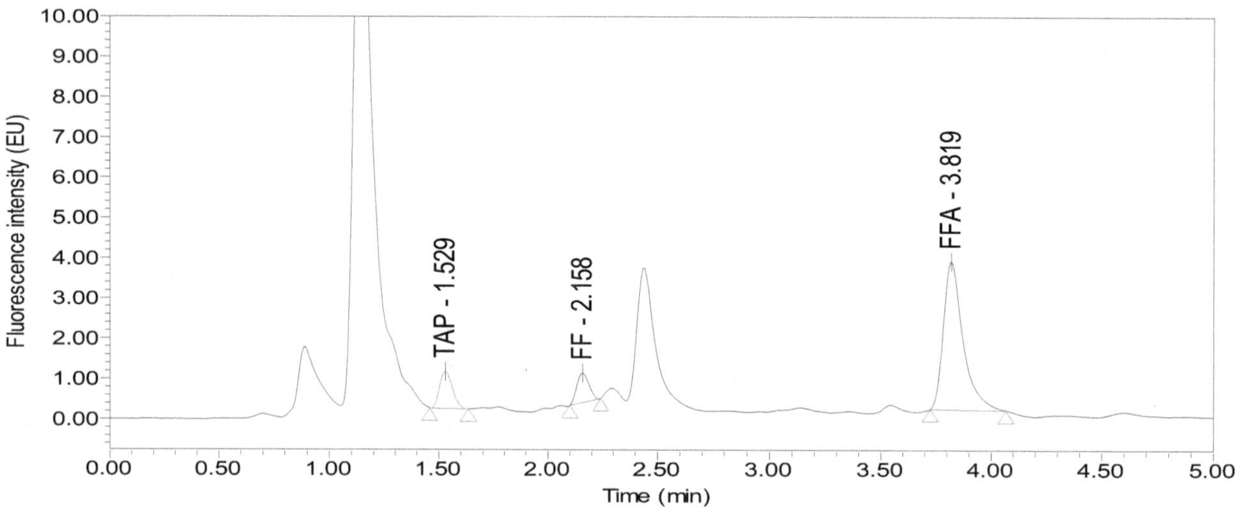

Figure 6. Chromatogram of the blank hen egg spiked with 25 μg/kg TAP and 50 μg/kg FF and FFA standards.

2.6. Real Sample Analysis

Our newly developed detection method was applied to the evaluation of real samples. A total of 150 commercial poultry egg samples (30 hen eggs, 30 duck eggs, 30 goose eggs, 30 pigeon eggs and 30 quail eggs) from a local supermarket were analyzed by the described method. Only hen eggs (34 and 20 μg/kg) and duck eggs (44 and 18 μg/kg) were found to contain FF and FFA residues, and none of the samples exceeded the MRL of 100 μg/kg (EU standard). From these data, we can evaluate the applicability and reliability of the newly developed method. Thus, this new UPLC-FLD method can be applied to the quantification of these drugs in poultry egg samples.

3. Materials and Methods

3.1. Chemicals and Reagents

TAP (99.0% purity), FF (99.0% purity) and FFA (99.8% purity) standards were obtained from Dr. Ehrenstorfer GmbH (Augsburg, Germany). HPLC-grade acetonitrile and triethylamine were purchased from EMD Millipore Company Inc. (Billerica, MA, USA) and Tedia Company Inc. (Fairfield,

OH, USA), respectively. Other reagents were of analytical grade and were supplied by Sinopharm Chemical Reagent Co. Ltd. (Shanghai, China).

3.2. *Standard and Working Solutions*

Stock solutions of TAP and FF at concentrations of 400.0 mg/L and of FFA at a concentration of 100.0 mg/L were prepared by dissolving TAP, FF, and FFA (initially dissolved in 1 mL of ultrapure water), respectively, in acetonitrile. Working standard solutions of TAP, FF, and FFA at different concentrations were prepared by diluting the stock solutions with acetonitrile-water (36:64, *v/v*). The stock solutions were stable for five months at −70 °C. Fresh working solutions were prepared by dilution of the stock solution before use.

3.3. *UPLC-FLD Instrumentation and Conditions*

A Waters ACQUITY UPLC System and a Waters fluorescence detector (Waters Corp., Milford, MA, USA) were used. The separation was achieved on an ACQUITY UPLC® BEH C_{18} (2.1 mm × 100 mm, 1.7 μm) chromatographic column. The column temperature was maintained at 30 °C. The injection volume was 10 μL. The analysis was carried out using acetonitrile and 0.005 mol/L NaH_2PO_4 solution containing 0.003 mol/L sodium dodecyl sulfate and 0.05% triethylamine with pH (FE20, METTLER TOLEDO, Shanghai, China) adjusted to 5.3 ± 0.1 with 85% phosphoric acid as the mobile phase (36:64, *v/v*) at a flow rate of 0.2 mL/min, and the excitation and emission wavelengths were 233 nm and 284 nm, respectively.

3.4. *Sample Preparation*

3.4.1. ASE Extraction

Homogeneous blank poultry egg samples (2.0 ± 0.02 g) were accurately weighed in a mortar and then ground with the appropriate amount of diatomite. The samples were then transferred into a 22 mL extraction tank, placed on the mechanical arm of the accelerated solvent extractor for extraction, and the extractant was acetonitrile:ammonia (98:2, *v/v*). The extraction method and parameters were as follows: extraction pressure of 1500 psi, extraction temperature of 80 °C, extraction time of 3 min, solvent flush of approximately 40% of the tank volume, and two cycles of static extraction. Finally, the extract was collected and left to stand.

3.4.2. Ultrasonic Extraction

Homogeneous blank poultry egg samples (2.0 ± 0.02 g) were accurately weighed into 50 mL polypropylene centrifuge tubes, and 1 mL of 30% acetonitrile solution was added. The solution was vortexed, and then 10 mL of acetonitrile:ammonia (98:2, *v/v*) was added. After ultrasonic extraction for 15 min via an ultrasonic cleaning machine (P300H Elma, Konstanz, Germany), the solution was centrifuged at 8000× *g* for 10 min in a desktop high-speed refrigerated centrifuge (5810R, Eppendorf, Hamburg, Germany), and the supernatant was transferred to a 50 mL propylene centrifuge tube. The extraction process was repeated a second time, and the two extracts were combined for further analysis.

3.4.3. Vortex Shock Extraction

Homogeneous blank poultry egg samples (2.0 ± 0.02 g) were accurately weighed into 50 mL polypropylene centrifuge tubes, and 1 mL of 30% acetonitrile solution was added. The solution was vortexed on a vortex oscillator (G560E, Scientific Industries Ltd., Bohemia, New York, NY, USA), 10 mL of acetonitrile:ammonia (98:2, *v/v*) was added, and the sample was vortexed for 2 min. The solution was centrifuged at 8000× *g* for 10 min and then transferred to 50 mL polypropylene centrifuge tubes. The extraction process was repeated a second time, and the two extracts were combined for further analysis.

3.4.4. Vortex Oscillating + Ultrasonic Extraction

Homogeneous blank poultry egg samples (2.0 ± 0.02 g) were accurately weighed into 50 mL polypropylene centrifuge tubes, and 1 mL of 30% acetonitrile solution was added. The samples were vortex mixed, 10 mL of acetonitrile:ammonia (98:2, *v/v*) was added, and they were vortex mixed for an additional 2 min. Then, they were subjected to ultrasonication for 15 min. After centrifugation at 8000× *g* for 10 min, the supernatant was transferred to a 50 mL polypropylene centrifuge tube. The extraction process was repeated a second time, and the two extracts were combined for further analysis.

3.4.5. Sample Purification

The collected extract was concentrated to near dryness in a centrifugal concentrator, and the residue was dissolved in 1 mL of acetonitrile and then defatted with acetonitrile-saturated n-hexane (hen, duck, and goose whole egg and yolk with 10 × 10 mL; egg white with 5 mL; and pigeon eggs and quail eggs with 8 × 8 mL). After vortexing for 1 min, the upper layer was left to rest for 5 min. The centrifuge tube was placed into the centrifugal concentrator to concentrate the lower liquid layer to dryness. The dried sample was reconstituted with 2.0 mL of the mobile phase, vortexed for 1 min, and centrifuged at 12,000× *g* for 15 min. The supernatant was passed through a 0.22 μm PVDF needle, and the filtrate was analyzed by UPLC-FLD.

3.5. Method Validation

This optimized method was validated according to the requirements defined by the EU and the FDA [41,42]. The selectivity of the method was estimated by preparing and analyzing 20 blank and spiked samples. The probable interferences from endogenous substances were assessed based on the chromatograms of blank and spiked poultry egg samples. The sensitivity of the method was assessed in terms of the LODs and LOQs. The LODs were defined by the concentration of each of the three analytes in the sample matrix that resulted in a signal-to-noise (S/N) ratio of 3:1. The LOQs were defined as the lowest concentration on the calibration curve of each of the three analytes giving a S/N ratio of 10:1.

The calibration curves were prepared based on the peak areas and the concentrations of the working solutions. A series of working standard solutions at concentrations of LOQ, 20.0, 50.0, 100.0, 150.0, 200.0, and 250.0 μg/kg for TAP; LOQ, 20.0, 50.0, 100.0, 200.0, 300.0, and 400.0 μg/kg for FF; and LOQ, 10.0, 20.0, 50.0, 100.0, 200.0, and 400.0 μg/kg for FFA were prepared by diluting the stock solutions with extract of blank sample matrix, and then these solutions were analyzed by the UPLC-FLD method.

The accuracy and precision of the method were evaluated by determining the recoveries of TAP, FF and FFA in poultry egg samples at concentrations of LOQ, 0.5 MRL, 1.0 MRL and 2.0 MRL. The recovery of the method was calculated by comparing the determined concentrations of samples to their theoretical concentrations.

4. Conclusions

In this study, an ASE method for the extraction of TAP, FF and FFA from poultry eggs was established. This extraction method has high extraction efficiencies and high recoveries (>80.0%), and the entire process is automated. This UPLC-FLD method for the quantitative determination of TAP, FF and FFA is accurate and sensitive. The parameters of this method were shown to meet the requirements of the Chinese Ministry of Agriculture, the EU and the FDA for the detection of veterinary drug residues.

Author Contributions: Conceptualization, K.X.; Data curation, B.W. and X.Z.; Formal analysis, G.Z., T.Z. and G.D. Funding acquisition, B.W., X.X. and K.X. Investigation, X.Z. Methodology, X.X. and K.X. Resources, Z.D. Software, Z.D. and G.D. Validation, Z.D., G.Z. and T.Z. Writing–original draft, B.W., X.X. and X.Z. Writing–review & editing, B.W. and X.X.

References

1. Lesnierowski, G.; Stangierski, J. What's new in chicken egg research and technology for human health promotion? - a review. *Trends Food Sci. Technol.* **2018**, *71*, 46–51. [CrossRef]
2. Ashraf, A.; Rahman, F.A.; Abdullah, N. Poultry feed in Malaysia: an insight into the Halalan Toyyiban issues. In Proceedings of the 3rd International Halal Conference (INHAC 2016), Selangor, Malaysia, 21–22 November 2016; Hashim, N.M., Shariff, N.N.M., Mahamood, S.F., Harun, H.M.F., Shahruddin, M.S., Bhari, A., Eds.; Springer: Singapore, 2018; pp. 511–531.
3. Tao, X.; He, Z.; Cao, X.; Jiang, H.; Li, H. Approaches for the determination of florfenicol and thiamphenicol in pork using a chemiluminescent ELISA. *Anal. Methods* **2015**, *7*, 8386–8392. [CrossRef]
4. Festing, M.F.; Diamanti, P.; Turton, J.A. Strain differences in haematological response to chloramphenicol succinate in mice: implications for toxicological research. *Food Chem. Toxicol.* **2001**, *39*, 375–383. [CrossRef]
5. Hanekamp, J.C.; Bast, A. Antibiotics exposure and health risks: chloramphenicol. *Environ. Toxicol. Pharmacol.* **2015**, *39*, 213–220. [CrossRef]
6. Hu, D.; Han, Z.; Li, C.; Lv, L.; Cheng, Z.; Liu, S. Florfenicol induces more severe hemotoxicity and immunotoxicity than equal doses of chloramphenicol and thiamphenicol in Kunming mice. *Immunopharmacol. Immunotoxicol.* **2016**, *38*, 472–485. [CrossRef]
7. Sadeghi, A.S.; Mohsenzadeh, M.; Abnous, K.; Taghdisi, S.M.; Ramezani, M. Development and characterization of DNA aptamers against florfenicol: fabrication of a sensitive fluorescent aptasensor for specific detection of florfenicol in milk. *Talanta* **2018**, *182*, 193–201. [CrossRef]
8. Sams, R.A. Florfenicol: chemistry and metabolism of a novel-broad-spectrum antibiotic. *Tieraerztliche Umschau* **1995**, *50*, 703–707.
9. Cornejo, J.; Pokrant, E.; Riquelme, R.; Briceno, C.; Maddaleno, A.; Araya-Jordan, C.; Martin, B.S. Single-laboratory validation of an LC-MS/MS method for determining florfenicol (FF) and florfenicol amine (FFA) residues in chicken feathers and application to a residue-depletion study. *Food Addit. Contam. Part A* **2017**, *34*, 469–476. [CrossRef]
10. Yang, F.; Yang, F.; Kong, T.; Wang, G.; Bai, D.; Liu, B. Pharmacokinetics of florfenicol and its metabolite florfenicol amine in crucian carp (*Carassius auratus*) at three temperatures after one single intramuscular injection. *J. Vet. Pharmacol. Ther.* **2018**, *41*, 739–745. [CrossRef]
11. The European Medicines Agency. *Commission Regulation (EU) No. 37/2010 of 22 December 2009 on Pharmacologically Active Substances and Their Classification Regarding Maximum Residue Limits in Foodstuffs of Animal Origin*; The European Commission: Brussels, Belgium, 2010.
12. U.S. Food and Drug Administration. *CFR-Code of Federal Regulations Title 21 Part 556 Tolerances for Residue of New Animal Drugs in Food*; U.S. Food and Drug Administration: Rockville, MD, USA, 2014.
13. Ministry of Agriculture of the People 's Republic of China. *Maxium Residue Level of Veterinary Drugs in Food of Animal Origin. Notice No. 235*; Ministry of Agriculture: Beijing, China, 2002.
14. Xie, K.; Jia, L.; Yao, Y.; Xu, D.; Chen, S.; Xie, X.; Pei, Y.; Bao, W.; Dai, G.; Wang, J.; et al. Simultaneous determination of thiamphenicol, florfenicol and florfenicol amine in eggs by reversed-phase high-performance liquid chromatography with fluorescence detection. *J. Chromatogr. B* **2011**, *879*, 2351–2354. [CrossRef]
15. Huang, S.; Gan, N.; Liu, H.; Zhou, Y.; Chen, Y.; Cao, Y. Simultaneous and specific enrichment of several amphenicol antibiotics residues in food based on novel aptamer functionalized magnetic adsorbents using HPLC-DAD. *J. Chromatogr. B Analyt. Technol. Biomed. Life Sci.* **2017**, *1060*, 247–254. [CrossRef]
16. Yang, J.; Sun, G.; Qian, M.; Huang, L.; Ke, X.; Yang, B. Development of a high-performance liquid chromatography method for the determination of florfenicol in animal feedstuffs. *J. Chromatogr. B Analyt. Technol. Biomed. Life Sci.* **2017**, *1068–1069*, 9–14. [CrossRef]
17. Yang, Q.H.; Xiao-Hui, A.I.; Rong, L.I. Simultaneous determination of chloramphenicol, thiamphnicol, florfenicol and forfenicol-amine in aquatic products by gas chromatographic method with solid phase extraction. *Chin. J. Anal. Lab.* **2015**, *34*, 533–537.
18. Peng, S.W.; Xin, L.J.; Mei, S.S. Determination of 3 chloramphenicols medicine residuesin animal-origin Chinese medicinal materials pheretima and hirudoin the retailer by gas chromatography. *Chin. J. Vet. Drug* **2017**, *51*, 65–69.
19. Chen, D.; Yu, J.; Tao, Y.; Pan, Y.; Xie, S.; Huang, L.; Peng, D.; Wang, X.; Wang, Y.; Liu, Z.; et al. Qualitative screening of veterinary anti-microbial agents in tissues, milk, and eggs of food-producing animals using

liquid chromatography coupled with tandem mass spectrometry. *J. Chromatogr. B. Analyt. Technol. Biomed. Life Sci.* **2016**, *1017–1018*, 82–88. [CrossRef]

20. Imran, M.; Habib, F.E.; Majeed, S.; Tawab, A.; Rauf, W.; Rahman, M.; Umer, M.; Iqbal, M. LC-MS/MS-based determination of chloramphenicol, thiamphenicol, florfenicol and florfenicol amine in poultry meat from the Punjab-Pakistan. *Food Addit. Contam. Part A Chem. Anal. Control. Expo. Risk Assess.* **2018**, *35*, 1530–1542. [CrossRef]
21. Marques, T.V.; Paschoal, J.A.R.; Barone, R.S.C.; Cyrino, J.E.P.; Rath, S. Depletion study and estimation of withdrawal periods for florfenicol and florfenicol amine in pacu (*Piaractus mesopotamicus*). *Aquac. Res.* **2018**, *49*, 111–119. [CrossRef]
22. Xie, X.; Wang, B.; Pang, M.; Zhao, X.; Xie, K.; Zhang, Y.; Wang, Y.; Guo, Y.; Liu, C.; Bu, X.; et al. Quantitative analysis of chloramphenicol, thiamphenicol, florfenicol and florfenicol amine in eggs via liquid chromatography-electrospray ionization tandem mass spectrometry. *Food Chem.* **2018**, *269*, 542–548. [CrossRef]
23. Shen, J.; Xia, X.; Jiang, H.; Li, C.; Li, J.; Li, X.; Ding, S. Determination of chloramphenicol, thiamphenicol, florfenicol, and florfenicol amine in poultry and porcine muscle and liver by gas chromatography-negative chemical ionization mass spectrometry. *J. Chromatogr. B. Analyt. Technol. Biomed. Life Sci.* **2009**, *877*, 1523–1529. [CrossRef]
24. Azzouz, A.; Ballesteros, E. Multiresidue method for the determination of pharmacologically active substances in egg and honey using a continuous solid-phase extraction system and gas chromatography-mass spectrometry. *Food Chem.* **2015**, *178*, 63–69. [CrossRef]
25. Amorello, D.; Barreca, S.; Gambacurta, S.; Gulotta, M.G.; Orecchio, S.; Pace, A. An analytical method for monitoring micro-traces of landfill leachate in groundwater using fluorescence excitation–emission matrix spectroscopy. *Anal. Methods* **2016**, *8*, 3475–3480. [CrossRef]
26. Fedeniuk, R.W.; Mizuno, M.; Neiser, C.; O'Byrne, C. Development of LC-MS/MS methodology for the detection/determination and confirmation of chloramphenicol, chloramphenicol 3-O-beta-d-glucuronide, florfenicol, florfenicol amine and thiamphenicol residues in bovine, equine and porcine liver. *J. Chromatogr. B. Analyt. Technol. Biomed. Life Sci.* **2015**, *991*, 68–78. [CrossRef]
27. Shin, D.; Kang, H.-S.; Jeong, J.; Kim, J.; Choe, W.J.; Lee, K.S.; Rhee, G.-S. Multi-residue determination of veterinary drugs in fishery products using liquid chromatography-tandem mass spectrometry. *Food Anal. Methods* **2018**, *11*, 1815–1831. [CrossRef]
28. Moretti, S.; Dusi, G.; Giusepponi, D.; Pellicciotti, S.; Rossi, R.; Saluti, G.; Cruciani, G.; Galarini, R. Screening and confirmatory method for multiclass determination of 62 antibiotics in meat. *J. Chromatogr. A* **2016**, *1429*, 175–188. [CrossRef]
29. Wang, K.; Lin, K.; Huang, X.; Chen, M. A simple and fast extraction method for the determination of multiclass antibiotics in eggs using LC-MS/MS. *J. Agric. Food Chem.* **2017**, *65*, 5064–5073. [CrossRef]
30. Orlando, E.A.; Costa Roque, A.G.; Losekann, M.E.; Simionato, A.V.C. UPLC-MS/MS determination of florfenicol and florfenicol amine antimicrobial residues in tilapia muscle. *J. Chromatogr. B Analyt. Technol. Biomed. Life Sci.* **2016**, *1035*, 8–15. [CrossRef]
31. Barreto, F.; Ribeiro, C.; Barcellos Hoff, R.; Costa, T.D. Determination of chloramphenicol, thiamphenicol, florfenicol and florfenicol amine in poultry, swine, bovine and fish by liquid chromatography-tandem mass spectrometry. *J. Chromatogr. A.* **2016**, *1449*, 48–53. [CrossRef]
32. Feng, J.B.; Huang, D.R.; Zhong, M.; Liu, P.; Dong, J.D. Pharmacokinetics of florfenicol and behaviour of its metabolite florfenicol amine in orange-spotted grouper (*Epinephelus coioides*) after oral administration. *J. Fish. Dis.* **2016**, *39*, 833–843. [CrossRef]
33. Yang, F.; Fang, Y.; Liu, Z.C. Determination of thiamphenicol, florfenicol and florfenicol amine residues in aquatic products by HPLC with fluorescence detection. *Chin. J. Vet. Drug* **2008**, *42*, 14–16.
34. Karami-Osboo, R.; Miri, R.; Javidnia, K.; Kobarfard, F. Simultaneous chloramphenicol and florfenicol determination by a validated DLLME-HPLC-UV method in pasteurized milk. *Iran. J. Pharm. Res.* **2016**, *15*, 361–368.
35. Zhai, Z.Y.; Xu, Y. Determination of thiamphenicol, florfenicol and florfenicol amine in lyophilized royal jelly powder by HPLC. *J. Prev. Med. Inf.* **2010**, *26*, 82–84.
36. Xie, K.Z.; Xu, D.; Chen, S.Q.; Xie, X.; Jia, L.F.; Huang, Y.P.; Guo, H.S.; Wang, J.Y.; Liu, Z.P. Simultaneous determination of residues of thiamphenicol forfenicol and forfenicol amine in chicken muscle by HPLC with

fluorescence detection. *Chin. J. Anal. Lab.* **2011**, *30*, 31–35.
37. Alechaga, E.; Moyano, E.; Galceran, M.T. Ultra-high performance liquid chromatography-tandem mass spectrometry for the analysis of phenicol drugs and florfenicol-amine in foods. *Analyst* **2012**, *137*, 2486–2494. [CrossRef]
38. Lehotay, S.J.; Lightfield, A.R. Simultaneous analysis of aminoglycosides with many other classes of drug residues in bovine tissues by ultrahigh-performance liquid chromatography-tandem mass spectrometry using an ion-pairing reagent added to final extracts. *Anal. Bioanal. Chem.* **2018**, *410*, 1095–1109. [CrossRef]
39. Yang, H.S.; Meng, Y.Z.; Meiqin, W.G. Simultaneous UHPLC-MS/MS determination of chloramphenicol and florfenicol in aquatic products with accelerated solvent extraction. *Phy. Chem. Test. Chem.* **2012**, *48*, 1353–1356.
40. Xiao, Z.; Song, R.; Rao, Z.; Wei, S.; Jia, Z.; Suo, D.; Fan, X. Development of a subcritical water extraction approach for trace analysis of chloramphenicol, thiamphenicol, florfenicol, and florfenicol amine in poultry tissues. *J. Chromatogr. A* **2015**, *1418*, 29–35. [CrossRef] [PubMed]
41. US Department of Health and Human Services, Food and Drug Administration, Center for Drug Evaluation and Research, Center for Veterinary Medicine. *Guidance for Industry: Bioanalytical Method Validation*; US Department of Health and Human Services: Washington, DC, USA, 2001.
42. The European Communities. Commission decision 2002/657/EC of 12 august 2002 implementing council directive 96/23/EC concerning the performance of analytical methods and the interpretation of results. *Off. J. Eur. Commun.* **2002**, *221*, 8–36.

3

Transfer of a Multiclass Method for over 60 Antibiotics in Food from High Resolution to Low Resolution Mass Spectrometry

Danilo Giusepponi [1], Fabiola Paoletti [1], Carolina Barola [1], Simone Moretti [1], Giorgio Saluti [1], Federica Ianni [2], Roccaldo Sardella [2] and Roberta Galarini [1,*]

1 Istituto Zooprofilattico Sperimentale dell'Umbria e delle Marche "Togo Rosati", 06126 Perugia, Italy
2 Department of Pharmaceutical Sciences, University of Perugia, 06123 Perugia, Italy
* Correspondence: r.galarini@izsum.it

Academic Editors: Alessandra Gentili and Chiara Fanali

Abstract: A multiclass method has been developed to screen and confirm a wide range of anti-microbial residues in muscle and milk, and validated using liquid-chromatography coupled to (low-resolution, LR) tandem mass spectrometry (LC-QqQ). Over sixty antibiotics, belonging to ten distinct families, were included in the method scope. The development process was rapidly concluded as a result of two previously implemented methods. This consisted of identical sample treatments, followed by liquid chromatography, and coupled with high-resolution (HR) mass spectrometry (LC-Q-Orbitrap). The validation study was performed in the range between 10–1500 $\mu g{\cdot}kg^{-1}$ for muscles and 2–333 $\mu g{\cdot}kg^{-1}$ for milk. The main performance characteristics were estimated and, then, compared to those previously obtained with HR technique. The validity of the method transfer was ascertained also through inter-laboratory studies.

Keywords: antibiotics; liquid chromatography mass spectrometry; milk; muscle; validation

---

1. Introduction

Antibiotics are widely used in livestock breeding to treat several diseases that appear in all the food producing animal species. To guarantee public health protection, the European Union requires member states to implement yearly monitoring plans to control the presence of antibiotic residues in food. Therefore, surveillance should be aimed particularly at controlling compliance with the maximum residue limits (MRLs), fixed in Table 1 of the Annex of Regulation (EC) No 37/2010 [1]. For several antibiotics, MRLs have been set in various matrices, such as eggs, fat, honey, kidney, liver, milk, and muscles and still, today, new MRLs are being fixed. In the early 2000s, the liquid chromatography coupled to tandem mass spectrometry technique (LC-QqQ) became essential in the routine analysis of single class of veterinary drug residues in food. Indeed triple quadrupole mass spectrometry analyzers were able to assure both greater sensitivity and selectivity than the traditional LC detectors, based on UV-Vis and fluorescence spectroscopy. In addition, for some important classes, such as aminoglycosides or avermectins, the need of a derivation step could be avoided. In the last ten years, the improvement of LC-QqQ systems allowed the realization of a further step in drug residue analysis, introducing procedures that are able to determine simultaneously more than one drug class [2–4]. As consequence, a remarkable effort has been made to progressively replace single-class with multiclass protocols, since this is a cost-effective way to improve the current residue control programs, thereby ensuring the determination of a wide number of compounds, with only few methods. Reviewing the main relevant published papers, some research groups recurred (Table 1). Among the control laboratories, the Official Food Control Authority of Zurich (Zurich, Switzerland), the RIKILT (Wageningen, Netherlands), the

European Union Reference Laboratory for Antimicrobial Residues in Food (EURL, Fougères, France), the National Institute for Agrarian and Veterinary Research (INIAV, Vila do Conde, Portugal), the Istituto Zooprofilattico Sperimentale dell'Umbria e delle Marche (IZSUM, Perugia, Italy), the Canadian Food Inspection Agency (Calgary, Canada), the Residue Analysis Laboratory of Laboratório Nacional Agropecuário (LANAGRO, Porto Alegre, Brazil), and the US Department of Agriculture (USDA, Wyndmoor, PA, USA) are mentioning.

Table 1. Overview of multiclass methods for the determination of veterinary drug residues in tissues and milk.

	N° of Veterinary Drugs	Matrix	Equipment	Reference	Laboratory/Centre [a]
1	18	Milk	LC-QqQ	Aguilera-Luiz et al. 2008 [5]	Almeria University (Spain)
2	39	Chicken muscle	LC-QqQ	Chico et al. 2008 [6]	Barcelona University (Spain)
3	>100	Muscle	LC-TOF	Kaufmann et al. 2008 [7]	OFCA-Zurich (Switerland)
4	>100	Milk	LC-TOF	Stolker et al. 2008 [8]	RIKILT (The Netherlands)
5	Ca 100	Meat and other food	LC-TOF	Peters et al. 2009 [9]	RIKILT (The Netherlands)
6	Ca 26	Animal tissues	LC-QqQ	Stubbings et al. 2009 [10]	FERA (UK)
7	58	Milk	LC-QqQ	Gaugain-Juhel et al. 2009 [11]	EURL (France)
8	21	Milk	LC-QqQ	Martinez-Vidal et al. 2010 [12]	Almeria University (Spain)
9	30	Milk	LC-Orbitrap, LC-Q-TOF, LC-QqQ	Romero-González et al. 2011 [5]	Almeria University (Spain)
10	>100	Meat and other food	LC-Orbitrap	Kaufmann et al. 2011 [13]	OFCA-Zurich (Switzerland)
11	>60	Meat	LC-LTQ-Orbitrap	Hurtaud-Pessel et al. 2011 [14]	EURL (France)
12	59	Milk and honey	LC-Q-TOF	Wang et al. 2012 [15]	CFIA-Calgary (Canada)
13	21	Meat	LC-QqQ	Bittencourt et al. 2012 [16]	LANAGRO (Brazil)
14	24	Milk and liver	LC-QqQ	Martins et al. 2014 [17]	LANAGRO (Brazil)
15	>100	Milk	LC-Q-Orbitrap	Kaufmann et al. 2014 [18]	OFCA-Zurich (Switzerland)
16	39	Liver	LC-QqQ	Freitas et al. 2015 [19]	INIAV (Portugal)
17	23	Liver	LC-QqQ	Martins et al. 2015 [20]	LANAGRO (Brazil)
18	>100	Milk	LC-Q-Orbitrap	Wang et al. 2015 [21]	CFIA-Calgary (Canada)
19	>100	Various food	LC-Q-TOF	Dasenaki et al. 2015 [22]	University of Athens (Greece)
20	76	Bovine muscle	LC-QqQ	Dasenaki et al. 2016 [23]	University of Athens (Greece)
21	62	Animal muscle	LC-Q-Orbitrap	Moretti et al. 2016 [24]	IZSUM (Italy)
22	62	Milk	LC-Q-Orbitrap	Moretti et al. 2016 [25]	IZSUM (Italy)
23	>120	Animal tissues	LC-QqQ/LC-Q-TOF	Anumol et al. 2017 [26]	USDA (USA)
24	174	Bovine tissues	LC-QqQ	Lehotay et al. 2018 [27]	USDA (USA)
25	44	Salmon	LC-Q-TOF	Gaspar et al. 2019 [28]	INIAV (Portugal)

[a] OFCA = Official Food Control Authority; CFIA:Canada Food Inspection Agency; FERA: The Food and Environment Research Agency; LANAGRO: Laboratório Nacional Agropecuário; INIAV: Instituto Nacional de Investigação Agrária e Veterinária; USDA: United States Department of Agriculture; IZSUM: Istituto Zooprofilattico Sperimentale dell'Umbria e delle Marche.

The universities of Barcelona (Spain), Almeria (Spain), and Athens (Greece) have been the most active in this analytical field. LC-QqQ techniques are the most consolidated and most common multiclass procedures for veterinary drugs. These techniques have been mainly developed using this type of equipment [5,6,10–12,16,17,19,20,23,27]. In 2008–2009, the Official Food Control Authority of Zurich and the Dutch RIKILT Institute proposed, for the first time, the application of high-resolution (HR) mass spectrometry, based on time-of-flight (TOF) technology [7–9]. About three years later, the same Laboratory of Zurich, and the research group of Almeria University developed multiclass procedures for veterinary drugs, respectively, in meat, and milk, using LC-Orbitrap technique, a new MS analyzer, that was commercialized in 2005 [13,29]. Later, the introduction of benchtop hybrid high-resolution mass spectrometers (mainly, Q-TOF and Q-Orbitrap) produced further advantages in terms of selectivity and accuracy and, accordingly, these kinds of equipment has been more commonly applied (Table 1) [14,15,18,21,22,24–26,28,29].

Based on all the above, multiclass methods are no longer innovative procedures, and there is interest in their wide diffusion. The possibility of easy implementation and sustainable daily management, independent from the available LC-MS equipment. The aim of this work was to discuss the transfer of previously developed multiclass methods for more than sixty antibiotics in meat and milk from an LC-Q-Orbitrap platform to an LC-QqQ one [24,25]. The performance characteristics of the new LC-QqQ methods were estimated by means of full validation studies carried out according to European Commission Decision 2002/657/EC [30]. Finally, a comparison between the two techniques was carried out in the light of their cost-effectiveness in routine analysis of veterinary drug residues.

2. Results and Discussion

2.1. Optimization of LC-MS/MS Conditions

The choice of analytes has been carried out using the most administered antibiotics in farm. Only the classes of aminoglycosides and colistins were excluded, as their high polarity hampers the chromatographic retention, based on the reversed-phase mechanism (C18 column). On the other hand, the addition of ion-pairing agents on the mobile phase produced remarkable ion suppression, with detrimental effects on all the other analytes [24]. The chromatographic conditions were optimized starting from the parameters set for the LC-Q-Orbitrap methods. In order to profitably increase analyte retention, the percentage of methanol (eluent B) was reduced from 5% down to 2% (by volume). According to a typical reversed-phase mechanism, this change allowed us to obtain retention times of about 0.5 min higher than the initial tested conditions (Figure S1).

The MS conditions were established without the infusion of the individual solutions of analytes, but by setting the transitions on the basis of the ion fragments previously studied [24]. As shown in Table 2, apart few exceptions, such as some beta-lactams ($[M + Na]^+$), sulfanilamide ($[M + H - NH_3]^+$), spiramycin, neospiramycin, cefquinome, tildipirosin, tilmicosin, tulathromycin marker, and tulathromycin ($[M + 2H]^{++}$), the selected precursor ion species were generally the protonated molecular ions ($[M + H]^+$). For macrolides, it is not uncommon for the choice of bi-charged ions to be used as a precursor, due to their favorable abundance among the formed charged species [31]. The sample preparation was exactly the same as that previously optimized by Moretti et al. [24,25]; however, two internal standards (ISs) were replaced, in order to either, decrease costs (metacycline instead of tetracycline-d6), or to improve the MS response (ceftiofur-d3 instead of cefadroxil-d4). In this context, the ISs were not used for quantification purposes, but only to perform the internal quality control by checking the success of the analytical operations, during the routine application of the procedure as well as to monitor the run-to-run differences in the retention times [8]. For this purpose, at the beginning of sample treatment, IS were added at 10 $\mu g \cdot kg^{-1}$ and, before the release of the results, the presence (S/N > 3) of all eight compounds must be verified. The analyte quantification was achieved by matrix-matched curves (external standardization), which corrected the concentration for the relevant recovery factor [32]. The LC-QqQ chromatograms of a blank muscle, and of the same spiked at 10 μg·kg^{-1}, are reported in Figures 1 and 2, respectively. Eight representative analytes are shown, starting from the polar metabolite of florfenicol (florfenicol amine, RT = 3.4 min) to the last eluting compound (rifaximin, RT = 20.7 min). The analogous chromatograms are shown also for milk (Figures 3 and 4).

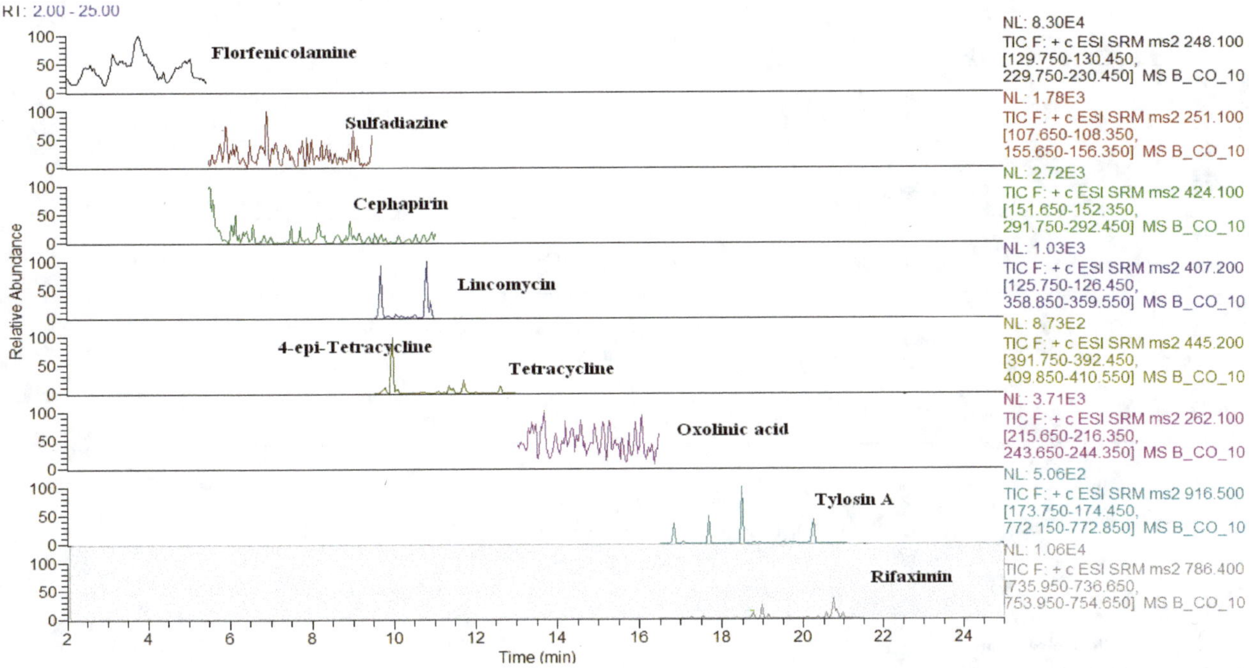

Figure 1. LC-QqQ chromatograms of a blank bovine muscle.

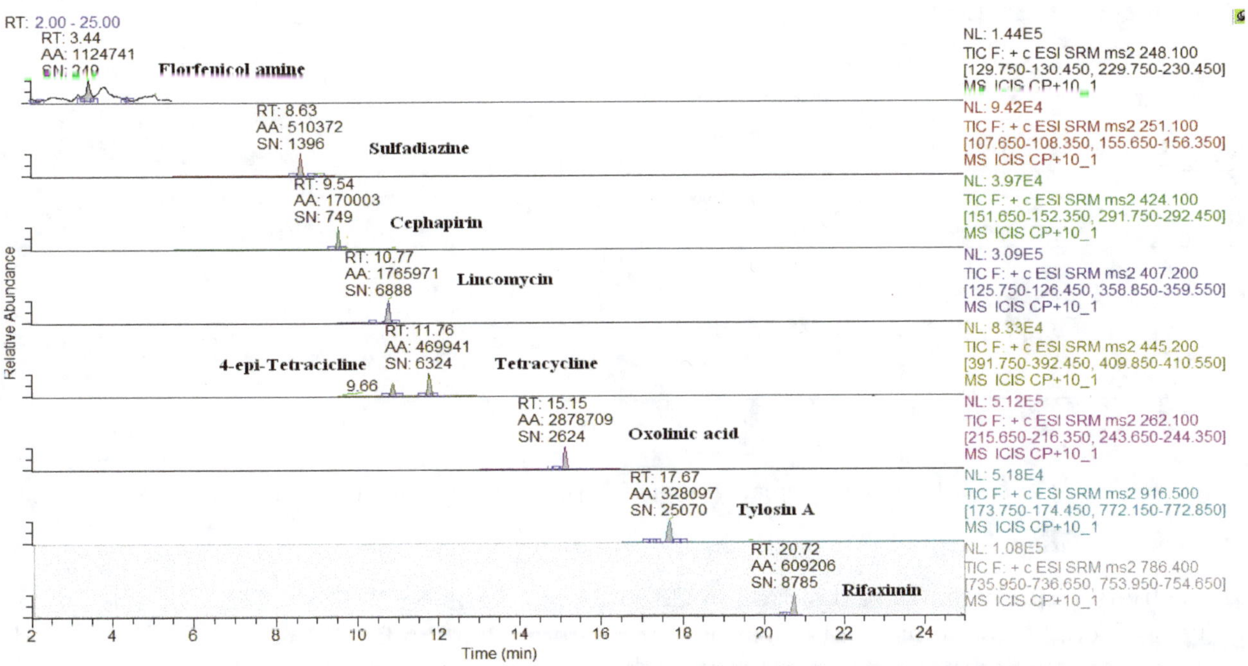

Figure 2. LC-QqQ chromatograms of a spiked bovine muscle (10 μg·kg^{-1}).

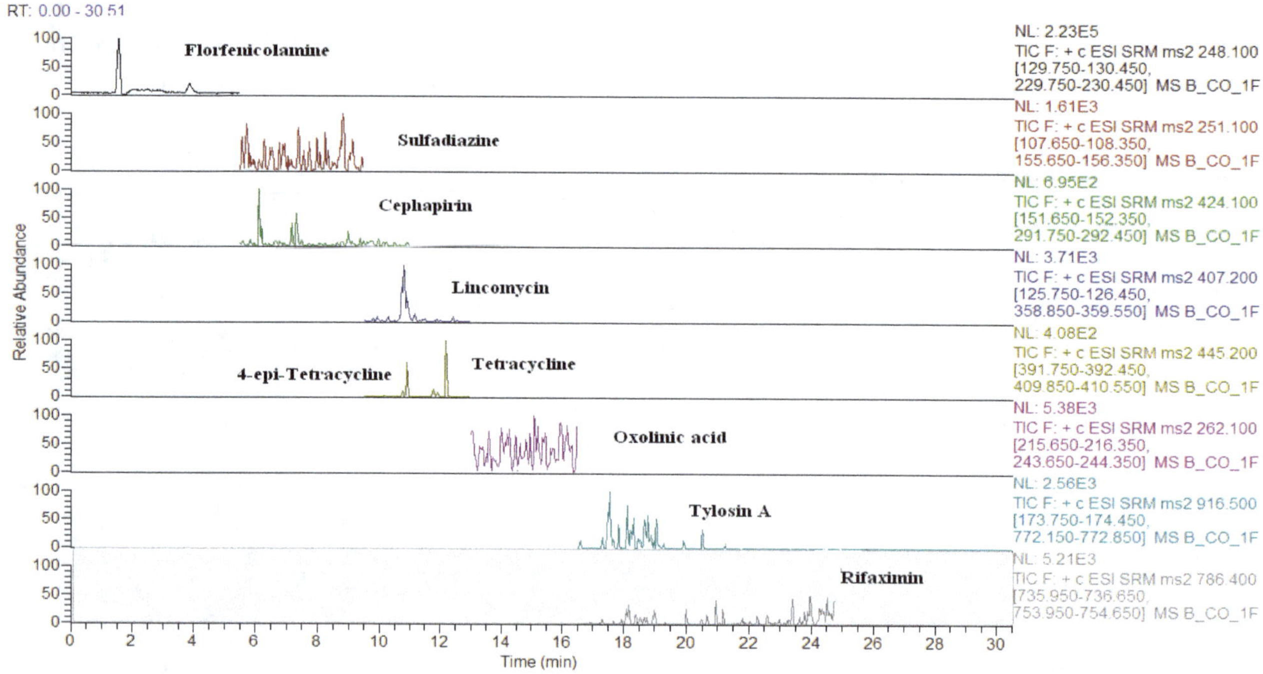

Figure 3. LC-QqQ chromatograms of a blank bovine milk.

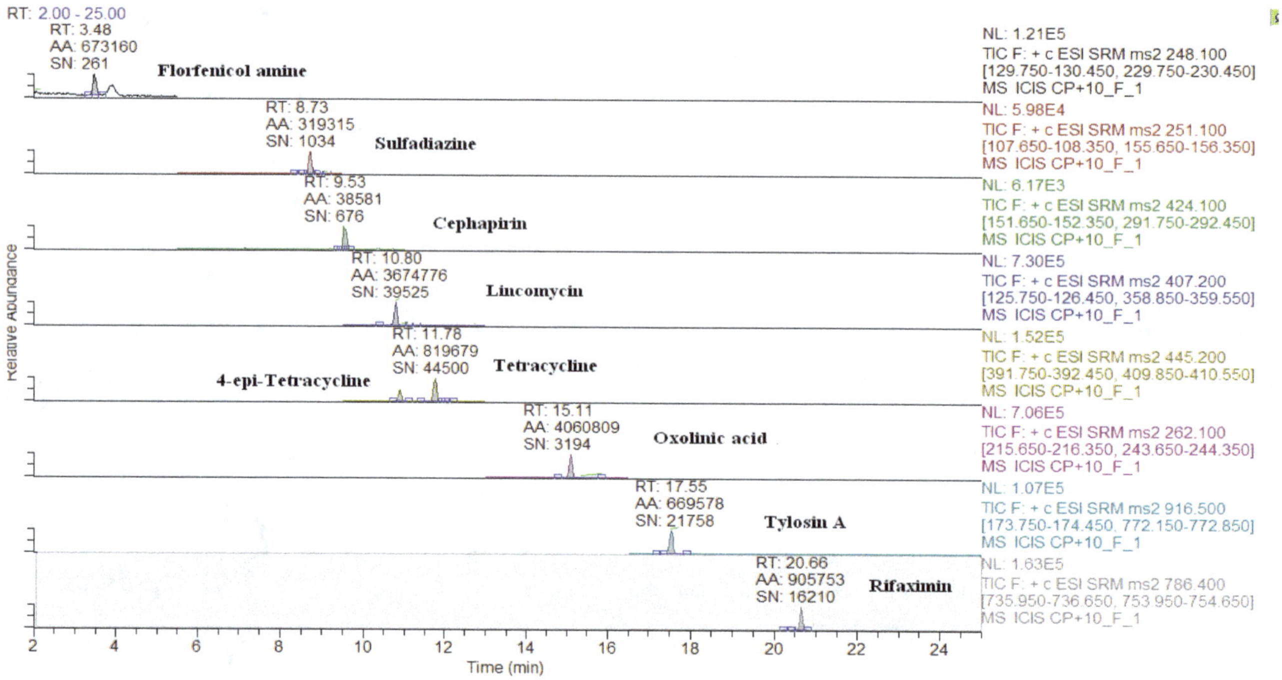

Figure 4. LC-QqQ chromatograms of a spiked bovine milk (10 μg·kg^{-1}).

2.2. Method Validation

Selectivity requirements are reported in Commission Decision 2002/657/EC [30]. The ion ratio of the two selected transitions (Table 2), and their relative retention times (<2.5%), were checked to confirm analyte identification. Linearity in the matrix was evaluated with five-points matrix-matched curves: 2, 10, 33, 100, and 150 μg·kg^{-1}. Therefore, levels higher than 150 μg·kg^{-1} had to be tested, and the final extract was diluted ten-fold or more, as reported in Tables S1 and S2. The linearity data are summarized in Table S3. For several analytes, the first calibration point (2 μg·kg^{-1}) had to be discarded, due to the scarce response. In other more critical cases (e.g., cefacetrile in meat/muscle, tildipirosin and tulathromycin markers in milk) additional points have been removed. Since Commission Decision 657/2002/EC [30] does not furnish precise criteria for evaluating linearity, the "Guidance document on analytical quality control and validation procedures for pesticide residues analysis in food and feed" was followed [33]. The percentage deviation of the back-calculated concentrations ($C_{measured}$) from the true concentrations (C_{true}) was calculated (1):

$$Deviation\ (\%) = \frac{(C_{measured} - C_{true})}{C_{true}} \cdot 100 \quad (1)$$

Table 2. Summary of the selected reactions transitions (SRM) monitored for the sixty-four targeted analytes.

N°	Analyte	Retention Time (min)	Adduct (*m/z*)	Precursor Ion (*m/z*)	Product Ions (*m/z*)	Collision Energy (eV)
1	Sulfaguanidine	2.85	$[M + H]^+$	215.1	92.0 156.0	15 20
2	Florfenicolamine	3.20	$[M + H]^+$	248.1	230.1 130.1	10 30
3	Sulfanilamide	3.30	$[M + H - NH_3]^+$	156.0	92.0 108.1	12 10
	Sulfanilamide-13C6	3.30	$[M + H - NH_3]^+$	162.0	98.1 114.1	13 13
4	Desacetylcephapyrin	6.80	$[M + H]^+$	382.1	152.0 226.0	30 20
5	Amoxicillin	8.30	$[M + H]^+$	366.1	349.1 114.0	10 20
6	Sulfadiazine	8.50	$[M + H]^+$	251.1	108.0 156.0	26 15
7	Sulfathiazole	9.20	$[M + H]^+$	256.0	92.1 156.0	28 15
8	Cephapyrin	9.45	$[M + H]^+$	424.1	292.1 152.0	20 30
9	Sulfapyridine	9.50	$[M + H]^+$	250.1	108.0 156.0	26 17
10	Tildipirosin	9.90	$[M+2H]^{++}$	367.7	281.2 98.1	20 18
11	Sulfamerazine	9.90	$[M + H]^+$	265.1	108.0 156.0	27 17
12	Cefquinome	10.00	$[M + 2H]^{++}$	265.1	134.2 199.1	20 20
13	Cefacetrile	10.15	$[M + Na]^+$	362.0	258.0 302.0	10 10
14	Cefalonium	10.50	$[M + H]^+$	459.1	337.0 152.0	10 20
15	Lincomycin	10.50	$[M + H]^+$	407.2	126.1 359.2	30 10
16	Tulathromycin marker	10.60	$[M + 2H]^{++}$	289.0	158.3 420.5	17 17
17	Thiamphenicol	10.60	$[M + H]^+$	356.0	308.0 229.0	20 20
18	Epitetracycline	10.60	$[M + H]^+$	445.2	410.2 392.1	20 30
19	Trimethoprim	10.70	$[M + H]^+$	291.1	261.1 230.1	30 30

Table 2. *Cont.*

N°	Analyte	Retention Time (min)	Adduct (m/z)	Precursor Ion (m/z)	Product Ions (m/z)	Collision Energy (eV)
20	Marbofloxacin	10.80	$[M + H]^+$	363.1	276.1 320.1	14 14
21	Sulfamethazine	11.10	$[M + H]^+$	279.1	92.1 124.1	31 28
	Sulfamethazine-^{13}C6	11.10	$[M + H]^+$	285.1	186.1	17
22	Epioxytetracycline	11.35	$[M + H]^+$	461.2	426.1 337.1	20 30
23	Norfloxacin[a]	11.50	$[M + H]^+$	320.1	231.2 282.1	39 29
24	Tetracycline	11.50	$[M + H]^+$	445.2	410.2 392.1	20 30
25	Cefalexin	11.70	$[M + H]^+$	348.1	158.0 174.1	10 20
26	Oxytetracycline	11.80	$[M + H]^+$	461.2	426.1 337.1	20 30
27	Ciprofloxacin	11.80	$[M + H]^+$	332.1	245.1 288.1	23 17
28	Enrofloxacin	11.90	$[M + H]^+$	360.2	245.0 316.1	26 19
	Enrofloxacin -d5	11.90	$[M + H]^+$	365.2	321.4	18
29	Tulathromycin	11.90	$[M + 2H]^{++}$	404.0	158.1 116.1	20 20
30	Danofloxacin	11.95	$[M + H]^+$	358.2	283.1 340.1	24 22
31	Cefazolin	12.00	$[M + H]^+$	455.0	323.1 156.0	10 20
32	Sulfamethoxazole	12.10	$[M + H]^+$	254.1	108.1 156.0	28 16
33	Difloxacin	12.30	$[M + H]^+$	400.1	299.1 356.1	28 18
34	Ampicillin	12.30	$[M + H]^+$	350.1	106.1 160.0	20 20
35	Sulfamonomethoxine	12.30	$[M + H]^+$	281.1	108.1 156.0	28 16
36	Florfenicol	12.40	$[M + H]^+$	358.0	241.0 340.0	20 10
	Florfenicol -d3	12.40	$[M + H]^+$	361.0	241.0	16
37	Cefoperazone	12.60	$[M + H]^+$	646.1	530.3 143.1	10 30
38	Sarafloxacin	12.60	$[M + H]^+$	386.1	342.1 299.1	18 26
39	Epichlortetracycline	12.85	$[M + H]^+$	479.1	444.1 154.0	20 30
40	Neospiramycin	13.40	$[M + 2H]^{++}$	350.2	160.1 174.1	10 20
41	Chlortetracycline	13.80	$[M + H]^+$	479.1	441.1 154.0	20 30
42	Spiramycin	14.05	$[M + 2H]^{++}$	422.3	702.4 174.1	10 20
	Spiramycin -d3	14.05	$[M + 2H]^{++}$	423.8	174.0	20
43	Sulfadimethoxine	14.40	$[M + H]^+$	311.1	108.1 156.0	29 21
44	Sulfaquinoxaline	14.80	$[M + H]^+$	301.1	92.1 156.0	30 21

Table 2. *Cont.*

N°	Analyte	Retention Time (min)	Adduct (*m/z*)	Precursor Ion (*m/z*)	Product Ions (*m/z*)	Collision Energy (eV)
45	Oxolinic Acid	15.00	$[M + H]^+$	262.1	216.0 244.0	29 18
46	Ceftiofur	15.10	$[M + H]^+$	524.0	241.0 126.0	20 30
	Ceftiofur-d3	15.10	$[M + H]^+$	527.0	244.1	15
	Metacycline	15.15	$[M + H]^+$	443.1	426.2	16
47	Gamithromycin	15.40	$[M + H]^+$	777.5	158.1 619.7	39 32
48	Tilmicosin	15.70	$[M + 2H]^{++}$	435.3	695.5 174.1	20 30
49	Doxycycline	15.50	$[M + H]^+$	445.2	428.1 321.1	10 35
50	Nalidixic Acid[a]	16.80	$[M + H]^+$	233.1	159.0 187.0	30 25
51	Tiamulin	17.10	$[M + H]^+$	494.3	192.1 119.0	20 30
52	Penicillin G	17.15	$[M + Na]^+$	357.1	198.1 182.0	20 20
	Penicillin G-d7	17.15	$[M + Na]^+$	364.0	205.2	13
53	Flumequine	17.20	$[M + H]^+$	262.1	202.0 244.0	33 19
54	Tylosina A	17.40	$[M + H]^+$	916.5	174.1 772.5	36 28
55	Erythromycin	17.60	$[M + H]^+$	734.5	576.4 158.1	20 30
56	3-O-Acetyltylosin	17.75	$[M + H]^+$	958.5	174.0 772.5	36 28
57	Oxacillin	18.20	$[M + Na]^+$	424.1	265.1 182.0	20 20
58	Penicillin V	18.20	$[M + Na]^+$	373.1	182.0 214.0	20 20
59	Cloxacillin	18.50	$[M + Na]^+$	458.1	299.0 182.0	20 20
60	Valnemulin	19.10	$[M + H]^+$	565.4	263.1 72.1	20 30
61	Dicloxacillin	19.20	$[M + Na]^+$	492.0	333.0 182.0	20 20
62	Nafcillin	19.30	$[M + H]^+$	415.1	171.1 199.1	34 13
63	Tilvalosin	19.45	$[M + H]^+$	1042.6	174.0 814.5	39 30
64	Rifaximin	20.60	$[M + H]^+$	786.4	754.3 736.3	20 30

[a] Acquired only in milk.

For each calibration point, it was verified that its value was not more than ± 20%. As an example, the calibration data of six analytes, belonging to different antibiotic families, are reported in Table S4 and Figures S2 and S3.

The overall recoveries and precisions data are listed in Table 3. For meat, seven validation levels were performed in the range 10–1500 $\mu g \cdot kg^{-1}$, whereas, in milk, five levels were investigated (10–333 $\mu g \cdot kg^{-1}$), with an additional concentration at 2 $\mu g \cdot kg^{-1}$ to check amoxicillin, ampicillin, and penicillin G accuracy at $\frac{1}{2}$ MRL, as required by the Commission Decision 2002/657/EC [30]. Moreover, in milk, two additional molecules were tested with respect to the original group of 62 compounds, that is, the two quinolones, nalidixic acid, and norfloxacin.

Table 3. Precision, recovery and matrix effect (muscle and milk).

Analyte [a,b]	Muscle				Milk			
	$CV_{r,pooled}$ (%)	$CV_{Rw, pooled}$ (%)	Rec (%)	ME [c] (%)	$CV_{r,pooled}$ (%)	$CV_{Rw, pooled}$ (%)	Rec (%)	ME [c] (%)
Sulfaguanidine	5.8	10	83	−14	5.0	14	61	**253**
Florfenicol Amine	3.2	6.1	85	**−22**	2.5	3.8	94	−13
Sulfanilamide	6.6	8.9	74	**−35**	5.8	12	66	**−21**
Desacetylcephapirin	7.6	7.3	76	−1	5.9	6.0	91	**−21**
Amoxicillin	4.5	6.3	64	−1	5.8	5.8	89	−10
Sulfadiazine	5.1	8.2	86	−18	4.8	8.6	74	−16
Sulfathiazole	5.2	8.0	83	−6	5.2	8.3	71	−15
Cephapirin	6.9	7.8	75	1	12	12	94	−5
Sulfapyridine	4.6	7.2	85	13	5.8	9.5	67	−13
Tildipirosin	6.2	10	72	−8	6.0	20	87	5
Cefquinome	6.1	7.2	97	−4	11	19	78	−16
Sulfamerazine	3.8	5.9	89	11	6.7	8.7	70	−13
Cefacetrile	12	13	80	**−26**	22	27	92	−5
Cefalonium	6.7	7.3	78	3	8.7	8.7	93	−6
Lincomycin	4.3	7.2	83	−12	2.3	4.4	92	**30**
Epitetracycline	6.2	9.2	66	**22**	4.7	7.6	96	5
Trimethoprim	3.4	7.8	90	−3	2.7	4.9	94	4
Thiamphenicol	10	11	87	−13	13	13	92	−13
Tulathromycin marker	5.6	8.6	81	−13	8.7	26	81	−18
Marbofloxacin	6.1	7.0	87	**−27**	4.3	4.9	96	−11
Sulfamethazine	4.1	7.2	85	−1	7.8	14	68	−13
Epioxytetracycline	8.8	13	62	**40**	13	15	90	−3
Norfloxacin	-	-	-	-	7.0	8.0	94	3
Tetracycline	6.3	9.4	71	**27**	4.5	5.0	91	**40**
Cefalexin	6.2	8.4	64	1	6.2	6.7	90	−7
Oxytetracycline	6.7	8.1	63	10	4.4	5.2	90	1
Ciprofloxacin	7.5	8.4	81	**−25**	6.1	7.5	95	−10
Enrofloxacin	4.7	6.6	93	−16	4.1	5.9	99	−1
Tulathromycin	8.3	16	69	10	7.5	13	94	5
Danofloxacin	5.7	7.2	90	−17	4.2	5.0	97	3
Cefazolin	8.4	9.0	83	−8	14	14	94	−11
Sulfamethoxazole	7.7	9.7	88	−16	6.7	6.7	84	**−24**
Difloxacin	5.0	7.4	93	−15	3.3	4.2	96	−13
Ampicillin	6.6	8.0	68	−2	9.4	10	88	−10
Sulfamonomethoxine	4.6	7.4	86	0	6.4	9.9	78	−19
Florfenicol	9.8	16	90	−17	13	13	94	**−28**
Cefoperazone	6.8	8.4	85	−13	16	18	103	**−25**
Sarafloxacin	5.8	7.4	86	**−26**	4.8	6.6	99	**−22**
Epichlorotetracycline	9.3	13	71	**65**	7.4	11	100	**22**
Neospiramycin	7.8	16	67	12	7.3	12	86	13
Chlortetracycline	5.8	7.7	69	**47**	5.7	7.2	92	**48**
Spiramycin	8.4	17	74	20	4.5	7.6	91	**21**
Sulfadimethoxine	4.5	8.1	88	−12	3.9	3.9	90	**−29**
Sulfaquinoxaline	6.1	7.7	86	**−26**	4.6	4.8	91	**−38**
Oxolinic Acid	4.9	6.7	97	4	2.6	6.6	96	−3
Ceftiofur	7.5	9.7	72	**−21**	6.7	7.1	95	**−24**
Gamithromycin	5.6	7.1	94	**55**	3.2	4.8	98	**26**
Tilmicosin	6.9	10	88	**49**	3.2	4.4	95	**38**
Doxycycline	7.0	9.5	69	1	6.6	7.4	95	10
Nalidixic Acid	-	-	-	-	3.6	6.0	94	−3
Penicillin G	7.9	9.0	85	**−21**	12	13	92	**37**
Tiamulin	9.3	13	88	**−21**	2.5	4.1	98	−5
Flumequine	4.3	7.3	95	−3	2.9	4.5	95	**52**
Tylosin A	7.7	15	85	**29**	2.5	4.8	96	**54**
Erythromycin A	4.9	8.6	89	**−27**	4.7	12	67	**−21**
3-O-Acetyltylosin	9.3	16	88	**42**	3.7	5.0	95	**53**
Oxacillin	6.0	11	83	**−29**	8.0	8.5	100	1
Penicillin V	7.9	11	85	**−28**	7.5	7.7	99	6

Table 3. *Cont.*

	Muscle				Milk			
Analyte [a,b]	$CV_{r,pooled}$ (%)	$CV_{Rw,\ pooled}$ (%)	Rec (%)	ME [c] (%)	$CV_{r,pooled}$ (%)	$CV_{Rw,\ pooled}$ (%)	Rec (%)	ME [c] (%)
Cloxacillin	8.2	11	84	**−27**	9.8	12	110	**−25**
Valnemulin	19	31	75	**−26**	2.7	5.4	108	**55**
Dicloxacillin	6.7	10	81	**−25**	13	13	99	−10
Nafcillin	5.0	8.2	84	**−7**	5.0	5.4	99	−19
Tylvalosin	9.0	19	93	**50**	5.4	6.9	101	**90**
Rifaximin	9.0	12	91	**33**	4.8	9.6	98	7

[a] Nalidixic acid and norfloxacin were not included in muscle method since these antibiotics were introduced later; [b] For valnemulin (muscle), cefacetrile (milk) and cefquinome (milk) the method can be used only for screening purposes (inadequate accuracy); [c] Values of matrix effect (ME) in bold are considered significant (>|20|%).

In this work, the classical validation scheme (0.5 MRL, MRL and 1.5 MRL), described in Commission Decision 2002/657/EC, was not applied. Paragraph 3.1.3. of the same decision allows the introduction of alternative models [30] and, since the validation studies of multiclass procedures have to consider dozens of MRLs, which can vary also in function of the animal species, the adoption of progressive validation levels, that are equal for all the analytes, was fully justified. On the other hand, when the Commission Decision was issued (2002), the development of multiclass procedures for the control of veterinary drug residues was not initiated. The spiking ranges were chosen, considering, on the one side, the reachable concentrations, and, on the other side, the relevant MRLs [1] in all the couplings analyte/matrix. Preliminary experiments demonstrated that, at levels lower than 10 $\mu g \cdot kg^{-1}$, the precision of several amphenicols, macrolides, beta-lactams, and tetracyclines became unsatisfactory.

The recovery factors (Rec) were established by comparing the peak area of each compound in the spiked samples against the peak area in the matrix matched standards, by which the antibiotics were added immediately prior to LC injection. The data in Table 3 summarize the average recoveries obtained in the whole validation range. In muscles, all recoveries were higher than 70%, except the majority of tetracyclines (62–69%), three beta-lactams (64–68%), and two macrolides (neospiramycin, 67% and tulathromycin, 69%). Analogously, in milk, recoveries higher than 70% were generally obtained, except for the more polar sulfonamides (61–70%) and one macrolide (erythromycin A, 67%). Since the quantification was performed with an external standardization, the raw results were always corrected for the relevant recovery factor, in order to correct the systematic error [32]. With regard to precision, the coefficients of variations (CVs) were calculated at each validation level both, repeatedly, and in intra-lab reproducibility conditions (CV_r, and CV_{wR}, respectively), by applying ANOVA. Moreover the CV_{wR} (and CV_r) were pooled to obtain an overall precision index, namely $CV_{wR,pooled}$ (2),

$$CV_{wR,pooled} = \sqrt{\frac{(n_1-1)CV^2_{wR1} + (n_2-1)CV^2_{wR2} + ...(n_n-1)CV^2_{wRn}}{(n_1-1)+(n_2-1)+...}} \tag{2}$$

where CV_{wR1}, CV_{wR2} ... CV_{wRn} were the coefficients of variation at the increasing levels 1, 2 ... n; n_1, n_2 ... n were the number of replicates at each level [34]. For a certain analyte, the $CV_{wR,pooled}$ give a single estimate of precision, which can be applied to calculate decision limits (CCα) and detection capabilities (CCβ) at whatever MRL value. Therefore, the $CV_{wR,pooled}$ can also be used to obtain CCα and CCβ, where new MRLs were fixed in Regulation 37/2010 [1], maximizing the cost-effectiveness of the validation study. The decision limit and detection capability were calculated as follows (equations 3 and 4):

$$CC\alpha_{MRL} = MRL + 1.64 \cdot CV_{wR,pooled} \cdot MRL \tag{3}$$

$$CC\beta_{MRL} = CC\alpha_{MRL} + 1.64 \cdot CV_{wR,pooled} \cdot CC\alpha_{MRL} \tag{4}$$

In Tables S4 and S5, the MRLs of the 64 antibiotics, in various food-producing animal species, and in bovine milk, together with the relevant CCαs and CCβs, are listed. The MRLs are those reported in the last consolidated text of Regulation 37/2010 [1,35]. In muscles, valnemulin demonstrated high imprecision ($CV_{wR,pooled}$ = 31%) and, therefore, for these compounds, the developed procedure could be only used for screening purposes (Table 3). For all the other antibiotics, $CV_{wR,pooled}$ was always lower than 20%. On the other hand, in milk, cefacetrile and tulathromycin marker demonstrated insufficient precision ($CV_{wR,pooled}$ > 22%). Matrix effects (ME%) listed in Table 3 were calculated as follows (equation 5),

$$ME(\%) = \frac{b_{MM}}{b_S} \times 100 \quad (5)$$

where $b_{MM,}$ and b_S were the slopes of matrix matched curves, and solvent standard curves, i.e., curves prepared in ammonium acetate 0.2 M, respectively. In the whole, although the sample purification was scarce, the matrix effects (suppression or enhancement) were limited and very few compounds demonstrated ME (%) higher than |50|%. This was probably because the long chromatographic run (30.5 min) allowed the distribution of matrix-interfering compounds and analytes, from preventing excessive bunching [26].

Since the signal-to-noise approach (S/N) is rather subjective [36], "operative" (fit for purpose) LOD and LOQ were fixed by examining the precision at each validation level (Table S7). All the compounds were detectable at the first concentration, i.e., 2 $\mu g{\cdot}kg^{-1}$ for amoxicillin, ampicillin, and penicillin G in milk and 10 $\mu g{\cdot}kg^{-1}$ for all the others. The only exception was cefacetrile, which is a scarcely ionizable molecule and, therefore, detectable and quantifiable from 33 $\mu g{\cdot}kg^{-1}$ in muscle (CV_r = 13%, CV_{wR} = 14%, recovery = 76%) and only detectable in milk (insufficient precision). A satisfactory accuracy for both amoxicillin and ampicillin was obtained at 2 $\mu g{\cdot}kg^{-1}$, whereas penicillin G demonstrated unsatisfactory precision at this level (CV_r = 23%, CV_{wR} = 26%) and, therefore, this beta-lactam could be quantified only starting from 10 $\mu g{\cdot}kg^{-1}$ (CV_r = 7.5%, CV_{wR} = 11%). Therefore, penicillin G, a fundamental drug for the treatment of sub-clinical mastitis in **lactating cows**, could not be quantified in milk at $\frac{1}{2}$ MRL (2 $\mu g{\cdot}kg^{-1}$) and the method was suitable only for screening purpose [30]. In Figure S4 the LC-QqQ chromatograms of these three compounds are shown in a blank milk sample and in the same sample spiked at 2 $\mu g{\cdot}kg^{-1}$.

2.3. *Comparison of LC-QqQ and LC-Q-Orbitrap Methods*

Comparing the recovery factors of the methods developed in meat, similar results were always obtained (differences < 15%), except for cefquinome, which demonstrated a higher recovery when LC-QqQ was applied (+ 26%: 97% LR vs. 71% HR) and valnemulin which, nevertheless, was not accurately quantifiable with the LR procedure, as discussed above. The recovery differences are visualized in Figure S5. Examining the data of the more polar cephalosporins (desacetylcephapirin, cephapirin, cefacetrile and cefalonium), cefquinome recovery appeared suspect (over-estimated) using QqQ technique. Since the sample preparation protocol was the same, this difference should be attributed to an instrumental technique. Interestingly, cefquinome co-eluted with sulfamerazine and, in addition, accidentally their precursor ion had the same nominal mass charge ratio i.e., *m/z* 265 (Table 2). In our laboratory, at the beginning of the development of multiclass methods for antibiotics in food (2014), two separate matrix-matched curves were prepared, one for beta-lactams and another for all other analytes [24]. This measure was precautionary in order to avoid the possible negative effects of methanol, contained in the intermediate solution of antibiotics other than beta-lactams (see Section 2.2) on beta-lactam stability, as described in the literature [37,38]. In saying that, the artefacts in the quantification of cefquinome and/or sulfamerazine could occur if "hidden" transitions were shared between these two compounds. This means that a transition, monitored only for one of the two analytes (Table 2), was shared by the other one, too. In order to verify this hypothesis, two individual solutions (50 ng·mL^{-1}) were separately injected, by simultaneously monitoring the four relevant SRMs (Figure S6). It was evident that sulfamerazine shared a "hidden" transition with cefquinome (m/z 265 > m/z 199—left side of Figure S6), but also viceversa (m/z 265 > m/z 156—right side of Figure S6). For sulfamerazine the fragment ion species at m/z 199 was formed by the rearrangement ion, by losing H_2SO_2 from the protonated molecule [39]. On the other hand, for cefquinome, the ion m/z 156 derived by the cleavage of C-C bond between oxime and the carbonyl group of C^7 amide ($[C_5H_6N_3OS]^+$) of the beta-lactam ring [40]. However, from a quantitative point of view, sulfamerazine could notably affect the peak area of cefquinome, but the contrary was much less, since sulfamerazine responded more significantly than cefquinome (about 4.9×10^6 vs. 1.1×10^6). Accordingly, when sulfamerazine was co-present, cefquinome concentration was significantly over-estimated. Later, observing that, in the adopted experimental conditions, beta-lactams were not deteriorated by methanol, the validation study in milk was performed by preparing only one matrix-matched calibration curve with all the analytes, including beta-lactams. In summary, although the development of LR procedures has been very fast and effective, due to previously-studied conditions [24,25], for one analyte, i.e., cefquinome, the choice of the precursor ion should be re-evaluated.

In Figure S7 the differences between LR and HR recovery factors in milk are shown. Tulathromycin marker and cefquinome had better recovery rates using HR detection, whereas valnemulin, cloxacillin, and rifaximin, showed better results using the LR system. As reported in our previous paper [25], LC-Q-Orbitrap suffers from "post interface ion suppression", which consists of instrument saturation when intense matrix-related compounds are present [41]. This phenomenon was more pronounced for the last eluting compounds, such as valnemulin, cloxacillin, and rifaximin, which explains the observed data in milk extracts. These have more interfering substances with respect to muscle [25]. Comparing the precision data (intra-laboratory reproducibility, $CV_{wR,pooled}$, see Tables S8 and S9), remarkable differences (≥15%) between the two techniques were observed only in milk and, again, tulathromycin marker and cefquinome revealed the worse performances when determined by LC-QqQ. $C_{wR,pooled}$ of tylosin, cloxacillin, and tylvalosin were about 10% higher when analyzed by LC-Q-Orbitrap (Figure S9).

According to accreditation rules [42], since 2014 our laboratory participated in Proficiency Test Schemes in meat and milk, by applying the LC-Q-Orbitrap methods in these products. Some of the stored test materials were then re-tested, by applying the new developed procedures. The results, together with the consensus values assigned by the Organizers, are listed in Table 4. Examining the acceptability ranges, satisfactory z-scores would have been always obtained, except in the case of amoxicillin in milk (sample code: MI1532-A1). This latter result was explainable with the well-known instability of penicillin antibiotics [43,44].

Table 4. Participation in Proficiency Test Schemes: comparison between methods.

Method		LC-Q-Orbitrap		LC-QqQ		
Sample Code/Year	Matrix	Analyte	Found Concentration (μg·kg^{-1})	Found Concentration (μg·kg^{-1})	Consensus Value (μg·kg^{-1})	Acceptability Range (μg·kg^{-1})
MI1432-A1/2014 [a]	Milk	Sulfamethazine	144	96	103	57–150
MI1432-A2/2014 [a]	Milk	Amoxicillin	5	ND	Not assigned	-
M1435-A1/2014 [a]	Pig muscle	Sulfamethazine	88	75	69	36–102
	Pig muscle	Sulfadimethoxine	32	23	27	12–42
M1433-A2/2014 [a]	Turkey muscle	Ciprofloxacin	5.5	5	5.6	3.2–8.1
	Turkey muscle	Enrofloxacin	173	152	160	92–227
MI1532-A1/2015 [a]	Milk	Amoxicillin	16	ND	14	5.6–22
MI1532-A2/2015 [a]	Milk	Sulfamethazine	165	131	134	75–191
MI1623-A1/2016 [a]	Milk	Flumequine	91	111	88	47–129
MI1623-A2/2016 [a]	Milk	Oxytetracycline	93	55	91	49–132
MI1715-A2/2017 [a]	Milk	Danofloxacin	91	80	74	39–109
484 (material C)/2018 [b]	Bovine Muscle	Marbofloxacin	178	193	170	100–240
484 (material C)/2018 [b]	Bovine muscle	Oxytetracycline	89	79	106	59–152
334/2019 [b]	Bovine muscle	Ciprofloxacin	10	10	NA [c]	-
334/2019 [b]	Bovine muscle	Enrofloxacin	82	84	NA [c]	-
544/2019 [b]	Bovine muscle	Tylosin A	54	87	NA [c]	-

[a] Test Veritas, Padova, Italy; [b] RIKILT, Wageningen, Netherlands; [c] The final Report is not yet available.

In summary, the main advantages and disadvantages of the two techniques are undoubtedly, the QqQ analyser, which only involves only a kind of acquisition, i.e., SRM mode. On the contrary, Q-Orbitrap forces more complex experiments, also because the detection of analytes at trace levels is complicated by the "post interface ion suppression" phenomenon. On this subject, we demonstrated that, in certain chromatographic regions, for milk it is not possible to reach the required limits using full-scan experiments, since the massive presence of interfering substances can drastically worsen the sensitivity [25]. With regard to the sample throughput, the sample preparation protocol is identical and, therefore, there is no great difference. Moreover, it must be highlighted that some performances of LR system are due to the obsolescence of the available equipment. For example, most likely, a more recent LR platform could reach comparable LODs with HR, i.e., lower than 10 $\mu g \cdot kg^{-1}$. All that said, the LC-QqQ technique is more suitable for routine laboratories, considering its user-friendliness and lesser cost (about three time lesser than LC-Q-Orbitrap).

3. Experimental

3.1. Chemical and Reagents

Acetonitrile (ACN) and methanol (LC-MS grade) were from Carlo Erba Reagents (Milan, Italy). Formic acid (50%) and *N,N′*-dimethylformamide (DMF) were purchased from Sigma-Aldrich (St. Louis, MO, USA). EDTA sodium salt dehydrate and ammonium acetate were provided by Sigma-Aldrich. Ultra-pure deionized water was generated by a Milli-Q purification apparatus (Millipore, Bedford, MA, USA). Amoxicillin, ampicillin, cloxacillin, dicloxacillin, nafcillin, oxacillin, penicillin G (benzylpenicillin), penicillin G-d7, penicillin V (phenoxymethylpenicillin), cefalonium, cefoperazone, cefquinome, ceftiofur, cephalexin, ciprofloxacin, danofloxacin, difloxacin, enrofloxacin, flumequine, marbofloxacin, nalidixic acid, norfloxacin, oxolinic acid, sarafloxacin, erithromycin A, spiramycin I, tylosin A, tilmicosin, sulfadiazine, sulfaguanidine, sulfadimethoxine, sulfamerazine, sulfamethazine (sulfadimidine), sulfamethoxazole, sulfanilamide, sulfapyridine, sulfaquinoxaline, sulfathiazole, trimethoprim, chlortetracycline, doxycycline, metacycline, oxytetracycline, tetracycline, florfenicol, florfenicol amine, thiamphenicol, lincomycin, rifaximin, tiamulin and valnemulin were obtained from Sigma-Aldrich. Sulfamonomethoxine was purchased from Dr. Ehrenstorfer (Augsburg, Germany); cefazolin, cefacetrile, ceftiofur-d3, cephapirin, desacetylcephapirin, florfenicol-d3, 3-*O*-acetyltylosin, gamithromycin, neospiramycin, spiramycin I-d3, tildipirosin, tulathromycin, tylvalosin, 4-epi-chlortetracycline, 4-epi-tetracycline, 4-epi-oxytetracycline, and tulathromycin marker (CP-60,300) were purchased from TRC Inc. (Toronto, Canada); sulfamethazine-$^{13}C6$, sulfanilamide-$^{13}C6$ and enrofloxacin-d5 were obtained from WITEGA (Berlin, Germany).

3.2. Standard Solutions

Individual stock standard solutions of 100 $\mu g \cdot mL^{-1}$ were prepared with methanol (amphenicols, lincosamides, macrolides, pleuromutilins, sulphonamides, tetracyclines and trimethoprim). The solubilization and storage conditions were previously studied [24]. Beta lactams were solubilized in H_2O/ACN 75/25 (*v*/*v*), except ceftiofur (DMF). Quinolones in MeOH/H_2O 80/20 (*v*/*v*), except ciprofloxacin, nalidixic acid, norfloxacin, enrofloxacin and oxolinic acid (DMF). Rifaximin was prepared in MeOH/H_2O 50/50 (*v*/*v*). The stock solutions were stored at −20 °C with variable storage times: From 1 month (cefquinome) to 24 months (sulphonamides). Working solutions (10, 1 and 0.1 $\mu g \cdot mL^{-1}$) were prepared from the relevant stock solutions diluting with H_2O/ACN 75/25 (*v*/*v*) for beta-lactams and with methanol for all the other antibiotics. The solutions of internal standards were prepared according to their native compound or class.

3.3. LC-MS-MS Conditions

LC-MS/MS measurements were performed by a Surveyor LC pump, coupled with a triple quadrupole mass spectrometer (TSQ Quantum Ultra, Thermo Fisher, San Jose, CA, USA), an electrospray source included, and operating in a positive ionization mode. Separation was achieved on a Poroshell 120 EC-C18 column (3.0 × 100 mm; 2.7 μm particle diameter), which was connected to a guard column Poroshell (2.1 × 5 mm), both from Agilent Technologies (Santa Clara, CA, USA). The flow rate used was 0.25 mL·min^{-1} and the column temperature set at 30 °C. Mobile phase A was an aqueous solution 0.1% (*v/v*) formic acid and eluent B methanol. The gradient profile started at 2% eluent B for 1 min and increased linearly up to 95% B in 19.5 min; this condition was maintained for 5 min before returning to initial condition in 1 min (2% B) and held for 4 min to equilibrate the column. The sample temperature was kept at 16 °C and the injection volume 10 μL. For MS detection, the parameters were the follows: Capillary temperature 300 °C, vaporizer 320 °C, spray voltage 3 kV, and a resolution setting of Q1 and Q3 *m/z* 0.7. Sheath gas and auxiliary gas (nitrogen) pressures were set at 35, and 15 arbitrary units, respectively. Collision gas (argon) pressure: 1.5 mtorr. The collision energies, that were associated with each transition, are listed in Table 2.

3.4. Sample Preparation

The sample preparation was described elsewhere [24,25]. Briefly, (1.50 ± 0.01) g of minced muscle or milk were weighed in a 50 mL Falcon tube. The sample was spiked with: (i) 15 μL of the solution of the two internal standards (ISs) of beta-lactams at 1 μg·mL^{-1} (ceftiofur-d3 and penicillin G-d7); (ii) 15 μL of the solution of the other six ISs (sulfanilamide-^{13}C6, sulfamethazine-^{13}C6, enrofloxacin-d5, florfenicol-d3, spiramycin-d3 and metacycline) at 1 μg mL^{-1}. For muscles, one hundred microliters of 0.1 M of EDTA was then added, and the sample was extracted with 3 mL of a mixture of acetonitrile/water 80/20 (*v/v*). Milk was extracted with one milliliter of 0.1 M of EDTA and 3 mL of acetonitrile. A second extraction with 3 mL of pure acetonitrile was performed for both matrices. After centrifugation, the reunited extracts were evaporated and solubilized in 1.5 mL of ammonium acetate 0.2 M. Ten μL was injected in the LC system.

3.5. Method Validation

The validation was carried out following the Commission Decision 2002/657/EC [30]. To test the selectivity, blank muscle samples, belonging to the main animal species (bovine, swine and poultry) and bovine milk samples from different origins, were analyzed. The linearity in the matrix has been evaluated in the range 2–150 μg·kg^{-1} (2, 10, 33, 100 and 150 μg·kg^{-1}). The matrix-matched solutions were prepared by adding the analytes immediately prior to LC injection. The curves were constructed, including at zero (blank), and by plotting the average peak area of analyte (three injections for each concentration point) against its concentration. An unweighted linear regression model was applied to the calibration data. The precision (repeatability and within-laboratory reproducibility), recovery (trueness), decision limit (CCα), and detection capability (CCβ) were studied following the experimental plans, described in Table S1 (muscle) and Table S2 (milk) of the Supplementary Material. A blank bovine muscle or milk was spiked at the beginning of the extraction procedure with the appropriate standard solutions. Four replicates (n = 4) at each level were carried out on the same day, along with the relevant matrix-matched calibration curve. Each series was repeated on three different days, and at varying times, operator, and calibration status of the LC-MS system. The spiking levels in muscles were; 10, 33, 100, 150, 333, 1000, and 1500 μg·kg^{-1} and 10, 33, 100, 150, and 333 μg kg^{-1} in milk. In milk, penicillin G, amoxicillin, and ampicillin were also tested at 2 μg·kg^{-1}. The Limits of Detection (LODs) and Quantification (LOQs) were estimated on the basis of the observed accuracy (recovery and precision) at the first validation concentration or, if necessary, at the second one.

4. Conclusions

The validity of the transfer, from LC-Q-Orbitrap to LC-QqQ of two multiclass methods for veterinary drugs in food, has been demonstrated. Using the LR platform, valnemulin in muscle, cefacetrile, and tulathromycin marker in milk did not reach acceptable precision. In addition, examining LC-QqQ results, a series of accidental events (co-elution, selected precursor ions, fragmentation pathway etc) produced an over-estimation of cefquinome in meat, due to the co-presence of sulfamerazine. If neglected, this phenomenon could give false positive results. Since the quantification process of the LC-Q-Orbitrap system was based on a different principle (peak area of the precursor ion measured with MS accuracy < 5 ppm), this drawback was not observed. However, in milk, LC-Q-Orbitrap achieved worse recoveries for some of the more non-polar analytes, eluting in the chromatographic zone in which the interfering substances were more abundant. The "post interface ion suppression", which is a specific phenomenon of Orbitrap mass analysers could explain this latter evidence. A further consideration concerns the chromatographic separation, which is fundamental when highly selective detectors are applied. A good separation of peaks, avoiding the overlapping and bunching of both analytes and endogenous substances, minimizes the risk of false positive results and reduces matrix effects. Finally, the described method transfer has been successfully performed on an obsolete LC-QqQ platform (fifteen year-old), encouraging the implementation of multiclass strategy also in routine laboratories with limited instrumental resources.

Supplementary Materials: The following are available online, Figure S1: LC-QqQ chromatograms of the three more polar analytes (sulfanilamide, sulfaguanidine and florfenicolamine) with different % MeOH at the beginning of the gradient: 2% (left) and 5% (right), Figure S2: Matrix-matched calibration curves of danofloxacin, doxycycline, oxacillin, spiramycin, sulfamethazine and tiamulin in meat, Figure S3: Matrix-matched calibration curves of of danofloxacin, doxycycline, oxacillin, spiramycin, sulfamethazine and tiamulin in milk, Figure S4: LC-QqQ chromatograms of a blank (a) and a milk sample spiked at 2 μg kg^{-1} (b) of the three penicillins with MRL at 4 μg kg^{-1}, Figure S5: Recovery differences between the LC-QqQ and LC-Q-Orbitrap methods in meat. Positive values (%) indicate better recovery of LC-QqQ procedure and vice versa, Figure S6: LC-QqQ chromatograms of individual solutions (50 ng mL^{-1}) of sulfamerazine (left) and cefquinome (right), Figure S7: Recovery differences between the LC-QqQ and LC-Q-Orbitrap methods in bovine milk. Positive values (%) indicate better recovery of LC-QqQ procedure and vice versa, Figure S8: Precision differences between the LC-QqQ and LC-Q-Orbitrap methods in meat. Positive values (%) indicate better precision of LC-QqQ procedure and vice versa, Figure S9: Precision differences between the LC-QqQ and LC-Q-Orbitrap methods in bovine milk. Positive values (%) indicate better precision of LC-QqQ procedure and vice versa, Table S1: Validation plan in bovine muscle, Table S2: Validation plan in milk, Table S3: Linearity studies in matrix (matrix-matched curves), Table S4: Calibration data (matrix-matched curves) of six representative analytes, Table S5: MRL (μg kg^{-1}), decision limits (μg kg^{-1}) and detection capabilities (μg kg^{-1}) of the 64 tested antibiotics in the main food-producing species, Table S6: MRL (μg kg^{-1}), decision limits (μg kg^{-1}) and detection capabilities (μg kg^{-1}) of the 64 antibiotics in aquaculture, rabbits, horses, Table S7: Estimated LODs and LOQs based on the observed accuracy at the first and at the second validation level.

Author Contributions: Conceptualization, D.G., S.M., G.S. and F.P.; methodology, G.S. and S.M.; validation, D.G., F.P. and C.B.; investigation, D.G., F.P., S.M, G.S. and C.B.; data curation, R.S. and F.I.; writing—original draft preparation, R.G.; writing—review and editing, R.G., R.S. and F.I.; supervision and funding acquisition, R.G.

References

1. European Communities Commission Regulation (EU) No 37/2010 of 22 December 2009 on pharmacologically active substances and their classification regarding maximum residue limits in foodstuffs of animal origin. *Off. J. Eur. Communities* **2010**, *L15*, 1–72. Available online: https://eur-lex.europa.eu/legal-content/EN/TXT/?uri=CELEX%3A32010R0037 (accessed on 30 May 2019).
2. Masiá, A.; Suarez-Varela, M.M.; Llopis-Gonzalez, A.; Picó, Y. Determination of pesticides and veterinary drug residues in food by liquid chromatography-mass spectrometry: A review. *Anal. Chim. Acta* **2016**, *936*, 40–61. [CrossRef] [PubMed]
3. Mainero Rocca, L.; Gentili, A.; Pérez-Fernández, V.; Tomai, P. Veterinary drugs residues: A review of the latest analytical research on sample preparation and LC-MS based methods. *Food Addit. Contam. - Part A Chem. Anal. Control. Expo. Risk Assess.* **2017**, *34*, 766–784. [CrossRef] [PubMed]
4. Rossi, R.; Saluti, G.; Moretti, S.; Diamanti, I.; Giusepponi, D.; Galarini, R. Multiclass methods for the analysis of antibiotic residues in milk by liquid chromatography coupled to mass spectrometry: A review. *Food Addit. Contam. - Part A Chem. Anal. Control. Expo. Risk Assess.* **2018**, *35*, 241–257. [CrossRef] [PubMed]

5. Aguilera-Luiz, M.M.; Vidal, J.L.M.; Romero-González, R.; Frenich, A.G. Multi-residue determination of veterinary drugs in milk by ultra-high-pressure liquid chromatography-tandem mass spectrometry. *J. Chromatogr. A* **2008**, *1205*, 10–16. [CrossRef] [PubMed]
6. Chico, J.; Rúbies, A.; Centrich, F.; Companyó, R.; Prat, M.D.; Granados, M. High-throughput multiclass method for antibiotic residue analysis by liquid chromatography-tandem mass spectrometry. *J. Chromatogr. A* **2008**, *1213*, 189–199. [CrossRef] [PubMed]
7. Kaufmann, A.; Butcher, P.; Maden, K.; Widmer, M. Quantitative multiresidue method for about 100 veterinary drugs in different meat matrices by sub 2-microm particulate high-performance liquid chromatography coupled to time of flight mass spectrometry. *J. Chromatogr. A* **2008**, *1194*, 66–79. [CrossRef] [PubMed]
8. Stolker, A.A.M.; Rutgers, P.; Oosterink, E.; Lasaroms, J.J.P.; Peters, R.J.B.; Van Rhijn, J.A.; Nielen, M.W.F. Comprehensive screening and quantification of veterinary drugs in milk using UPLC-ToF-MS. *Anal. Bioanal. Chem.* **2008**, *391*, 2309–2322. [CrossRef]
9. Peters, R.J.B.B.; Bolck, Y.J.C.C.; Rutgers, P.; Stolker, A.A.M.M.; Nielen, M.W.F.F. Multi-residue screening of veterinary drugs in egg, fish and meat using high-resolution liquid chromatography accurate mass time-of-flight mass spectrometry. *J. Chromatogr. A* **2009**, *1216*, 8206–8216. [CrossRef]
10. Stubbings, G.; Bigwood, T. The development and validation of a multiclass liquid chromatography tandem mass spectrometry (LC-MS/MS) procedure for the determination of veterinary drug residues in animal tissue using a QuEChERS (QUick, Easy, CHeap, Effective, Rugged and Safe) approac. *Anal. Chim. Acta* **2009**, *637*, 68–78. [CrossRef]
11. Gaugain-Juhel, M.; Delépine, B.; Gautier, S.; Fourmond, M.P.; Gaudin, V.; Hurtaud-Pessel, D.; Verdon, E.; Sanders, P. Validation of a liquid chromatography-tandem mass spectrometry screening method to monitor 58 antibiotics in milk: A qualitative approach. *Food Addit. Contam. - Part A Chem. Anal. Control. Expo. Risk Assess.* **2009**, *26*, 1459–1471. [CrossRef] [PubMed]
12. Martínez Vidal, J.L.; Frenich, A.G.; Aguilera-Luiz, M.M.; Romero-González, R. Development of fast screening methods for the analysis of veterinary drug residues in milk by liquid chromatography-triple quadrupole mass spectrometry. *Analytical and Bioanalytical Chemistry.* **2010**, *397*, 2777–2790. [CrossRef] [PubMed]
13. Kaufmann, A.; Butcher, P.; Maden, K.; Walker, S.; Widmer, M. Development of an improved high resolution mass spectrometry based multi-residue method for veterinary drugs in various food matrices. *Anal. Chim. Acta* **2011**, *700*, 86–94. [CrossRef] [PubMed]
14. Hurtaud-Pessel, D.; Jagadeshwar-Reddy, T.; Verdon, E. Development of a new screening method for the detection of antibiotic residues in muscle tissues using liquid chromatography and high resolution mass spectrometry with a LC-LTQ-Orbitrap instrument. *Food Addit. Contam. - Part A Chem. Anal. Control. Expo. Risk Assess.* **2011**, *28*, 1340–1351. [CrossRef] [PubMed]
15. Wang, J.; Leung, D. The challenges of developing a generic extraction procedure to analyze multi-class veterinary drug residues in milk and honey using ultra-high pressure liquid chromatography quadrupole time-of-flight mass spectrometry. *Drug Test. Anal.* **2012**, *1*, 103–111. [CrossRef] [PubMed]
16. Bittencourt, M.S.; Martins, M.T.; de Albuquerque, F.G.S.; Barreto, F.; Hoff, R. High-throughput multiclass screening method for antibiotic residue analysis in meat using liquid chromatography-tandem mass spectrometry: A novel minimum sample preparation procedure. *Food Addit. Contam. - Part A Chem. Anal. Control. Expo. Risk Assess.* **2012**, *29*, 508–516. [CrossRef] [PubMed]
17. Martins, M.T.; Melo, J.; Barreto, F.; Barcellos Hoff, R.; Jank, L.; Soares Bittencourt, M.; Bazzan Arsand, J.; Scherman Schapoval, E.E. A simple, fast and cheap non-SPE screening method for antibacterial residue analysis in milk and liver using liquid chromatography-tandem mass spectrometry. *Talanta* **2014**, *129*, 374–383. [CrossRef] [PubMed]
18. Kaufmann, A.; Butcher, P.; Maden, K.; Walker, S.; Widmer, M. Multi-residue quantification of veterinary drugs in milk with a novel extraction and cleanup technique: Salting out supported liquid extraction (SOSLE). *Anal. Chim. Acta* **2014**, *820*, 56–68. [CrossRef]
19. Freitas, A.; Barbosa, J.; Ramos, F. Multidetection of antibiotics in liver tissue by ultra-high-pressure-liquid-chromatography-tandem mass spectrometry. *J. Chromatogr. B Anal. Technol. Biomed. Life Sci.* **2015**, *976-977*, 49–54. [CrossRef]
20. Martins, M.T.; Barreto, F.; Hoff, R.B.; Jank, L.; Arsand, J.B.; Feijó, T.C.; Schapoval, E.E.S. Determination of quinolones and fluoroquinolones, tetracyclines and sulfonamides in bovine, swine and poultry liver using LC-MS/MS. *Food Addit. Contam. - Part A Chem. Anal. Control. Expo. Risk Assess.* **2015**, *32*, 333–341. [CrossRef]

21. Wang, J.; Leung, D.; Chow, W.; Chang, J.; Wong, J.W. Development and Validation of a Multiclass Method for Analysis of Veterinary Drug Residues in Milk Using Ultrahigh Performance Liquid Chromatography Electrospray Ionization Quadrupole Orbitrap Mass Spectrometry. *J. Agric. Food Chem.* **2015**, *63*, 9175–9187. [CrossRef] [PubMed]
22. Dasenaki, M.E.; Thomaidis, N.S. Multi-residue determination of 115 veterinary drugs and pharmaceutical residues in milk powder, butter, fish tissue and eggs using liquid chromatography-tandem mass spectrometry. *Anal. Chim. Acta* **2015**, *880*, 103–121. [CrossRef] [PubMed]
23. Dasenaki, M.E.; Michali, C.S.; Thomaidis, N.S. Analysis of 76 veterinary pharmaceuticals from 13 classes including aminoglycosides in bovine muscle by hydrophilic interaction liquid chromatography–tandem mass spectrometry. *J. Chromatogr. A* **2016**, *1452*, 67–80. [CrossRef] [PubMed]
24. Moretti, S.; Dusi, G.; Giusepponi, D.; Pellicciotti, S.; Rossi, R.; Saluti, G.; Cruciani, G.; Galarini, R. Screening and confirmatory method for multiclass determination of 62 antibiotics in meat. *J. Chromatogr. A* **2016**, *1429*, 175–188. [CrossRef] [PubMed]
25. Moretti, S.; Cruciani, G.; Romanelli, S.; Rossi, R.; Saluti, G.; Galarini, R. Multiclass method for the determination of 62 antibiotics in milk. *J. Mass Spectrom.* **2016**, *51*, 792–804. [CrossRef] [PubMed]
26. Anumol, T.; Lehotay, S.J.; Stevens, J.; Zweigenbaum, J. Comparison of veterinary drug residue results in animal tissues by ultrahigh-performance liquid chromatography coupled to triple quadrupole or quadrupole–time-of-flight tandem mass spectrometry after different sample preparation methods, including use of. *Anal. Bioanal. Chem.* **2017**, *409*, 2639–2653. [CrossRef] [PubMed]
27. Lehotay, S.J.; Lightfield, A.R. Simultaneous analysis of aminoglycosides with many other classes of drug residues in bovine tissues by ultrahigh-performance liquid chromatography–tandem mass spectrometry using an ion-pairing reagent added to final extracts. *Anal. Bioanal. Chem.* **2018**, *2018. 410*, 1095–1109. [CrossRef]
28. Gaspar, A.F.; Santos, L.; Rosa, J.; Leston, S.; Barbosa, J.; Vila Pouca, A.S.; Freitas, A.; Ramos, F. Development and validation of a multi-residue and multi-class screening method of 44 antibiotics in salmon (Salmo salar) using ultra-high-performance liquid chromatography/time-of-flight mass spectrometry: Application to farmed salmon. *J. Chromatogr. B Anal. Technol. Biomed. Life Sci.* **2019**, *1118–1119*, 78–84. [CrossRef]
29. Romero-González, R.; Aguilera-Luiz, M.M.; Plaza-Bolaños, P.; Frenich, A.G.; Vidal, J.L.M. Food contaminant analysis at high resolution mass spectrometry: Application for the determination of veterinary drugs in milk. *J. Chromatogr. A* **2011**, *1218*, 9353–9365. [CrossRef]
30. Commission Decision (2002/657/EC) of 12 August 2002 implementing Council Directive 96/23/EC concerning the performance of analytical methods and the interpretation of results. *Off. J. Eur. Communities* **2002**, *L221*, 8–36.
31. Dickson, L.C. Performance characterization of a quantitative liquid chromatography-tandem mass spectrometric method for 12 macrolide and lincosamide antibiotics in salmon, shrimp and tilapia. *J. Chromatogr. B Anal. Technol. Biomed. Life Sci.* **2014**, *967*, 203–210. [CrossRef] [PubMed]
32. SANCO/2004/2726-rev 4-December 2008 Guidelines for the Implementation of Decision 2002/657/EC. European Commission Health & Consumer Protection Directorate-General. 2008. Available online: https://ec.europa.eu/food/sites/food/files/safety/docs/cs_vet-med-residues_cons_2004-2726rev4_en.pdf (accessed on 30 May 2019).
33. SANTE/11813/2017. Guidance document on analytical quality control and validation procedures for pesticide residues analysis in food and feed. In European Commission Health & Consumer Protection Directorate-General. 2017. Available online: https://ec.europa.eu/food/sites/food/files/plant/docs/pesticides_mrl_guidelines_wrkdoc_2017-11813.pdfwebsite (accessed on 30 May 2019).
34. Barvick, V.J.; Wllison, S.R.L. Part (d): Protocol for uncertainty evaluation from validation data. In *VAM Project 3.2.1. Development and Harmonisation of Measurement Uncertainty Principles*; LGC: Teddington, UK, 2000.
35. European Communities Commission Regulation (EU) No 37/2010 of 22 December 2009 on pharmacologically active substances and their classification regarding maximum residue limits in foodstuffs of animal origin. Consolidated test. 3 March 2019. Available online: https://eur-lex.europa.eu/collection/eu-law/consleg.html (accessed on 30 May 2019).
36. Galarini, R.; Moretti, S.; Saluti, G. Quality Assurance and Validation General Considerations and Trends. In *Chromatographic Analysis of the Environment Mass Spectrometry Based Approaches, Fourth Edition*; Leo, M.L., Nollet, D.A.L., Eds.; CRC Press: Boca Raton, FL, USA, 2017; ISBN 9781315316208.

37. Grujic, S.; Vasiljevic, T.; Lausevic, M.; Ast, T. Study on the formation of an amoxicillin adduct with methanol using electrospray ion trap tandem mass spectrometry. *Rapid Commun. Mass Spectrom.* **2008**, *22*, 67–74. [CrossRef] [PubMed]
38. Mastovska, K.; Lightfield, A.R. Streamlining methodology for the multiresidue analysis of beta-lactam antibiotics in bovine kidney using liquid chromatography-tandem mass spectrometry. *J. Chromatogr. A* **2008**, *1202*, 118–123. [CrossRef] [PubMed]
39. Asteggiante, L.G.; Nunez, A.; Lehotay, S.J.; Lightfield, A.R. Structural characterization of product ions by electrospray ionization and quadrupole time-of-flight mass spectrometry to support regulatory analysis of veterinary drug residues in foods. *Rapid Commun. Mass Spectrom.* **2014**, *28*, 1061–1081. [CrossRef] [PubMed]
40. Niessen, W.M.A.; Correa, R.A.C. *Interpretation of MS-MS Mass Spectra of Drugs and Pesticides*; John Wilewy and Sons: Hoboken, NJ, USA, 2017.
41. Kaufmann, A.; Walker, S. Extension of the Q Orbitrap intrascan dynamic range by using a dedicated customized scan. *Rapid. Commun. Mass Spectrom.* **2016**, *30*, 1087–1095. [CrossRef] [PubMed]
42. *General requirements for the competence of testing and calibration laboratories*, Second ed.; International Standard Organization: Geneva, Switzerland, 2005.
43. Berendsen, B.J.A.A.; Elbers, I.J.W.W.; Stolker, A.A.M.M. Determination of the stability of antibiotics in matrix and reference solutions using a straightforward procedure applying mass spectrometric detection. *Food Addit. Contam. - Part A Chem. Anal. Control. Expo. Risk Assess.* **2011**, *28*, 1657–1666. [CrossRef]
44. Gaugain, M.; Chotard, M.P.; Verdon, E. Stability study for 53 antibiotics in solution and in fortified biological matrixes by LC/MS/MS. *J. AOAC Int.* **2013**, *96*, 471–480. [CrossRef]

4

Ion Mobility Spectrometry in Food Analysis: Principles, Current Applications and Future Trends

Maykel Hernández-Mesa [1,2,3,*], David Ropartz [2], Ana M. García-Campaña [1], Hélène Rogniaux [2], Gaud Dervilly-Pinel [3] and Bruno Le Bizec [3]

1 Department of Analytical Chemistry, Faculty of Sciences, University of Granada, Campus Fuentenueva s/n, E-18071 Granada, Spain
2 INRA, UR1268 Biopolymers Interactions Assemblies, F-44316 Nantes, France
3 Laboratoire d'Etude des Résidus et Contaminants dans les Aliments (LABERCA), Oniris, INRA UMR 1329, Route de Gachet-CS 50707, F-44307 Nantes CEDEX 3, France
* Correspondence: maykelhm@ugr.es

Academic Editors: Alessandra Gentili and Chiara Fanali

Abstract: In the last decade, ion mobility spectrometry (IMS) has reemerged as an analytical separation technique, especially due to the commercialization of ion mobility mass spectrometers. Its applicability has been extended beyond classical applications such as the determination of chemical warfare agents and nowadays it is widely used for the characterization of biomolecules (e.g., proteins, glycans, lipids, etc.) and, more recently, of small molecules (e.g., metabolites, xenobiotics, etc.). Following this trend, the interest in this technique is growing among researchers from different fields including food science. Several advantages are attributed to IMS when integrated in traditional liquid chromatography (LC) and gas chromatography (GC) mass spectrometry (MS) workflows: (1) it improves method selectivity by providing an additional separation dimension that allows the separation of isobaric and isomeric compounds; (2) it increases method sensitivity by isolating the compounds of interest from background noise; (3) and it provides complementary information to mass spectra and retention time, the so-called collision cross section (CCS), so compounds can be identified with more confidence, either in targeted or non-targeted approaches. In this context, the number of applications focused on food analysis has increased exponentially in the last few years. This review provides an overview of the current status of IMS technology and its applicability in different areas of food analysis (i.e., food composition, process control, authentication, adulteration and safety).

Keywords: food quality; IMS; food composition; food process control; food authentication; food adulteration; food safety

1. Introduction

In the current context of food trade globalization and due to the recognized impact of the diet on human health, food analysis has become more important than ever. Food analysis has today gone beyond the traditional analysis of the major components of food and is more complex and broader. In addition to the nutritional value of foodstuffs (i.e., carbohydrates, proteins, lipids, vitamins, minerals and water), food analysis has been focused on food safety for a long time, mainly in the determination of residues of pesticides and veterinary drugs. Nowadays, food safety analysis encompasses a wide variety of compounds including natural contaminants (e.g., toxins) or anthropogenic contaminants such as persistent organic pollutants (POPs) [1]. Understanding the interactions of food with the environment and consequences (i.e., large-scale production, organic production, environmental contamination, etc. and including the control of food processes, packaging, etc.), as well as its effects on consumers (e.g., investigation of bioactive compounds), is also gaining great importance. Other issues such as food authentication (i.e., quality, origin, etc.), adulteration and fraud detection have also acquired great relevance in the field of food chemistry in the last few years [2].

Food analysis involves the determination of a wide range of compounds with different chemical nature. Chromatographic techniques (i.e., liquid chromatography (LC) and gas chromatography (GC)) coupled to mass spectrometry (MS) are the gold standard for this purpose because they allow the analysis of molecules with different polarity and volatility as well as provide mass spectra for compound identification. However, LC–MS and GC–MS methods still face several challenges related to the complexity of food matrices, the presence of compounds at different concentration levels (from pg/kg and pg/L to mg/kg and mg/L levels), and the existence of isobars and isomers that are not separated in the chromatographic dimension and cannot be distinguished by MS.

Consequently, the development of more advanced analytical strategies is required for the analysis of food composition, including nutritive and bioactive components, as well as to guarantee food safety and avoid food fraud. Within this framework, ion mobility spectrometry (IMS) has been recently introduced in the food chemistry field in order to improve LC–MS and GC–MS workflows. Despite the fact that IMS can be used alone as analytical tool, it can also be coupled to other analytical separation techniques such as LC, GC, capillary electrophoresis (CE), or supercritical fluid chromatography (SFC) as well as to MS, enhancing their performance characteristics in terms of sensitivity, peak capacity, and compound identification [3,4]. Its coupling with front-end separation techniques and with MS has emerged as a useful approach to extend the current boundaries of analytical methods in food science, and it can be anticipated that it will rapidly be growing in this field.

The implementation of IMS within the food analytical field is quite new [5–8], and is still barely known by many researchers in this scientific area. Despite IMS fundamentals have been developed since the beginning of the 20th century [9], it has not been until the recent commercialization of hyphenated ion mobility-mass spectrometry (IM–MS) instruments when this technique has really caught the attention of researchers from multiple fields, including food science. As a result, the number of publications on IMS applications in food analysis has rapidly increased over the last few years (Figure 1). Within this context, this review provides a general overview of IMS principles and presents the current state of the art of this technology for food analysis purposes.

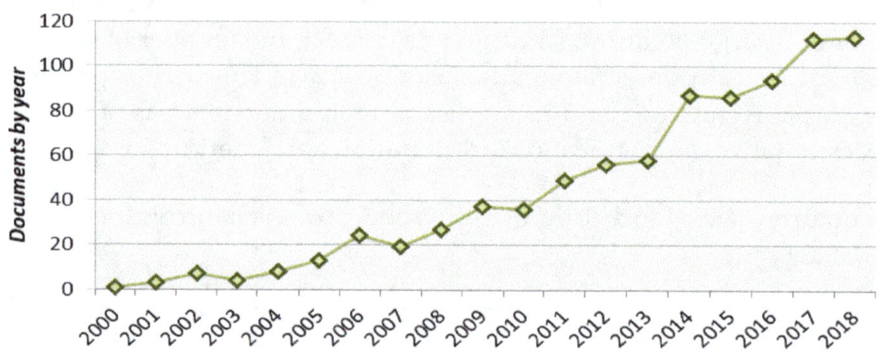

Figure 1. Search of literature related to ion mobility spectrometry (IMS) in food analysis from 2000 to 2018 on Scopus database. The terms "ion mobility spectrometry" and "food" have been included in the search topic.

2. Overview of Ion Mobility Spectrometry (IMS) Technique and Potential

IMS is an electrophoretic separation technique in which ionized compounds are separated in a neutral gas phase at atmospheric or near to atmospheric pressure. Therefore, separation takes places under an electric field (E) and is the result of the difference in mobility (K) of ions in the drift cell. K refers to the time (t_d) required by the ions to traverse the length (l) of the mobility cell and is related to the electric field according to Equation (1). In this equation, v_d represents the steady-state net ion/gas relative velocity [10].

$$K = \frac{l}{t_d\,E} = \frac{v_d}{E} \tag{1}$$

This physical property depends on several experimental conditions such as temperature (T), pressure (p) and gas number density (N). The reduced mobility (K_0), which refers to standard conditions (Equation (2); $N_0 = 2.687 \times 10^{25}\ m^{-3}$, $p_0 = 760$ Torr, $T_0 = 273.16$ K), is typically reported instead of K in order to allow comparison between studies independently of the experimental conditions.

$$K_0 = K\,\frac{N}{N_0} = K\frac{p}{p_0}\frac{T_0}{T} \tag{2}$$

Although the 'momentum transfer collision integral' (Ω), commonly referred to as the collision cross section (CCS), is reported as the response resulting from ion mobility measurements [11], t_d or K are actually the variables that are measured when performing IMS experiments. Nevertheless, both parameters can be easily correlated according to Mason–Schamp equation (Equation (3)) when the separation occurs at low electric fields [12].

$$K = \frac{3}{16}\frac{ze}{N}\left(\frac{2\pi}{\mu k_B T}\right)^{1/2}\frac{1}{\Omega} \tag{3}$$

where z and e represent the absolute charge of the ion and the elementary charge, respectively; μ encompasses the reduced mass of the ion–neutral drift gas pair (i.e., $\mu = mM/(m + M)$; m and M are the ion and gas-particle masses, respectively); and k_B is the Boltzmann constant.

CCS represents the averaged momentum transfer impact area of the ion and is a molecular parameter related to ions size, shape and charge state. Therefore, CCS is widely used for structural elucidation since it provides knowledge about the three dimensional (3D) conformation of ions in the gas phase [13,14]. The correlation existing between this parameter and the mass-to-charge ratio (*m/z*) is not negligible. However, it also gives additional information to retention index (e.g., retention time, electrophoretic mobility, migration time, etc.), mass spectra, fragmentation and isotopic patterns, etc., for peak annotation in analytical workflows, especially in omics approaches (e.g., metabolomics, lipidomics) and, ultimately, for compound identification [15,16].

2.1. IMS Instrumentation

Nowadays, there is a wide variety of stand-alone IMS and IM–MS instruments on the market, but they are based on different technologies that offer different advantages [4,14,17]. They can be classified in time-dispersive, space-dispersive and trapping (i.e., ion confinement and release) technologies [18]. Drift tube ion mobility spectrometry (DTIMS) and travelling tube ion mobility spectrometry (TWIMS) are two different types of time-dispersive forms. For example, DTIMS is currently commercialized by Agilent (i.e., 6560 Ion Mobility LC/quadrupole-time of flight (Q–ToF)), Excellims Corporation (i.e., high-performance ion mobility spectrometry (HPIMS) systems including MA3100 and RA4100 HPIMS–MS instruments), Gesellschaft für Analytische Sensorsysteme mbH (G.A.S.) (i.e., GC–IMS systems) and TOFWerk (i.e., IMS-ToF), whereas TWIMS is available from Waters Corporation (i.e., Synapt G2-Si and Vion IMS QTof). High-field asymmetric waveform ion mobility spectrometry (FAIMS) and differential ion mobility spectrometry (DIMS or DMS) belong to space-dispersive techniques. Both FAIMS and DMS are based on the same principles of operation and mainly differ in the geometry of the cell. FAIMS cells are curved whereas DMS cells present a planar geometry and, consequently, they can lead to different analytical properties [19]. FAIMS systems are currently available from Owlstone Medical (i.e., ultraFAIMS) and Thermo Fisher Scientific (i.e., Thermo Scientific FAIMS Pro interface), whereas DMS is commercialized by SCIEX (i.e., SelexION). Until now, trapped ion mobility spectrometry (TIMS), which represents a type of trapping technology, is the only IMS system of its class that is currently commercially available (i.e., timsTOF from Bruker Daltonics).

Regarding CCS-related measurements, they can only be carried out using IMS instruments that operate at low electric fields (e.g., DTIMS) because the reduced mobility is independent of the electric field under this condition. K_0 becomes dependent of the reduced field strength (E/N) at high electric fields (E/N > 4–10 Townsends (Td); 1 Td = 10^{-21} V m^2) and Equation (3) is no longer applicable [10]. In this sense, only primary DTIMS methods can be applied to obtain CCS values directly, whereas secondary DTIMS, TWIMS and TIMS methods require system calibration using reference compounds of known CCS. Only those compounds characterized in terms of CCS by primary DTIMS methods can be used as calibrants [20]. The same results should be obtained for the CCS of a specific ion independently of the IMS platform employed if secondary methods operate under the same experimental conditions (i.e., temperature, drift gas, E/N, etc.) as primary methods. Since this is not always possible, CCS is a method-dependent value [20]. Specific annotation ($^{\text{IMS form}}CCS_{\text{drift gas}}$, IMS form: DT for DTIMS, TW for TWIMS, TIMS for TIMS; and drift gas: N_2, CO_2, He, etc.) is currently accepted to indicate the type of IMS technology and drift gas used for CCS measurements [11].

In DTIMS and TWIMS methods, ions travel through the drift cell against the buffer gas describing a similar path, so they are separated according to their mobility in the drift cell (Figure 2). Compact molecules collide less frequently with the molecules of drift gas (normally N_2 or He, but also CO_2), so they present a higher mobility (i.e., a smaller CCS) and cross the cell faster than elongated molecules.

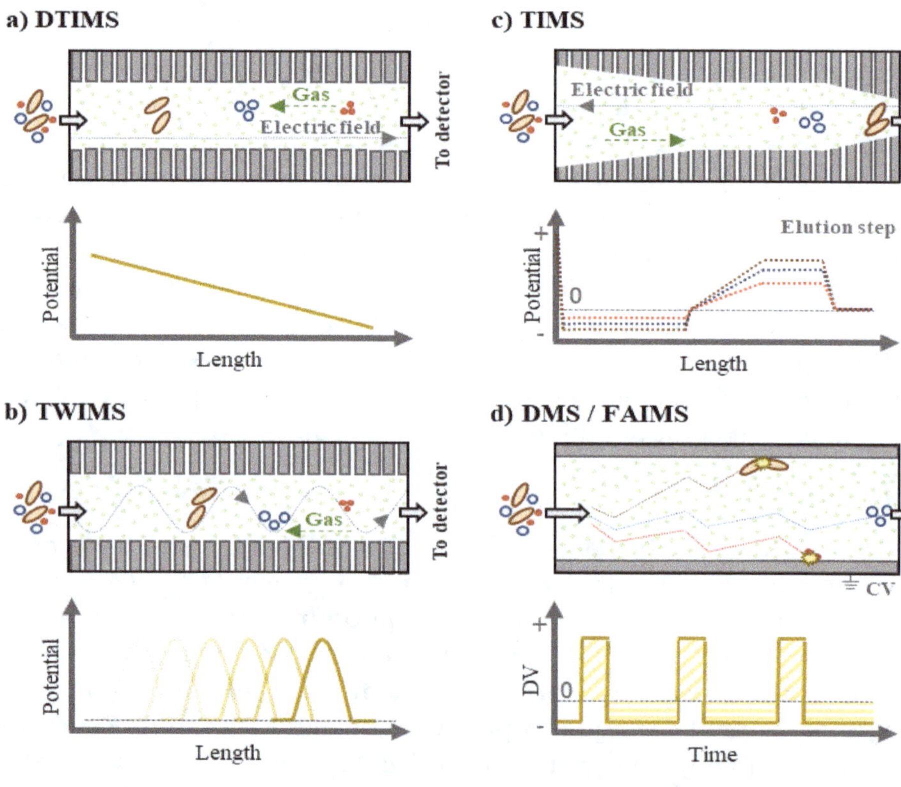

Figure 2. Schematic representation of commercially available IMS forms.

DTIMS is the former and simplest form of IMS, and DTIMS cells consist of a series of piled electrodes that generate a weak uniform electric field. In general, E/N varies between 1 to 15 Td [20]. DTIMS works as a primary method when stepped field experiments are carried out. In this case, the arrival time (t_A) is measured at multiple fixed drift voltages, and each t_A value represents the sum of drift time (t_d) of ions and the time elapsing between the moment when ions exit the drift tube and their detection (t_0). Subsequently, the reduced mobility can be directly obtained by applying Equation (4) and the CCS is further calculated by Mason–Schamp equation, Equation (3) [21].

$$t_A = t_d + t_0 = \left(\frac{l^2}{K_0} \times \frac{T_0 p}{T p_0}\right) \times \frac{1}{\Delta V} = \text{slope} \times \frac{1}{\Delta V} + t_0 \tag{4}$$

In general, the term 'arrival time distribution' (ATD) should be used because IMS measurements are based on a population of ions rather than on a single ion and they do not reach one point of the instrument at the same time. ATD gives information about how homogeneous is the population of ions, and it should be reported in addition to t_A [20]. Furthermore, it must be noted that stepped field methods are limited to the K_0/CCS characterization of molecules because the analysis of complex samples is not feasible with them. Consequently, DTIMS usually works in single-field mode in which a single linear voltage is applied along the drift tube. In this condition, DTIMS operates as a secondary method and CCS calibration is required to measure this molecular descriptor [22].

Unlike in DTIMS, a dynamic electric field is applied in TWIMS to separate the ions [17]. TWIMS systems consist of a stacked ring ion guide where a pulsed differential voltage is applied to each electrode. As a result, a wave of electric potential travels along the drift tube and propels the ions axially. Ions are subjected to a varying voltage with a maximum E/N of 160 Td [10], and their separation depends on the speed and magnitude of the voltage wave. Voltage waves move faster than the ions, so they roll over the wave and require a succession of them to reach the exit of the mobility cell. In addition, in order to prevent the ions being pushed towards the drift tube wall, radio-frequency (RF) voltages of opposite phases are periodically applied to adjacent electrodes causing the radial confinement of ions [4,23].

On the other hand, TIMS instruments consist of three regions of electrodes: entrance funnel, TIMS tunnel, and exit funnel (Figure 2). Unlike in DTIMS and TWIMS where an electric field is applied to push the ions through the cell, in TIMS systems, an electric field (i.e., 45–150 Td) is applied to trap the ions and they are only dragged through the drift tube towards the detector by the gas [24]. Initially, ions are accumulated for a fixed period of time and directed to the entrance funnel. Subsequently, they are released into the TIMS funnel traversing an axial electric field gradient (EFG). When the velocity of the buffer gas (v_g) is equaled by the opposite steady state drift velocity of the ion (v_d), ions reach a stationary state and are trapped in a moving column of gas. Finally, the EFG is decreased and ions are released towards the exit funnel. Large ions with lower K are eluted before more compact ions since the dependence of the drift time on K_0 is opposite to DTIMS [25].

Finally, FAIMS and DMS instruments do not separate ions in function of their mobility in a neutral gas as in DTIMS, TWIMS or TIMS, but rather by the ratio of low-field to high-field mobilities [17]. Consequently, CCS values cannot be obtained by these systems. In FAIMS/DMS systems, a time-dependent electric field is applied between two parallel electrodes. Low and high electric fields of opposite polarities are simultaneously applied (i.e., high electric fields > 30 Td) [26], taking into account that the product of the voltage (commonly referred as dispersive voltage, DV) and time for each condition must be the same as shown in Figure 2 [19]. Low electric fields are applied for longer periods than high electric fields. Moreover, a compensation voltage (CV) is applied to one of the electrodes with the aim of avoiding that ions collide against them. As a result, the trajectory of ions is altered ensuring that they migrate to the exit of the cell. In this sense, FAIMS and DMS act as selective instruments because the CV is an analyte-dependent parameter and they are not able to scan the complete CV range during the transition of ions through the cell. Unlike in other IMS forms where all the ions are normally detected, in FAIMS/DMS technology, only those analytes related to the selected CVs reach the detector.

More detailed information about the operation and physical principles of the different IMS forms can be found in specialized literature [17,19,25,27,28]. Furthermore, a guide about how IMS experiments must be reported, including CCS measurements, has been recently published [20]. It is recommended to follow this guide for the communication of IMS-related results because it will probably set the basis for future guidelines and standards in ion mobility.

2.2. Collision Cross Section (CCS) for Structural Elucidation

As previously mentioned, IMS also gives access to the CCS characteristic which, in certain cases, provides additional information to mass spectra and retention index. It can be highly valuable for a higher confidence in the determination of residues and contaminants in food safety or for achieving a more complete fingerprint of food products. In the last few years, several CCS databases have been built in an attempt to use this characteristic as an identification parameter [11,15]. In general, there is still much controversy about the added value provided by this parameter in comparison to mass spectra. It cannot be denied that CCS and *m/z* are not fully orthogonal parameters since a close correlation exists between them, as observed in Figure 3B. Nevertheless, slight differences existing between the CCS of molecules with similar or equal *m/z* (i.e., isobars and isomers) can be enough to distinguish them (differences of at least 1.5–1.8% (2.0–6.5 Å^2) in apparent CCS values for accurate CCS determination [29]).

Regarding the application of CCS databases, an error threshold of ±2% is currently accepted for CCS measurements in comparison to CCS values reported by them (Equation (5)). However, the threshold of ±2% could potentially be reduced, which will give more confidence to the results [30]. Even so, this threshold for CCS measurements can still result in being more effective than isotopic pattern or fragmentation criteria to reduce the number of false positive results in automated screening workflows and avoid the requirement of post manual verification or confirmation analysis [31].

$$\% \text{ CCS error} = \frac{\left(\text{CCS}^{\text{measured}} - \text{CCS}^{\text{database}}\right)}{\text{CCS}^{\text{database}}} \times 100 \tag{5}$$

2.3. *IMS Hyphenation*

IMS separations typically take place in the millisecond range, so they can be easily carried out after traditional chromatographic or electrophoretic separations (i.e., LC, GC, SFC, and CE), which occur in the second range. The selection of the chromatographic/electrophoretic technique is obviously influenced by the nature of analytes (e.g., SFC for polar compounds, CE for polar and ionic compounds, GC for volatiles). Moreover, MS separations last microseconds so IM-MS hyphenation is also possible. IMS is normally coupled to time-of-flight (ToF)-MS due to its high acquisition rate. Indeed, as indicated in Section 2.1, several IMS systems combined with ToF-MS technology are already available on the market as integrated instruments.

GC was one of the first analytical techniques to be coupled to IMS, which has been mainly used as a detector in GC–IMS configurations [3,32]. GC–IMS hyphenation has been widely applied to the analysis of volatile compounds in food samples because stand-alone IMS systems generally provide low resolution. Both analytical techniques present a high degree of orthogonality, so selectivity and peak capacity are improved. However, LC–IM–MS platforms are currently becoming very popular and are used for numerous applications [4]. In this context, the integration of IMS in LC–MS workflows introduces a third separation dimension that improves peak capacity and allows the separation of isobars and isomers [33]. The implementation of IMS in LC–MS workflows increase peak capacity at least 2 or 3-fold [34,35], although this improvement ultimately depends on IMS resolution and the analytical application. Moreover, chromatographic peaks are extracted from background noise which provides cleaned-up chromatograms and mass spectra, leading to an improvement of the limits of detection (LODs) and sensitivity. Mass spectral data can also be interpreted more easily. Undoubtedly, these characteristics have contributed to the implementation of IMS technology in the food science field where samples present a high complexity and some compounds need to be determined at low concentration levels (i.e., ppb and ppt ranges).

From a technical point of view, limited dynamic range and ion loss have been traditionally attributed to LC–IM–MS in comparison to LC–MS methods [36]. However, several technological advances have been accomplished in the last years and, for example, some recent studies indicate that hyphenation with an ion mobility device does not affect the linear dynamic range [37]. Furthermore, it is important to remark that each IMS form can improve the performance of LC–MS methods in one way or another, but they are not exempt of limitations [4,14]. Therefore, the choice of the IMS technology will depend on the purpose of the intended application. In general, FAIMS and DMS provide higher orthogonality to MS than DTIMS and TWIMS but they act as signal filters according to the CV selected. Consequently, information about the sample is lost because the number of detected compounds is limited.

Finally, independent of whether stand-alone IMS, IM–MS or LC–IM–MS platforms are used, samples need to be ionized prior to entering into the mobility cell. Nowadays, electrospray ionization (ESI) is the gold standard for IMS analyses, although ^{63}Ni radioactive ionization sources are widely used in portable IMS instruments. Other ionization sources such as photoionization, corona discharge (CD) and pulse glow discharge (PGD) have also been used in food applications, but to a lesser extent [8]. Sample ionization is a process to be controlled because, as well as ion transportation and storage, it influences the conformation of ions and, consequently, their mobility. In addition, different adducts and ions with different charges states and sites (e.g., protomers) can be detected for the same compound [20], and their formation during the ionization can be altered by experimental conditions (i.e., ionization source and related parameters, solvents composition, etc.). In the context of food analysis, specifically in food safety, the identification of protomers is of high relevance since it could justify why some compounds lead to non-compliance with confirmation criteria based on ion fragmentation ratios and are not detected in screening analyses [38,39]. In addition, if it is applicable, a different t_d, K or CCS could potentially be obtained for each ion which provides more information for peak annotation. But it also hinders data treatment due to the number of species or 'features' that are detected. In this sense, data treatment generally remains as the main bottleneck of hyphenated LC–IM–MS systems, and a higher development is still required to successfully integrate IMS data in current LC–MS workflows.

3. Applications of IMS in Food Analysis

The application of IMS in food science is still in its early years, especially those approaches involving IM–MS or LC–IM–MS hyphenation. Despite IMS, and specifically IM-MS, has found its main application in omics approaches (i.e., proteomics and metabolomics, including lipidomics and glycomics) [40–42], this technique has barely been exploited for the specific analysis of composition and nutritional value of food, and their health benefits and risks (i.e., foodomics). Nevertheless, it is expected that its use will be extended rapidly to foodomics where a wide range of food metabolites need to be characterized for understanding their effects on human health. For example, the effects of coffee consumption on lipids profile (i.e., 853 lipid species from 14 lipid classes) have been investigated by DMS-triple quadrupole (QqQ)/linear ion trap (LIT)–MS [43]. This lipidomics approach has suggested that coffee intake alters glycerophospholipid metabolism and supports previous studies about the health benefits of drinking coffee.

The interest for this technology is growing and an increasing number of IMS-based methods have already been developed for food analysis, especially for food safety and food authentication purposes. A selection of the most recent applications of IMS in food science, including food composition, process, authentication, adulteration, and safety, is presented below. This section intends to provide a general vision of the potential of this technique for food analysis rather than being a comprehensive summary of all related articles that have been reported in the last years.

In general, IMS applications in food science cover from the determination of a predetermined number of compounds (i.e., targeted analysis) to the analysis of a large non-predetermined set of molecules (i.e., non-targeted analysis). As will be shown in the following sections, IMS has mainly been applied to targeted or semi-targeted analysis until now. However, it is expected that it will be widely used in non-targeted approaches, such as in food fingerprinting, where a large number of compounds are detected and the performance characteristics of analytical methods need to be improved (i.e., requirement of higher resolving power, tools to support molecular identification, simplification of mass spectra, etc.). On the other hand, current IMS applications in food analysis can be classified in three types: 1) those using stand-alone IMS instruments in which their intrinsic low selectivity is not a limitation, 2) approaches where IMS is coupled to a chromatography technique, usually GC, and mainly acts as a detector, and 3) applications based on LC–IM–MS workflows (alternatively

GC–IM–MS) in which the potential of IMS is fully exploited [i.e., extra separation dimension that improves selectivity and sensitivity (separation of isomers and isobars, and isolation of compounds of interest form background noise), compound selection based on their CVs (in FAIMS and DMS), CCS as an additional molecular descriptor (except in FAIMS and DMS)].

3.1. Food Composition

From a chemical composition point of view, food consists of a wide variety of compounds that can be divided in macronutrients (i.e., proteins, carbohydrates, lipids and vitamins) and micronutrients (e.g., polyphenols, etc.). Nowadays, IMS applications in food composition cover a wide variety of substances (e.g., lipids, peptides, phenolic compounds, terpenes, etc.) [44–47], including allergens [48], and samples (e.g., flours, olive oil, mushrooms, etc.) [48–50]. Due to the complexity of food matrices, stand-alone IMS has rarely been investigated in food composition studies, although it has found a wider application in food process analysis where few and specific compounds are determined, as will be shown in Section 3.2. As an example, stand-alone IMS has been used for the analysis of seven alcohol sweeteners in chewing gum, identifying the presence of sorbitol [51]. Other peaks were also observed in the ion mobility spectrum which were attributed to gum base components. However, this fact highlights the requirement of IM–MS for proper identification. Furthermore, more analytical information is obtained when IMS is combined with other analytical techniques such as LC or GC. Chromatographic and IMS separations are not correlated [44,46] (Figure 3), so their coupling improves peak capacity and enables the detection of a larger number of compounds. This advantage is widely used for food fingerprinting at low cost [50,52,53], without applying MS which is an expensive technology. In this case, if complete food characterization is required, standards of those substances that are potentially present in the sample must also be characterized in terms of retention time and mobility, drift time, or CCS. Consequently, peaks resulting from sample analysis can be tentatively identified (MS is mandatory for identity confirmation).

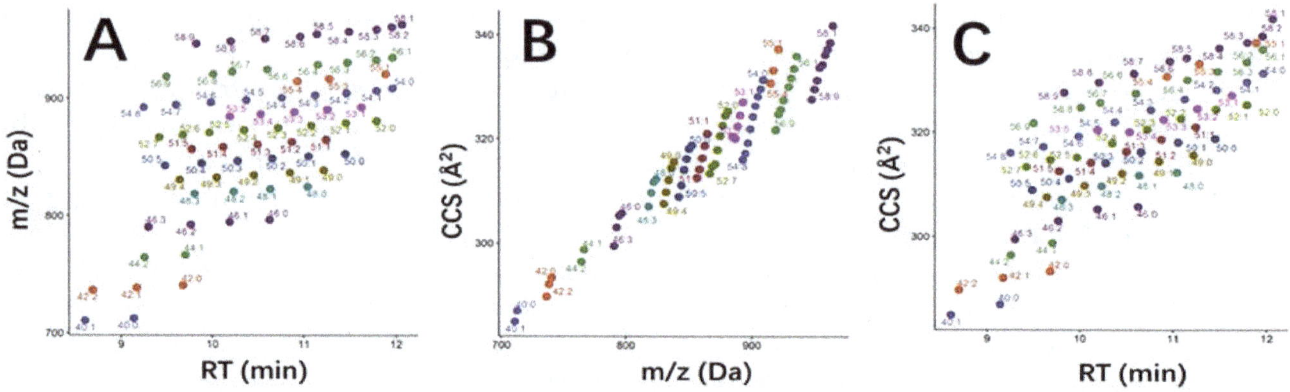

Figure 3. Selected triacylglycerides found in milk and comparison of the measured orthogonal parameters: (**A**) m/z vs. retention time (RT), (**B**) collision cross section (CCS) vs. m/z, and (**C**) CCS vs. RT. Figure reprinted with permission from [44]. Copyright (2018) American Chemical Society.

In this context, the volatile organic compounds (VOCs) fraction of olive oils has been widely studied by GC–IMS, and headspace (HS) analysis has been usually carried out for this purpose [49,52,53]. In general, IMS offers high sensitivity allowing the detection of olive oil volatiles in the range of 0.01–0.05 ppm, which is usually below the odor threshold of many compounds in extra virgin olive oils (EVOO) [53]. Several aldehydes (e.g., hexanal, 2-hexenal), ketones (e.g., 1-penten-3-ol), alcohols (e.g., 1-pentanol, 1-hexanol, etc.) and esters (e.g., hexylacetate), which are related to desirable attributes of olive oils, are present in their VOCs fraction. Their targeted analysis by HS–GC–DTIMS combined with chemometrics has been shown to be an effective strategy to characterize olive oils and distinguish them according to their quality (i.e., extra virgin, virgin and lampante) [49]. As alternative to HS analysis, olive oils samples can be submitted to laser desorption–GC–IMS analysis for the simultaneous characterization of VOCs fraction and detection of semi- and non-volatile compounds (e.g., (*E*,*Z*)-2,4-dodecadiene, 1-dodecene, (*E*)-2-hexenal, phenol, ß-pinene, benzaldehyde, acetic acid, limonene, 1-hexanol, nonanal, 1-heptanol, and octanal [54]. In comparison to HS, a higher number of signals were detected when laser desorption was applied to the analysis of oil samples. Despite the analysis of certain markers providing useful information about olive oils quality, statistical models for olive oils classification are improved when oil fingerprinting (i.e., semi-targeted or non-targeted analysis) is carried out [52,53]. This is because food fingerprinting gives a more complete picture of samples composition than the analysis of specific compounds. In the case of using GC–IMS methods, this implies that samples are differentiated according to biomarkers not identified, but characterized analytically. In addition to olive oils, HS–GC–DTIMS has also been applied to characterize the volatile fingerprint of fresh and dried *Tricholoma matsutake* Singer samples [50] and green tea aromas [55].

Unlike GC–IMS, LC–IMS coupling has not been evaluated so far for food composition analysis according to the literature. Nevertheless, the determination of phenolic acids in seedling roots by high-performance liquid chromatography (HPLC)–ESI–DTIMS has been recently reported [56], and this application can be extended to food composition analysis. Several compounds co-eluted in the chromatographic dimension (i.e., vanillic acid and caffeic acid; ferulic acid and sinapic acid; benzoic acid and salicylic acid), but they were separated in the ion mobility dimension. It demonstrated the potential of IMS for improving analytical separations at low analysis times (< 30 ms).

A new trend is to integrate IMS in LC–MS workflows for achieving the separation of isomers and isobars that are not resolved in the chromatographic dimension. For example, milk oligosaccharide isomers lacto-N-hexaose (LNH) and lacto-N-neo-hexaose (LNnH) are difficult to separate by LC and tandem MS provides similar fragmentation patterns in both positive and negative mode. However, both carbohydrates are baseline resolved by IMS under negative ionization conditions [57]. In this case, the statement 'IMS separates ions rather than molecules' is clearly exemplified. Deprotonated forms of LNH and LNnH were separated because they presented different CCS, but it was not the case for their sodium adducts. Therefore, ionization conditions have a special relevance for IMS results and, if there are not ionization constraints (i.e., ion suppression, low ionization for targeted analytes), they must be investigated when performing IMS experiments.

In general, IMS is considered as a third separation dimension in LC–MS workflows, but it can also act as a fourth dimension if a two-dimensional (2D) chromatographic separation is performed prior to IM–MS analysis. Food matrices are highly complex and one single LC or GC separation can be insufficient to provide a satisfactory degree of resolving power for their characterization. Advanced approaches such as LC × LC and GC × GC have been developed to enhance peak capacity which is increased even more when IMS is integrated in LC × LC–MS workflows, as shown in the analysis of phenolic compounds in chestnut (i.e., ellagitannins and gallotannins), grape seed (i.e., procyanidins), rooibos tea and red wine (i.e., flavonoid and non-flavonid phenolics) [58]. TWIMS improved the practical peak capacity from 7- to 17-fold (depending on the sample) in comparison to a 2D–hydrophilic interaction liquid chromatography (HILIC)-reverse phase (RP)–LC–ToF–MS method. Regarding the orthogonality degree of separation methods, lower orthogonality was found for a HILIC–TWIMS combination (52%), whereas the orthogonality existing between RP-LC and TWIMS was surprisingly higher (73%) than the orthogonality resulted from HILIC–RP–LC combination (67%). 2D–LC–IMS–MS approaches clearly reveal a higher complexity of food samples and improve the probabilities of distinguishing between isobaric and isomeric compounds [59]. For example, 2D–ultra-high performance liquid chromatography (UHPLC)–TWIMS–ToF–MS has been applied to the characterization of ginsenosides in white and red ginsengs which are diet supplements, and from the 201 compounds identified, 10 pairs of co-eluting isobaric ginsenosides were resolved only by IMS [60]. On the other hand, 2D–GC–IMS–MS methods have not been reported for the analysis of food samples, but the potential of this strategy has already been shown for the analysis of volatiles in medical herbs [61]. Consequently, similar approaches can be applied to the determination of the VOCs fraction of foodstuff.

Finally, IMS provides structural information of the ionized compounds (i.e., CCS), giving a more complete overview of food composition [44,62]. In the context of food composition analysis, $^{DT}CCSN_2$ libraries have been developed for the characterization of phenolic acids in wine [46] and lipids in bovine milk [44]. In this last case, lipid identification rates were increased when CCS, in addition to retention time and accurate *m/z*, was considered as molecular descriptor. Machine-learning algorithms were developed for the classification of 429 lipids according to their family class (i.e., triacylglycerides, diacylglycerides, phosphatidylcholines, and sphingomyelins) and carbon number. In general, a satisfactory classification rate (84.01%) was achieved when retention time and *m/z* were selected as analytical parameters. Classification accuracy was increased to 91.78% when CCS was included in the model, mainly due to the separation of isomeric species. This model was further improved to include the unsaturation level and, as a proof of concept, finally applied to the classification of 2087 bovine milk lipidomics data. As a result, 429 lipids, which were previously identified, were accurately classified whereas 179 unknown lipids were also annotated confidently. The identification of steviol glycosides represents another example of the applicability of CCS values for food characterization [62]. Steviol glycosides encompass a wide range of isomeric compounds and related molecules that can co-elute or result in similar fragment ions, impacting identification certainty. In order to improve the characterization of food products containing steviol sweeteners, a library of $^{TW}CCSN_2$ values was developed and applied to the fingerprinting of 55 food commodities. $^{TW}CCSN_2$ values contributed to the identification of several isomeric pairs such as rebaudioside E and rebaudioside A ($[M-H]^-$ *m/z* 965.4230, CCS = 289.2 and 298.8 Å^2) and stevioside and rebaudioside B ($[M-H]^-$ *m/z* 803.3701, CCS = 269.6 and 261.2 Å^2), without requiring ion fragmentation for identification, which cannot provide relevant information at low analyte concentrations.

3.2. Food Process Control

Food process control requires rapid response analytical tools, mainly intended as the analysis of many volatile compounds, in order to monitor industrial processes in near-real time and make rapid decisions if needed. In this sense, VOCs fraction can be related to storage conditions and production process, being also an indicator of the shelf life of food products. Consequently, volatiles are usually analyzed as part of food process and quality control. Stand-alone IMS represents a good option for this purpose because it allows the quick detection of volatile compounds (~20 s) at low cost, which is also attractive for industrial companies.

IMS has been widely applied to control the freshness of food products with special attention to the determination of biogenic amines [7,8], which are usually associated with fermentation processes and food degradation. Stand-alone IMS devices, which typically consist of DTIMS systems, are commercialized with different ionization sources. Therefore, the instrument to be used depends on the nature of analytes and their ionization characteristics. This is not a disadvantage for the analysis of biogenic amines by IMS because similar LODs can be reached (i.e., 0.1–1.2 ppm expressed as vapor in air for trimethylamine, putrescine and cadaverine), independent of the ionization source (i.e., photo-ionization, corona discharge, and radioactive ion sources) [63]. Stand-alone IMS methods have been reported for the determination of histamine in tuna fish [64], trimethylamine in seafood [65], and the simultaneous analysis of histamine, putrescine, cadaverine, tyramine in canned fish [66]. Stand-alone IMS applications have also been developed to detect the presence of off-flavors and contaminants generated during food processing and storage such as 2,4,6-trichloroanisole [67], and furfural and hydroxymethylfurfural toxicants [68].

Due to the low selectivity of stand-alone IMS instruments, the combination of HS–GC with IMS is recommended for the simultaneous analysis of several volatile compounds in liquid and solid samples in food quality control [5]. Despite chromatographic separations requiring several minutes, HS–GC–IMS can be less time-consuming and more environmentally friendly than other analytical and physico-chemical approaches currently used in food process control [69]. HS–GC–IMS methods have been applied to study lipid oxidation processes experienced by roasted peanuts [70] and EVOO [69] during storage.

In addition to the selective analysis of specific volatile substances, the profiling of VOCs fraction by IMS approaches can provide more complete information about the freshness and storage conditions of food products such as fish [71] and eggs [72]. VOCs profiling by GC–IMS usually requires chemometrics for data treatment in order to classify the samples, although the majority of compounds remain unidentified. It does not represent a limitation for sample classification, but compound identification is essential to understand fermentation and other decay processes. For example, up to 35 potential markers related to freshness were detected in the VOC fraction of eggs, but only a few compounds (i.e., butyl acetate, heptanal, dimethyl disulfide, dimethyl trisulfide and 1-butanol) were identified (Figure 4) [72]. Nevertheless, eggs were correctly classified according to their freshness (from 0 to 5 days at room temperature) by a principal component analysis (PCA) model. A supervised orthogonal partial least square discriminant analysis (OPLS–DA) model was finally developed to maximize the separation of both groups (i.e., fresh eggs vs. non-fresh eggs).

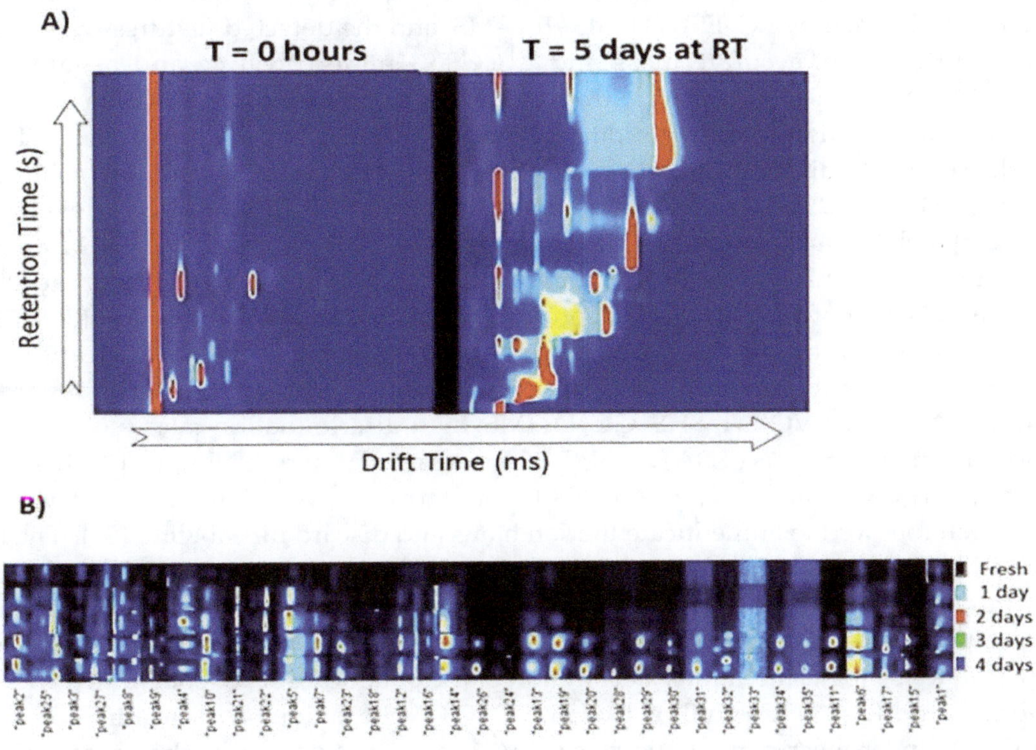

Figure 4. (**A**) Ion mobility spectrum of an egg product at T = 0 h (**left**) and after 5 days at room temperature (**right**). The red line identifies the reaction ion peak (RIP) position. (**B**) Global overview of the spots or 'features' identified in one egg product at different time points (from 0 to 4 days). Figure reprinted with permission from [72]. Copyright (2019) Elsevier.

Volatile compounds are not only related to the shelf life of food products and their degradation. The monitoring of VOCs fraction can also be decisive in food process and quality. For example, beer fermentation is conditioned by the presence of diacetyl and 2,3-pentadione whose concentrations must be reduced below the human odor threshold. In comparison to traditional methods, the use of GC–IMS for their determination allows decreasing the analysis time from 3 h to 10 min [5]. GC-IMS has also been applied to the characterization of VOCs fraction during the fermentation of lychee beverages [73] and wine [74]. One of the advantages of GC–IMS over other analytical tools is that it enables online monitoring and control of bioprocesses [74]. However, GC separations still require several minutes, which is a disadvantage of real-time decisions. By contrast, IM–MS analyses take a few milliseconds allowing the direct monitoring of VOCs evolution, which involves time- and cost-saving on sample preparation as well as avoids process interruption. Corona discharge–DTIMS–ToF–MS has been proposed for the on-line monitoring of VOCs formation in coffee roasting processes [75]. More than 150 VOCs were observed during the roasting of Brazilian *Coffea arabica* beans, and several alkyl pyrazines, fatty acids and other organic acids were identified. Some of them were isomers and isobars, so MS was not enough to identify them. Nevertheless, the integration of DTIMS in the analytical workflow allowed their separation according to their $^{DT}CCSN_2$ (e.g., unsaturated fatty acids presented smaller $^{DT}CCSN_2$), facilitating their identification.

3.3. Food Authentication

Chemical fingerprinting of food products and chemometrics are widely used for food authentication and ultimately to identify food fraud [76]. Within this framework, metabolomics fingerprinting is a promising strategy for food authentication that combines both approaches. In this sense, it is very efficient and has already shown clear benefits over traditional methods [77]. At the moment, only one study has been reported about the applicability of LC–IM–MS in metabolomics fingerprinting for food authentication [78]. The metabolic fingerprint of 42 red wines from the Republic of Macedonia was obtained by LC–ESI–DTIMS–ToF–MS, and the detected features were characterized in terms of retention time, accurate mass and $^{DT}CCSN_2$ under positive and negative ionization conditions. $^{DT}CCSN_2$ values added an extra-identification point for the putative identification of several phenolic compounds and other grape reaction products. After data treatment by PCA, Vranec wines were clearly distinguished from other varieties such as Cabernet Sauvignon or Merlot wines.

In general, IMS-based methods for food authentication have mainly been developed for the fingerprinting of volatile and semi-volatile compounds, usually involving GC–IMS hyphenation. GC–IMS methods in combination with chemometrics have been successfully applied to the authentication assessment of oils [79,80], meat products [81,82], wines [83] and honey [84]. For example, EVOO are 'designation of origin' (DO) products of high value, so their authentication is required not only to detect any fraud related to oil quality but also to identify mislabeling regarding origin. Based on the analysis of the VOCs by HS–GC–DTIMS, high discrimination rates (i.e., 98% and 92% by PCA-linear discriminant analysis (LDA) and PCA-k-nearest neighbors (kNN), respectively) have been reached to effectively discriminate between EVOO from Italy and Spain [80]. A similar HS–GC–DTIMS approach has been followed to authenticate Iberian hams and discard mislabeling [81]. The monitoring of specific non-identified markers by HD–GC–DTIMS and the application of orthogonal projections to latent structures discriminant analysis allowed to distinguish hams with a discrimination rate of 100%. As a third example, a HS–GC–DTIMS method has also been applied to identify the botanical source of different types of honey [84], since faulty declaration of the botanical source constitutes one of the main frauds on honey products. VOC profiling by HS–GC–DTIMS provided a discrimination rate (>98.6% according to PCA-linear-discriminant analysis (LDA)) as high as proton nuclear magnetic resonance (^{1}H-NMR) spectroscopy, which has been widely investigated for honey authentication in the last years.

Despite the efficiency showed by GC–IMS for the authentication of food products, the information resulting from GC–IMS analyses is very limited since MS is not included in the analytical workflow and markers cannot be identified. Only those compounds whose standards are available in the laboratory can be identified according to their retention and drift time, whereas the majority of compounds in the sample remain unknown. Nevertheless, discrimination of food products of different geographical origin, nature, quality, etc. can be achieved by GC–IMS–chemometrics methods based on unidentified markers. These strategies are usually recommended as screening approaches since molecular identities are generally required for confirmation.

3.4. Food Adulteration

The detection of food adulteration represents a particular case of food authentication in which substances are intentionally added to food products to decrease and mask their quality or valuable ingredients are removed in order to obtain higher economic profit. Consequently, similar IMS-based approaches as those shown in Section 3.3 have been proposed for the identification of this type of food fraud. HS–GC–IMS in combination with chemometrics has been evaluated for detecting the adulteration of winter honey (derived from *Schefflera actinophylla* (Endl.) *Harms* and wild *Eurya* spp. (Theaceae)) with cheaper *Sapium* honey [85], canola oil with other vegetable oils (i.e., sunflower, soybean, and peanut oils) [86], and crude palm oil with process byproducts of lower quality (i.e., palm fiber oil and sludge palm oil) [87]. As previously mentioned, GC–IMS methods should mainly be used as screening approaches because MS results essential for confirmation. However, screening methods are required for rapid analysis at lower cost and are of special interest for carrying out in situ analyses. Thus, only a few suspicious samples are sent to the laboratory for confirmation. In this context, stand-alone in combination with chemometrics has been investigated for the rapid detection of EVOO [88], sesame oil [89], and flaxseed oil [90] adulteration with vegetable oils of lower quality. Therefore, and despite the lack of selectivity of IMS, these applications demonstrate its efficiency to initially discriminate between allegedly adulterated samples and samples that are in compliance with their quality and labeling.

3.5. Chemical Food Safety

Chemical food safety covers the determination of a wide range of residues and contaminants (i.e., pesticides, veterinary drugs, toxins, environmental contaminants, etc.) in feed and food-related matrices; being one of the food science fields where IMS has found a wider application. Within this framework, the analysis of pesticides in a great variety of vegetables (i.e., apples, tomatoes, cucumbers, etc.), juices, oils, animal feed and water samples has been the topic most investigated [8]. IMS has been shown to be a solution for answering current society's concerns such as the presence of glyphosate in drinking water (as well as in other food products). The determination of glyphosate is quite difficult due to its ionic character, low volatility and low molecular weight. However, the LOD achieved by IMS for drinking water analysis (i.e., 10 μg/L) is comparable to those reported by HPLC and GC–MS (i.e., 0.02–50 μg/L) or ion chromatography (i.e., 15.4 μg/L) methods [91]. Despite of the potential of IMS for food safety applications, other food contaminants and residues rather than pesticides have been scarcely studied by IMS [8].

The analysis of residues and contaminants in food products faces the same issues as other areas of food analysis (i.e., high complexity of food matrices requiring high selective analytical techniques, presence of isomers and isobars, and requirement of detecting compounds at very low concentration levels) but, in this particular case, it must also comply several regulations according to the compounds analyzed. From a legal perspective, the employment of MS is practically mandatory in order to reach enough identification points (IP), so IM–MS related methods are of major interest for chemical food safety applications. However, IMS-based methods do not follow current guidelines and regulations concerning analytical methods intended to the determination of pesticides (e.g., SANTE/11813/2017 [92]), veterinary drugs (e.g., Regulation 2002/657/EC [93] or mycotoxins (e.g., SANTE/12089/2016 [94]). Indeed, because this particular technique has not been considered in the corresponding documents and no IP is allocated to CCS determination yet, this impairs its use as official identification criteria. For instance, scientists currently working is the revision of decision 2002/657/EC are debating the value of this new parameter as a criterion for identification and, if necessary, the value of the IP awarded [(acceded on 12 June 2019), [95]].

Under this context, stand-alone IMS and GC–IMS methods have been widely proposed for chemical food safety analyses [8], but they are outside the current frame of food law enforcement. Moreover, despite the use of stand-alone IMS has been shown to be an effective strategy for the rapid and in situ detection of residues and contaminants [96], its selectivity is limited and samples must normally be submitted to selective sample treatments prior analysis. In this sense, molecularly imprinted solid-phase extraction (MIPSE) and immunoaffinity chromatography (IAC) have been commonly employed for the selective extraction of the analytes of interest and to avoid matrix background.

For example, IAC has been applied to the analysis of fungicides in strawberry juices and wines [97] and mycotoxins (i.e., aflatoxins B1 and B2) in pistachios [98], whereas MISPE has been use to study the uptake and translocation of a neonicotinoid pesticide (i.e., imidacloprid) in chili and tomato plants [99]. Regarding GC–IMS methods, IMS has been useful to separate pesticides that were not resolved in the chromatographic dimension [100,101], although in general it merely acts as detector [32].

In chemical food safety, tandem mass spectrometry (MS/MS) and high-resolution mass spectrometry (HRMS) are usually employed with the aim of achieving the unequivocal identification of compounds. The combination of IMS and HRMS or MS/MS has commonly been applied to clean up mass spectra by removing matrix interferences and signal background. As a result, the task of mass spectra interpretation is reduced and compounds at lower concentrations are detected more easily. FAIMS and DMS are ion filters and this characteristic has been exploited to reduce the signal background in the determination of the mycotoxin zearalenone and its metabolites in cornmeal [102]. LODs were consequently improved up to 25 and 42.5-fold when applying FAIMS–ESI–MS instead of ESI–MS/MS or ESI–MS, respectively. Despite the characteristic of ion filtering is usually assigned to FAIMS/DMS technology, DTIMS and TWIMS are also able to isolate analyte signals from chemical background, improving concentration sensitivity [30,103]. In the context of chemical food safety, signal to noise ratio (S/N) of LC–ESI–ToF–MS methods intended to monitoring steroids and their metabolites in livestock were increased between 2 and 7 fold by the integration of TWIMS in the analytical workflow (Figure 5) [30].

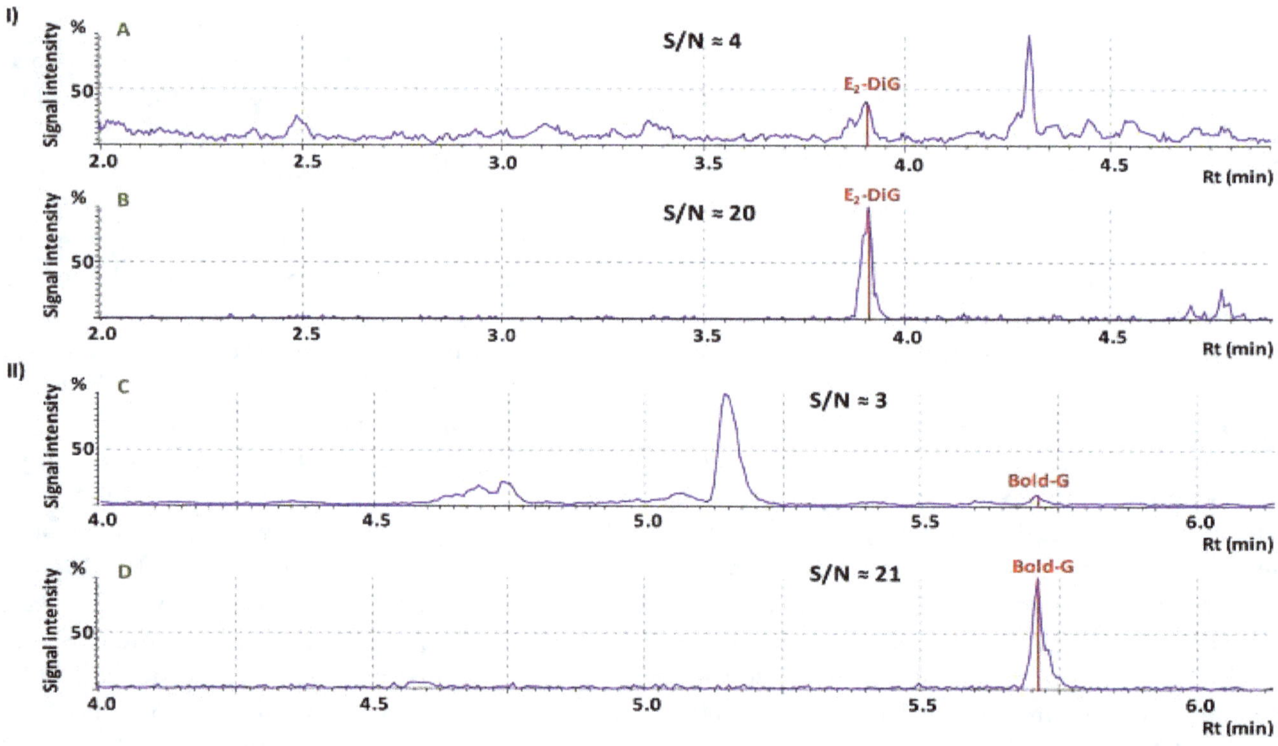

Figure 5. Extracted ion chromatograms (EICs) resulted from the analysis of: I) estradiol diglucuronide (E_2-DiG; 2 µg mL^{-1}; $[M + Na]^+$), and II) boldenone glucuronide (Bold-G; 0.2 µg mL^{-1}; $[M - H]^-$) in adult bovine urine samples. The following filters were applied for signal processing of related total ion chromatograms: A) *m/z* 647, B) *m/z* 647 and drift time range between 11.3 and 11.7 ms, C) *m/z* 461, D) *m/z* 461 and drift time range between 4.9 and 5.2 ms. Figure adapted from [30], which is licensed under CC BY-NC-ND 4.0 (changes: example III has been removed).

On the other hand, IMS–ESI–MS methods provide rapid throughput as observed in the separation of isomeric perfluoroalkyl substances (PFAS) by DMS–trap quadrupole (LTQ)–MS [104]. Nevertheless, food samples are very complex and matrix compounds can cause ion suppression when applying

ESI ionization. This issue has a great impact on signal sensitivity so it must be avoided for achieving robust and confident analyses. Thus, samples must normally be submitted to LC (or GC, SFC, CE) separation after specific sample treatments and prior to IM–MS analysis in order to overcome this effect. For example, FAIMS coupled to LTQ–Orbitrap has been shown to be a useful tool to separate paralytic shellfish toxins epimeric pairs [105], but their analysis in shellfish tissue extracts required a previous HILIC separation due to severe ionization suppression from matrix components. In general, 3D–LC–IM–MS separation is usually recommended when IMS is applied within the framework of food safety.

In this sense, LC and IMS are complementary separation techniques because they are based on different separation principles. Saxitoxins, which are a class of marine neurotoxins present in shellfish, comprise a group of isomers that are not differentiated by MS. Some diastereomers can be separated by HILIC but it does not distinguish between non-sulphated saxitoxins analogues. These molecules can be separated by TWIMS which makes HILIC–TWIMS–MS the best solution for the analysis of these group of substances in one single run [106]. Other examples of isobars and isomers separation by IMS include veterinary drugs such as ractopamine/isoxuprine ($[M + H]^+$; m/z 302.1751, $^{TW}CCSN_2 = 171.9$ Å^2 and m/z 302.2025, $^{TW}CCSN_2 = 173.6$ Å^2, respectively) by TWIMS [107] and isomeric environmental contaminants such as 2,2′,4,5′,6-pentachlorobiphenyl (PCB-103) and 3,3′,4,4′,5-pentachlorobiphenyl (PCB-126) ($[M - Cl + O]^-$; m/z 304.9095, $^{DT}CCSN_2 = 160.7$ and 164.4 Å^2, respectively) by DTIMS [108]. Furthermore, if the separation of isobars and isomers cannot be achieved under standard conditions in the drift cell (i.e., only containing the drift gas), organic solvents (e.g., acetonitrile, methanol, isopropanol, etc.) can be added into the buffer gas in DMS and FAIMS systems in order to improve ion mobility separation [109]. This approach has been followed for the separation of the neurotoxin β-*N*-methylamino-L-alanine (BMAA) and its isomer β-amino-N-methylalanine (BAMA) in mussel tissues [110]. Both isomers were only resolved by HILIC–DMS–QqQ/LIT–MS when 0.35% acetonitrile was added in the DMS carrier gas. DMS also improved the low sensitivity traditionally observed for the analysis of BMAA by LC–MS methods.

Finally, as previously mentioned, IMS also provides additional analytical information to support molecular assignment. For example, drift time has been used to identify 100 pesticides in different vegetables and fruits by LC–TWIMS–ToF–MS [111]. This analytical property is not influenced by the matrix but drift times are instrument-dependent, so these data cannot be extrapolated to other IMS platforms. For this reason, the use of the CCS as the information provided by IMS has been extended. Consequently, CCS databases can be created and applied to support the determination of compounds, reducing the number of false negative/positives found in classical LC–MS workflows [112], and increasing the detection rates at low concentration levels of residues [39]. In the current context of Regulation 2002/657/EC revision, it has emerged the proposal to include the CCS as IP for the identification of veterinary drugs in food [95]. Scientists and experts are currently debating the value of grating IPs to the CCS and, if so, to determine its contribution in the 4 and 5 IPs required for the confirmation of substances with maximum residues limits (MRLs) and non-authorized substances, respectively. However, several concerns have been raised such as the lack of CCS databases and the tolerance accepted for CCS measurements. While, until a few years ago, CCS had been barely used as verification parameter in screening approaches due to the lack of databases, this situation has now recently changed and several CCS databases for pesticides [112,113], veterinary drugs [114–116], mycotoxins [117] and other contaminants [108], have been reported in the last five years.

Finally, recent applications of LC–ESI–TWIMS–ToF–MS methods, which include the CCS as signal filter in data processing, report the analysis of mycotoxins in cereals [117] and pesticides in green tea powder, fresh garlic, leek, fresh herb chives and rye [112], fish feed [31], and in 20 food products (e.g., strawberries, honey, chia seed, etc.) that belong to six different commodities according to SANTE/11813/2017 [39]. Other LC-TWIMS-MS applications in the context of food safety encompass the identification and structural characterization of residues and contaminants, either for evaluating exposure [118] or for studying their biotransformation [119].

4. Current Perspectives of Ion Mobility Spectrometry

Significant progress has been made in IMS technology in the last years, especially in IM–MS hyphenation. This fact has been directly reflected in the number of IMS-related applications that are currently published in various scientific fields including food science, and the number of related papers will keep growing due to technological developments in this analytical technique. Advances are expected in three different senses: IMS coupling with MS technologies offering higher resolution than ToF-MS (i.e., Fourier transform ion cyclotron resonance (FTICR) and Orbitrap–MS), improvement in IMS resolution, and implementation of the CCS as a parameter to support molecular assignment. In this section, only the last two topics will be discussed in more detail because the first is more related to improvements in MS acquisition rates than to IMS evolution. Several prototypes based on DTIMS/TWIMS–FTICR/Orbitrap–MS hyphenation have already been proposed; so further developments are expected [41]. These developments will certainly open up new possibilities for enhanced food fingerprinting, compounds identification and, ultimately, the discovery of substances, including contaminants, present in food but not detected until now.

4.1. Improvement in Peak Resolution

The resolving power (R_p) of IMS is typically expressed in the scale of CCS for IMS-platform comparison, although it excludes FAIMS/DMS systems. R_p is calculated according to Equation (6) where ΔCCS represents the full width of the peak at half its maximum height (FWHM).

$$R_p = \frac{CCS}{\Delta CCS} \tag{6}$$

As a consequence of the R_p of IMS, the integration of this analytical technique in LC–MS workflows increase their peak capacity. Thus, it can be expected that LC-IM-MS will quickly replace LC–MS methods in many applications, including food analysis. Current IMS instrumentation provide a R_p up to 300, so molecules presenting CCS differences (Equation (7)) as small as 0.5% can be separated [120]. However, commercial systems are not able to routinely offer this type of performance. TIMS is the only capable to provide a R_p higher than 200 [41], which is considered the lower limit of ultra-high resolution (UHR)IMS [23]. R_p of DTIMS systems is around 100 whereas TWIMS instruments currently reach a maximum R_p of 40 [14,23]. Therefore, the development of UHRIMS and its implementation in routine analysis obviously represent some of the main challenges of IMS, but it will also contribute to extend the use of this technique.

$$\Delta CCS_{A,B}(\%) = \frac{CCS_B - CCS_A}{\text{average } CCS_{A,B}} \times 100 \tag{7}$$

As a result of the improvements on R_p expected in UHRIMS technology, IMS separations will be able to provide similar peak capacities as LC at lower analysis times, which will be revolutionary from an analytical point of view and transformative for several applications [121]. In general, limitations on the applied field involve that R_p can only be increased by extending the drift path, although it presents some practical constraints [17,23]. In this sense, two IMS approaches (i.e., cyclic-TWIMS and structure for lossless ion manipulations (SLIM)–TWIMS) have successfully overcome these drawbacks and, consequently, are gaining attention due to their enhanced R_p. In addition, a third alternative was previously explored, namely the high resolution ion cyclotron mobility spectrometry (cyclic-DTIMS) [122], which is a circular 180.88 cm-length drift tube consisting of four quarter-circle drift tubes with ion funnels in between to re-focus the ions. Ions undergo multiple passes through the drift path in order to improve their separation. Successive improved versions of this technology have been shown to reach a R_p in excess of 1000 (as a consequence of 100 transits or cycles and involving a drift length of over 180 m) [123]. Nevertheless, no further research has been reported about cyclic-DTIMS, probably due to the long measurement times required and the significant loss of ions at long drift times of more than a second [23].

In cyclic-TWIMS, the drift cell is arranged in a circle configuration with pre- and post-store cells for ejecting the ions in and out [124,125]. As a result, ions can be submitted to several passes enhancing their separation. The resolution of multiple passes is given by Equation (8) where A is the single pass resolution (~65, 98 cm pass length), n represents the number of passes, and z is the ion charge state.

$$R_P = A(nz)^{1/2} \tag{8}$$

One single separation can be enough to obtain a general overview of samples and select the ATD regions of interest that will be submitted to multiple passes. The selection of ATD regions is required in order to avoid that the fastest ions trap the slowest ones, causing a 'wrapping effect'. Due to its enhanced R_p, cyclic-TWIMS provides new insights of ions conformation not previously described, as observed for the three protomeric species of the veterinary drug danofloxacin (Figure 6) [126]. As discussed above, it is relevant to characterize protomers in the food safety context because they do not fragment in the same way and results could be non-compliant with confirmation criteria based on ion fragmentation ratios. This technology was recently launched to the market by Waters Corporation during the last American Society for Mass Spectrometry (ASMS) meeting (2019).

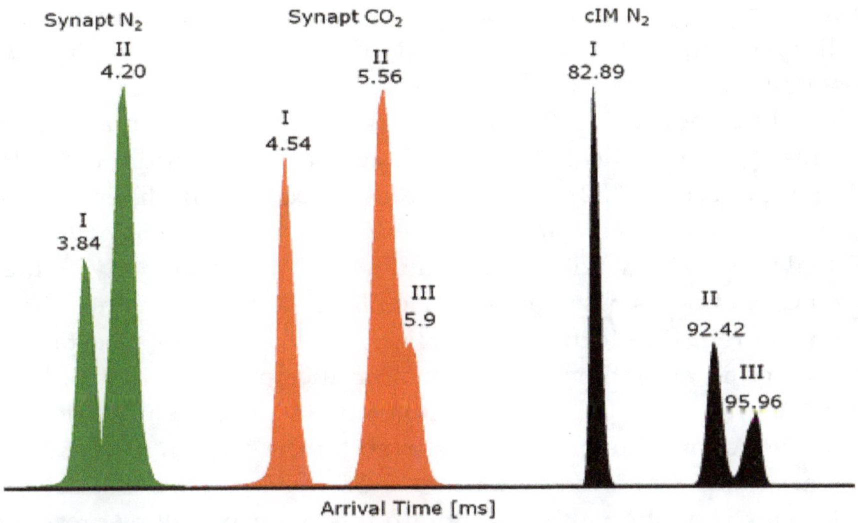

Figure 6. Danofloxacin IMS protomer separation using the Synapt (N_2 and CO_2 IMS gas) and cyclic-IMS (N_2 IMS gas, 5 passes) systems. Figure reprinted with permission from [126]. Copyright (2019) Wiley.

SLIM–TWIMS technology is not yet commercially available, but it is also catching the attention due to its high R_p resulted from its long drift path (~337, 540 m; 40 passes, 13.5-m path) [121]. As occurs with cyclic-IMS, SLIM–IMS can be based on DTIMS or TWIMS separations. However, TWIMS technology is usually applied because it allows longer path lengths than DTIMS, which presents voltage limitations [23]. SLIM–TWIMS systems consist of two planar surfaces fabricated using printed circuit boards and containing an array of electrodes. Ions travel through a serpentine path and are confined due to the application of DC and RF potentials [127]. SLIM–TWIMS offers a longer drift path than cyclic-TWIMS and, as a consequence, a larger range of mobilities can be monitored simultaneously. Although SLIM–TWIMS approaches have not been directly applied to food analysis, it has been shown to improved separations of lipid and peptide isomers [128] as well as glycans that differ in CCS by as little as 0.2% [129], which can have potential applicability in food composition analysis and food fingerprinting.

4.2. Implementation of CCS in Current Analytical Workflows

The CCS parameter is closely correlated to *m/z*, but its use as molecular descriptor in traditional analytical workflows has been shown to be effective in improving data processing by reducing false positive/negative results [112,113]. Despite its potential to support compound identification, the lack of CCS libraries is considered to be the major drawback for reaching this goal and implementing the CCS parameter in current analytical workflows [8]. In an effort to overcome this issue, several CCS databases have been reported in the last years [15], although they are not enough considering the number of molecules that remain uncharacterized. In order to extend the knowledge provided by CCS databases created experimentally, other strategies such as computational modeling and, more recently, machine-learning based prediction have also been investigated [130]. Machine-learning approaches are very effective for CCS prediction and are becoming very popular for generating CCS values of molecules on a large scale because their application only takes a few seconds/minutes whereas computational modeling is computationally intensive. For example, a machine-learning model based on artificial neural networks (ANNs) has been developed for the CCS prediction of small molecules selecting 205 compounds for model training, verification and blind test sets (ratio 68:16:16) [131]. This model was subsequently applied to the analysis of ten pesticides in spinach samples, and deviations between the observed and predicted CCS values for the protonated ions of pesticides were smaller than 5.3%.

CCS databases are typically based on one-platform measurements and, taking into account that CCSs are conditional values, the implementation of current libraries in other IMS platforms already introduces uncertainty into CCS measurements [20]. Theoretically, CCS values are platform-independent but they are influenced by several experimental parameters (i.e., E/N, temperature, and buffer gas) as indicated above. Specifically, the nature of buffer gas has a great influence on the CCS of molecules [132], and CCS values measured in different drift gases cannot be directly compared, as shown for a wide variety of pesticides whose CCSs have been obtained in He, CO_2, N_2O and SF_6 by DTIMS [133]. Within this framework, it may actually be more appropriate to develop standardized CCS libraries based on multiple platform measurements. Inter-platform studies should also consider different IMS forms (i.e., DTIMS, TWIMS, TIMS) to establish if standardized CCSs can be implemented in all of them or whether, on the contrary, ^{DT}CCS, ^{TW}CCS and ^{TIMS}CCS values must be reported and used. In general, similar CCS values are provided by these three IMS technologies when using the same buffer gas as observed for the sodium adduct of 25-hydroxyvitamin D3 [134]. However, another recent study about the CCS of 35 pharmaceuticals, 64 pesticides, and 25 metabolites of pesticides has shown that, although the majority of ions present differences lower than 1.1% between ^{DT}CCS and ^{TW}CCS values, both values cannot be compared in all cases [135]. Deviations up to 6.2% between ^{DT}CCS and ^{TW}CCS values were observed for several ions.

In the same vein, there is still a knowledge gap in the use of CCS databases and the variability associated with CCS measurements. Recent studies have shown high repeatability over time [30,112], and negligible impact of sample matrix on CCS values [39,113]. Despite this, the use of this parameter for specific analytical applications regulated by guidelines requires a comprehensive assessment of reproducibility across different laboratories and instrument types. Inter-laboratory reproducibility has barely been evaluated, and only few studies have tackled this issue [22,112,136]. This information is highly relevant to establish confidence intervals for CCS measurements and thresholds for their comparison with CCS values in databases. An inter-laboratory study involving four DTIMS-platforms have recently reported absolute bias of 0.54% to the standardized stepped field $^{DT}CCS_{N_2}$ values on the reference system [22]. Under this context, the current threshold of 2% accepted for CCS measurements seems to be too wide and could be reduced. As a result, the number of possibilities for peak assignment will be decreased when using the CCS for compound identification. Based on current evidence, this threshold could potentially be reduced to at least 1–1.5% [30,112,113]. Looking ahead to the implementation of the CCS parameter to support compound identification, a score system based on deviations from standardized CCS values should also be developed. Scoring systems based on mass spectra are typically used for the putative identification of molecules, so a similar approach could be suitable for the integration of CCS values in data processing workflows.

In addition to the standardization of CCS libraries and establishment of thresholds for CCS measurements, there is a third issue that requires further development. CCS calibration must be carried out in DTIMS (in single-field mode), TWIMS and TIMS systems before CCS measurements, and must be performed under the same operational conditions applied to obtain the CCS of analytes. Until now, there is neither consensus on the application of standardized calibration protocols nor on the application for this purpose of primary standards, which also serve for preparing reference materials [20]. A protocol has been reported for TWIMS calibration and operation [137], but different calibration procedures and calibrants are applied in other IMS systems. Under this context, 'Agilent tunemix calibration standard' (i.e., mixture of hexakis-fluoropropoxyphosphazines) [22], poly-DL-alanine [107], and 'major mix IMS/ToF calibration kit' from Waters (i.e., mixture of small molecules such as caffeine or sulfadimethoxine, and including poly-DL-alanine) [116] are currently widely used as CCS calibrants. Other alternative CCS calibrants such as the dextran have also been investigated for obtaining the CCS of carbohydrates [138]. Therefore, more studies about CCS calibration are still needed since, in the case of some classes of compounds, the accuracy of CCS measurements depends on the chemical nature of the calibrant [139]. Within this framework, both standardized calibration procedures and primary standards of different chemical class seem to be crucial in order to increase confidence in CCS comparison and measurements. The development of reference materials and standardized calibration protocols will contribute to the implementation of the CCS, but it calls for a great effort from IMS community including IMS suppliers.

5. Conclusions and Perspectives in Food Analysis

The implementation of IMS in food science is quite new but its applicability is growing exponentially. Nowadays, there is a wide range of methods intended for food composition, process control, authentication, adulteration and safety analysis, covering a great variety of molecules (i.e., macronutrients such as proteins, lipids, carbohydrates, vitamins; micronutrients such as polyphenols; process and biodegradation products such as biogenic amines; residues such as pesticides and veterinary drugs; or contaminants such as toxins and environmental pollutants), and involving different IMS forms (i.e., DTIMS, TWIMS, TIMS, FAIMS and DMS).

Stand-alone IMS instruments are portable and provide a quick response (< 30 ms), so they have been shown to be very efficient for in situ analysis and real-monitoring, such as in food process control where rapid decisions have to be made. It can be expected that miniaturized and portable IMS instruments will be implemented for food analysis as already occurs in the analysis of chemical warfare agents and drugs in airports, courts, etc. Potential applications are in situ monitoring of ripening processes of crops, depletion of residues after veterinary treatments in farms, etc.

In many applications, stand-alone IMS approaches offer low selectivity, so they are limited to the determination of specific compounds and usually require exhaustive sample treatments prior analysis due to the complexity of food matrices. In order to improve selectivity, IMS is commonly coupled to a chromatographic technique, especially GC if MS detection is not used. GC–IMS methods are very popular for the analysis of VOC fraction of food. This approach is widely used in food authentication where HS analysis is typically carried out. HS–GC–IMS in combination with chemometrics is broadly applied for food fingerprinting and product discrimination according to their quality, origin, etc. However, this strategy barely exploits the potential of the IMS technique (since it usually acts as simple detector) and is limited by the low number of compounds that are identified (i.e., only substances with standards available in the laboratory are identified based on their retention and drift times). The development of GC–IM–MS strategies will give more knowledge about VOCs composition, overcoming the current boundaries of GC–IMS approaches for food characterization and authentication. Consequently, for example, food fermentation and decay processes will be better understood since mass spectra will be obtained.

The recent commercialization of hyphenated IM–MS instruments, usually as part of LC–IM–MS platforms, has been the reason why IMS is becoming very popular in food science. IMS is extending the current boundaries of LC–IM–MS methods by introducing an extra separation dimension that allows the separation of isobars and isomers (not always separated in the chromatographic dimension and undistinguished by MS) and the isolation of analytes of interest from chemical background (so improving S/N and enhancing sensitivity). Although the correlation existing between IMS and MS has been a topic of discussion for a long time, the applications included in this review show that a higher number of compounds are detected by LC–IMS–MS in comparison to traditional LC–MS workflows, which gives a more complete picture for food characterization. Moreover, the integration of IMS provides cleaned-up chromatograms and mass spectra facilitating data interpretation.

In addition, IMS gives additional information to retention index and mass spectra of molecules (i.e., K_0, drift time, CV and/or CCS) which can ultimately be used for compound identification. Within this framework, the CCS of molecules has acquired great relevance. Despite this parameter is correlated to *m/z*, it has been shown to be useful to distinguish compounds with different chemical nature, but also to differentiate among close chemical molecules. From a regulatory point of view, the application of the CCS parameter for the identification of residues and contaminants in food is currently under discussion. Nowadays, the development of open-access CCS databases (based on inter-laboratory measurements for proposing normalized reference values), the requirement of standardized calibration procedures and calibrants for CCS measurements, and the establishment of thresholds for CCS measurements, are viewed as the main challenges to tackle for the implementation of CCS in food analysis. Based on current knowledge and trends, CCS will definitely be included in current analytical workflows to support food characterization and, probably, legally accepted as complementary information to confirm the presence or absence of residues and contaminants in food products.

Technological developments experienced by IMS, and especially by IM–MS hyphenation, are going to have a great impact on the implementation of this technique in the food science field. Enhancements in R_p, currently lead by cyclic-TWIMS and SLIM–TWIMS technologies, will contribute to overcome the challenges arise from the complexity of food matrices (i.e., high number of compounds with different chemical nature at different concentration levels). Under this context, they will enable the discovery of unknown food components with bioactive properties which are normally at low concentration levels and are typically masked by isobaric and isomeric compounds or by food major components. In addition, more complete fingerprints of food products will be achieved, providing more detailed information for food authentication and detection of food adulteration.

Finally, the integration of IMS in current LC–MS (or GC–MS, CE–MS, SFC–MS) workflows involves certain technical challenges for operators already working in food analysis and for whom this technique is generally unknown. The addition of a third dimension also involves more complex data. This issue hinders data interpretation, especially in the case of food fingerprinting where large datasets are generated. Therefore, the development of simple benchtop LC–IM–MS platforms and user-friendly software for IMS operation and data treatment will be crucial to the success of the implementation of IMS in food analysis.

Author Contributions: M.H.-M. conceptualized and wrote the manuscript. A.M.G.-C., G.D.-P. and B.L.B. provided scientific knowledge about food science and its related challenges. D.R., H.R., G.D.-P., and B.L.B. contributed to develop the ideas about the potential of IMS for food analysis. D.R., H.R., A.M.G.-C., G.D.-P. and B.L.B. reviewed and edited the manuscript. H.R. and A.M.G.-C. provided scientific and administrative support to M.H.-M.'s fellowship project. G.D.-P. and B.L.B. initiated and supervised the project. All authors edited, read and approved the final manuscript.

References

1. Lehotay, S.J.; Chen, Y. Hits and misses in research trends to monitor contaminants in foods. *Anal. Bioanal. Chem.* **2018**, *410*, 5331–5351. [CrossRef] [PubMed]
2. Medina, S.; Pereira, J.A.; Silva, P.; Perestrelo, R.; Câmara, J.S. Food fingerprints—A valuable tool to monitor food authenticity and safety. *Food Chem.* **2019**, *278*, 144–162. [CrossRef] [PubMed]

3. Zheng, X.; Wojcik, R.; Zhang, X.; Ibrahim, Y.M.; Burnum-Johnson, K.E.; Orton, D.J.; Monroe, M.E.; Smith, R.D.; Baker, E.S. Coupling front-end separations, ion mobility spectrometry, and mass spectrometry for enhanced multidimensional biological and environmental analyses. *Annu. Rev. Anal. Chem.* **2017**, *10*, 71–92. [CrossRef] [PubMed]
4. D'Atri, V.; Causon, T.; Hernandez-Alba, O.; Mutabazi, A.; Veuthey, J.-L.; Cianferani, S.; Guillarme, D. Adding a new separation dimension to MS and LC-MS: What is the utility of ion mobility spectrometry? *J. Sep. Sci.* **2018**, *41*, 20–67. [CrossRef] [PubMed]
5. Vautz, W.; Zimmermann, D.; Hartmann, M.; Baumbach, J.I.; Nolte, J.; Jung, J. Ion mobility spectrometry for food quality and safety. *Food Addit. Contam.* **2006**, *23*, 1064–1073. [CrossRef] [PubMed]
6. Arce, L.; Valcárcel, M. The role of ion mobility spectrometry to support the food protected designation of origin. In *Comprehensive Analytical Chemistry—Food Protected Designation of Origin: Methodologies and Applications*; de la Guardia, M., González, A., Eds.; Elsevier: Amsterdam, The Netherlands, 2013; Volume 60, pp. 221–249.
7. Karpas, Z. Applications of ion mobility spectrometry (IMS) in the field of foodomics. *Food Res. Int.* **2013**, *54*, 1146–1151. [CrossRef]
8. Hernández-Mesa, M.; Escorrou, A.; Monteau, F.; Le Bizec, B.; Dervilly-Pinel, G. Current applications and perspectives of ion mobility spectrometry to answer chemical food safety issues. *TrAC Trends Anal. Chem.* **2017**, *94*, 39–53. [CrossRef]
9. Uetrecht, C.; Rose, R.J.; van Duijn, E.; Lorenzen, K.; Heck, A.J.R. Ion mobility mass spectrometry of proteins and protein assemblies. *Chem. Soc. Rev.* **2010**, *39*, 1633–1655. [CrossRef]
10. Gabelica, V.; Marklund, E. Fundamentals of ion mobility spectrometry. *Curr. Opin. Chem. Biol.* **2018**, *42*, 51–59. [CrossRef]
11. May, J.C.; Morris, C.B.; McLean, J.A. Ion mobility collision cross section compendium. *Anal. Chem.* **2017**, *89*, 1032–1044. [CrossRef]
12. Revercomb, H.E.; Mason, E.A. Theory of plasma chromatography/gaseous electrophoresis—A review. *Anal. Chem.* **1975**, *47*, 970–983. [CrossRef]
13. Creaser, C.S.; Griffiths, J.R.; Bramwell, C.J.; Noreen, S.; Hill, C.A.; Thomas, C.L.P. Ion mobility spectrometry: A review. Part 1. Structural analysis by mobility measurement. *Analyst* **2004**, *129*, 984–994. [CrossRef]
14. Lanucara, F.; Holman, S.W.; Gray, C.J.; Eyers, C.E. The power of ion mobility-mass spectrometry for structural characterization and the study of conformational dynamics. *Nat. Chem.* **2014**, *6*, 281–294. [CrossRef] [PubMed]
15. Picache, J.A.; Rose, B.S.; Balinski, A.; Leaptrot, K.L.; Sherrod, S.D.; May, J.C.; McLean, J.A. Collision cross section compendium to annotate and predict multi-omic compound identities. *Chem. Sci.* **2019**, *10*, 983–993. [CrossRef] [PubMed]
16. Mairinger, T.; Causon, T.J.; Hann, S. The potential of ion mobility-mass spectrometry for non-targeted metabolomics. *Curr. Opin. Chem. Biol.* **2018**, *42*, 9–15. [CrossRef] [PubMed]
17. Ewing, M.A.; Glover, M.S.; Clemmer, D.E. Hybrid ion mobility and mass spectrometry as a separation tool. *J. Chromatogr. A* **2016**, *1439*, 3–25. [CrossRef] [PubMed]
18. May, J.C.; McLean, J.A. Ion mobility-mass spectrometry: Time-dispersive instrumentation. *Anal. Chem.* **2015**, *87*, 1422–1438. [CrossRef] [PubMed]
19. Schneider, B.B.; Nazarov, E.G.; Londry, F.; Vouros, P.; Covey, T.R. Differential mobility spectrometry/mass spectrometry history, theory, design optimization, simulations, and applications. *Mass Spectrom. Rev.* **2016**, *35*, 687–737. [CrossRef] [PubMed]
20. Gabelica, V.; Shvartsburg, A.A.; Afonso, C.; Barran, P.; Benesch, J.L.P.; Bleiholder, C.; Bowers, M.T.; Bilbao, A.; Bush, M.F.; Campbell, J.L.; et al. Recommendations for reporting ion mobility mass spectrometry measurements. *Mass Spectrom. Rev.* **2019**, *38*, 291–320. [CrossRef]
21. Nichols, C.M.; May, J.C.; Sherrod, S.D.; McLean, J.A. Automated flow injection method for the high precision determination of drift tube ion mobility collision cross sections. *Analyst* **2018**, *143*, 1556–1559. [CrossRef]
22. Stow, S.M.; Causon, T.J.; Zheng, X.; Kurulugama, R.T.; Mairinger, T.; May, J.C.; Rennie, E.E.; Baker, E.S.; Smith, R.D.; McLean, J.A.; et al. An interlaboratory evaluation of drift tube ion mobility-mass spectrometry collision cross section measurements. *Anal. Chem.* **2017**, *89*, 9048–9055. [CrossRef] [PubMed]
23. Kirk, A.T.; Bohnhorst, A.; Raddatz, C.-R.; Allers, M.; Zimmermann, S. Ultra-high-resolution ion mobility spectrometry—Current instrumentation, limitations, and future developments. *Anal. Bioanal. Chem.* **2019**

in press. [CrossRef] [PubMed]

24. Ridgeway, M.E.; Lubeck, M.; Jordens, J.; Mann, M.; Park, M.A. Trapped ion mobility spectrometry: A short review. *Int. J. Mass Spectrom.* **2018**, *425*, 22–35. [CrossRef]
25. Michelmann, K.; Silveira, J.A.; Ridgeway, M.E.; Park, M.A. Fundamentals of trapped ion mobility spectrometry. *J. Am. Soc. Mass Spectrom.* **2015**, *26*, 14–24. [CrossRef]
26. Cumeras, R.; Figueras, E.; Davis, C.E.; Baumbach, J.I.; Gràcia, I. Review on ion mobility spectrometry. Part 1: Current instrumentation. *Analyst* **2015**, *140*, 1376–1390. [CrossRef] [PubMed]
27. Eiceman, G.A.; Karpas, Z. *Ion. Mobility Spectrometry*, 2nd ed.; CRC Press, Taylor & Francis: Boca Raton, FL, USA, 2005.
28. Shvartsburg, A.A.; Smith, R.D. Fundamentals of traveling wave ion mobility spectrometry. *Anal. Chem.* **2008**, *80*, 9689–9699. [CrossRef]
29. Causon, T.J.; Hann, S. Theoretical evaluation of peak capacity improvements by use of liquid chromatography combined with drift tube ion mobility-mass spectrometry. *J. Chromatogr. A* **2015**, *1416*, 47–56. [CrossRef] [PubMed]
30. Hernández-Mesa, M.; Monteau, F.; Le Bizec, B.; Dervilly-Pinel, G. Potential of ion mobility-mass spectrometry for both targeted and non-targeted analysis of phase II steroid metabolites in urine. *Anal. Chim. Acta X* **2019**, *1*, 100006. [CrossRef]
31. Regueiro, J.; Negreira, N.; Hannisdal, R.; Berntssen, M.H.G. Targeted approach for qualitative screening of pesticides in salmon feed by liquid chromatography coupled to traveling-wave ion mobility/quadrupole time-of-flight mass spectrometry. *Food Control.* **2017**, *78*, 116–125. [CrossRef]
32. Kanu, A.B.; Hill, H.H., Jr. Ion mobility spectrometry detection for gas chromatography. *J. Chromatogr. A* **2008**, *1177*, 12–27. [CrossRef]
33. Lapthorn, C.; Pullen, F.; Chowdhry, B.Z. Ion mobility spectrometry-mass spectrometry (IMS-MS) of small molecules: Separating and assigning structures to ions. *Mass Spectrom. Rev.* **2013**, *32*, 43–71. [CrossRef] [PubMed]
34. Haynes, S.E.; Polasky, D.A.; Dixit, S.M.; Majmudar, J.D.; Neeson, K.; Ruotolo, B.T.; Martin, B.R. Variable-velocity traveling-wave ion mobility separation enhancing peak capacity for data-independent acquisition proteomics. *Anal. Chem.* **2017**, *89*, 5669–5672. [CrossRef] [PubMed]
35. Arthur, K.L.; Turner, M.A.; Reynolds, J.C.; Creaser, C.S. Increasing peak capacity in nontargeted omics applications by combining full scan field asymmetric waveform ion mobility spectrometry with liquid chromatography-mass spectrometry. *Anal. Chem.* **2017**, *89*, 3452–3459. [CrossRef] [PubMed]
36. Hill, H.H.; Simpson, G. Capabilities and limitations of ion mobility spectrometry for field screening applications. *Field Anal. Chem. Technol.* **1997**, *1*, 119–134. [CrossRef]
37. Kaufmann, A.; Walker, S. Comparison of linear intrascan and interscan dynamic ranges of Orbitrap and ion-mobility time-of-flight mass spectrometers. *Rapid Commun. Mass Spectrom.* **2017**, *31*, 1915–1926. [CrossRef] [PubMed]
38. Kaufmann, A.; Butcher, P.; Maden, K.; Widmer, M.; Giles, K.; Uría, D. Are liquid chromatography/ electrospray tandem quadrupole fragmentation rations unequivocal confirmation criteria? *Rapid Commun. Mass Spectrom.* **2009**, *23*, 958–998. [CrossRef] [PubMed]
39. Bauer, A.; Kuballa, J.; Rohn, S.; Jantzen, E.; Luetjohann, J. Evaluation and validation of an ion mobility quadrupole time-of-flight mass spectrometry pesticide screening approach. *J. Sep. Sci.* **2018**, *41*, 2178–2187. [CrossRef]
40. Tu, J.; Zhou, Z.; Li, T.; Zhu, Z.-J. The emerging role of ion mobility-mass spectrometry in lipidomics to facilitate lipid separation and identification. *TrAC Trends Anal. Chem.* **2019**, *116*, 332–339. [CrossRef]
41. Chouinard, C.D.; Nagy, G.; Smith, R.D.; Baker, E.S. Ion mobility-mass spectrometry in metabolomic, lipidomic, and proteomic analyses. *Compr. Anal. Chem.* **2019**, *83*, 123–159. [CrossRef]
42. Manz, C.; Pagel, K. Glycan analysis by ion mobility-mass spectrometry and gas-phase spectroscopy. *Curr. Opin. Chem. Biol.* **2018**, *42*, 16–24. [CrossRef]
43. Kuang, A.; Erlund, I.; Herder, C.; Westerhuis, J.A.; Tuomilehto, J.; Cornelis, M.C. Lipidomic response to coffee consumption. *Nutrients* **2018**, *10*, 1851. [CrossRef] [PubMed]
44. Blaženović, I.; Shen, T.; Mehta, S.S.; Kind, T.; Ji, J.; Piparo, M.; Cacciola, F.; Mondello, L.; Fiehn, O. Increasing compound identification rates in untargeted lipidomics research with liquid chromatography drift time-ion mobility mass spectrometry. *Anal. Chem.* **2018** *90*, 10758–10764. [CrossRef] [PubMed]

45. López-Morales, C.A.; Vázquez-Leyva, S.; Vallejo-Castillo, L.; Carballo-Uicab, G.; Muñoz-García, L.; Herbet-Pucheta, J.E.; Zepeda-Vallejo, L.G.; Velasco-Velázquez, M.; Pavón, L.; Pérez-Tapia, S.M.; et al. Determination of peptide profile consistency and safety of collagen hydrolysates as quality attributes. *J. Food Sci.* **2019**, 430–439. [CrossRef] [PubMed]
46. Causon, T.J.; Došen, M.; Reznicek, G.; Hann, S. Workflow development for the analysis of phenolic compounds in wine using liquid chromatography combined with drift-tube ion mobility-mass spectrometry. *LC-GC N. Am.* **2016**, *34*, 854–867.
47. Rodríguez-Maecker, R.; Vyhmeister, E.; Meisen, S.; Martinez Rosales, A.; Kuklya, A.; Telgheder, U. Identification of terpenes and essential oils by means of static headspace gas chromatography-ion mobility spectrometry. *Anal. Bioanal. Chem.* **2017**, *409*, 6595–6603. [CrossRef] [PubMed]
48. Alves, T.O.; D'Almeida, C.T.S.; Victorio, V.C.M.; Souza, G.H.M.F.; Cameron, L.C.; Ferreira, M.S.L. Immunogenic and allergenic profile of wheat flours from different technological qualities revealed by ion mobility mass spectrometry. *J. Food Compos. Anal.* **2018**, *73*, 67–75. [CrossRef]
49. Garrido-Delgado, R.; Dobao-Prieto, M.M.; Arce, L.; Valcárcel, M. Determination of volatile compounds by GC-IMS to assign the quality of virgin olive oil. *Food Chem.* **2015**, *187*, 572–579. [CrossRef]
50. Guo, Y.; Chen, D.; Dong, Y.; Ju, H.; Wu, C.; Lin, S. Characteristic volatiles fingerprints and changes of volatile compounds in fresh and dried *Tricholoma matsutake* Singer by HS-GC-IMS and HS-SPME-GC-MS. *J. Chromatogr. B* **2018**, *1099*, 46–55. [CrossRef]
51. Browne, C.A.; Forbes, T.P.; Sisco, E. Detection and identification of sugar alcohol sweeteners by ion mobility spectrometry. *Anal. Methods* **2016**, *8*, 5611–5618. [CrossRef]
52. Contreras, M.D.M.; Jurado-Campos, N.; Arce, L.; Arroyo-Manzanares, N. A robustness study of calibration models for olive oil classification: Targeted and non-targeted fingerprint approaches based on GC-IMS. *Food Chem.* **2019**, *288*, 315–324. [CrossRef]
53. Gerhardt, N.; Schwolow, S.; Rohn, S.; Pérez-Cacho, P.R.; Galán-Soldevilla, H.; Arce, L.; Weller, P. Quality assessment of olive oils based on temperature-ramped HS-GC-IMS and sensory evaluation: Comparison of different processing approaches by LDA, kNN, and SVM. *Food Chem.* **2019**, *278*, 720–728. [CrossRef] [PubMed]
54. Liedtke, S.; Seifert, L.; Ahlmann, N.; Hariharan, C.; Franzke, J.; Vautz, W. Coupling laser desorption with gas chromatography and ion mobility spectrometry for improved olive oil characterization. *Food Chem.* **2018**, *255*, 323–331. [CrossRef] [PubMed]
55. Li, J.; Yuan, H.; Yao, Y.; Hua, J.; Yang, Y.; Dong, C.; Deng, Y.; Wang, J.; Li, H.; Jiang, Y.; et al. Rapid volatiles fingerprinting by dopant-assisted positive photoionization ion mobility spectrometry for discrimination and characterization of Green Tea aromas. *Talanta* **2019**, *191*, 39–45. [CrossRef] [PubMed]
56. Lu, Y.; Guo, J.; Yu, J.; Guo, J.; Jia, X.; Liu, W.; Tian, P. Two-dimensional analysis of phenolic acids in seedling roots by high performance liquid chromatography electrospray ionization-ion mobility spectrometry. *Anal. Methods* **2019**, *11*, 610–617. [CrossRef]
57. Struwe, W.B.; Baldauf, C.; Hofmann, J.; Rudd, P.M.; Pagel, K. Ion mobility separation of deprotonated oligosaccharide isomers—Evidence for gas-phase charge migration. *Chem. Commum.* **2016**, 12353–12356. [CrossRef] [PubMed]
58. Venter, P.; Muller, M.; Vestner, J.; Stander, M.A.; Tredoux, A.G.J.; Pasch, H.; De Villiers, A. Comprehensive three-dimensional LC × LC × ion mobility spectrometry separation combined with high-resolution MS for the analysis of complex samples. *Anal. Chem.* **2018**, *90*, 11643–11650. [CrossRef] [PubMed]
59. Stephan, S.; Jakob, C.; Hippler, J.; Schmitz, O.J. A novel four-dimensional analytical approach for analysis of complex samples. *Anal. Bioanal. Chem.* **2016**, *408*, 3751–3759. [CrossRef]
60. Zhang, H.; Jiang, J.M.; Zheng, D.; Yuan, M.; Wang, Z.Y.; Zhang, H.M.; Zheng, C.W.; Xiao, L.B.; Xu, H.X. A multidimensional analytical approach based on time-decoupled online comprehensive two-dimensional liquid chromatography coupled with ion mobility quadrupole time-of-flight mass spectrometry for the analysis of ginsenosides from white and red ginsengs. *J. Pharm. Biomed. Anal.* **2019**, *163*, 24–33. [CrossRef]
61. Lipok, C.; Hippler, J.; Schmitz, O.J. A four dimensional separation method based on continuous heart-cutting gas chromatography with ion mobility and high resolution mass spectrometry. *J. Chromatogr. A* **2018**, *1536*, 50–57. [CrossRef]
62. McCullagh, M.; Douce, D.; Hoeck, V.; Goscinny, S. Exploring the complexity of steviol glycosides analysis using ion mobility mass spectrometry. *Anal. Chem.* **2018**, *90*, 4585–4595. [CrossRef]

63. Karpas, Z.; Guamán, A.V.; Pardo, A.; Marco, S. Comparison of the performance of three ion mobility spectrometers for measurement of biogenic amines. *Anal. Chim. Acta* **2013**, *758*, 122–129. [CrossRef] [PubMed]
64. Cohen, G.; Rudnik, D.D.; Laloush, M.; Yakir, D.; Karpas, Z. A novel method for determination of histamine in tuna fish by ion mobility spectrometry. *Food Anal. Methods* **2015**, *8*, 2376–2382. [CrossRef]
65. Cheng, S.; Li, H.; Jiang, D.; Chen, C.; Zhang, T.; Li, Y.; Wang, H.; Zhou, Q.; Li, H.; Tan, M. Sensitive detection of trimethylamine based on dopant-assisted positive photoionization ion mobility spectrometry. *Talanta* **2017**, *162*, 398–402. [CrossRef] [PubMed]
66. Parchami, R.; Kamalabadi, M.; Alizadeh, N. Determination of biogenic amines in canned fish samples using head-space solid phase microextraction based on nanostructured polypyrrole fiber coupled to modified ionization región ion mobility spectrometry. *J. Chromatogr. A* **2017**, *1481*, 37–43. [CrossRef] [PubMed]
67. Zarpas, Z.; Guamán, A.V.; Calvo, D.; Pardo, A.; Marco, S. The potential of ion mobility spectrometry (IMS) for detection of 2,4,6-trichloroanisole (2,4,6-TCA) in wine. *Talanta* **2012**, *93*, 200–205. [CrossRef]
68. Kamalabadi, M.; Ghaemi, E.; Mohammadi, A.; Alizadeh, N. Determination of furfural and hydroxymethylfurfural from baby formula using headspace microextraction base don nanostrutured polypyrrole fiber coupled with ion mobility spectrometry. *Food Chem.* **2015**, *181*, 72–77. [CrossRef]
69. Garrido-Delgado, R.; Dobao-Prieto, M.M.; Arce, L.; Aguilar, J.; Cumplido, J.L.; Valcárcel, M. Ion mobility spectrometry versus classical physico-chemical analysis for assessing the shelf life of extra virgin olive oil according to container type and storage conditions. *J. Agric. Food Chem.* **2015**, *63*, 2179–2188. [CrossRef]
70. Tzschoppe, M.; Haase, H.; Höhnisch, M.; Jaros, D.; Rohm, H. Using ion mobility spectrometry for screening the autoxidation of peanuts. *Food Control.* **2016**, *64*, 17–21. [CrossRef]
71. Raatikainen, O.; Reinikainen, V.; Minkkinen, P.; Ritvanen, T.; Muje, P.; Pursiainen, J.; Hiltunen, T.; Hyvönen, P.; Von Wright, A.; Reinikainen, S.-P. Multivariate modelling of fish freshness index based on ion mobility spectrometry measurements. *Anal. Chim. Acta* **2005**, *544*, 128–134. [CrossRef]
72. Cavanna, D.; Zanardi, S.; Dall'Asta, C.; Suman, M. Ion mobility spectrometry coupled to gas chromatography: A rapid tool to assess eggs freshness. *Food Chem.* **2019**, *271*, 691–696. [CrossRef]
73. Tang, Z.-S.; Zeng, X.-A.; Brennan, M.A.; Han, Z.; Niu, D.; Huo, Y. Characterization of aroma profile and characteristic aromas during lychee wine fermentation. *J. Food Process. Preserv.* **2019**, in press. [CrossRef]
74. Halbfeld, C.; Ebert, B.E.; Blank, L.M. Multi-capillary column-ion mobility spectrometry of volatile metabolites emitted by *Saccharomyces Cerevisiae*. *Metabolites* **2014**, *4*, 751–774. [CrossRef] [PubMed]
75. Gloess, A.N.; Yeretzian, C.; Knochenmuss, R.; Groessl, M. On-line analysis of coffee roasting with ion mobility spectrometry-mass spectrometry (IMS-MS). *Int. J. Mass Spectrom.* **2018**, *424*, 49–57. [CrossRef]
76. Danezis, G.P.; Tsagkaris, A.S.; Camin, F.; Brusic, V.; Georgiou, C.A. Food authentication: Techniques, trends & emerging approaches. *TrAC Trends Anal. Chem.* **2016**, *85*, 123–132. [CrossRef]
77. Cubero-Leon, E.; Peñalver, R.; Maquet, A. Review on metabolomics for food authentication. *Food Res. Int.* **2014**, *60*, 95–107. [CrossRef]
78. Causon, T.J.; Ivanova-Petropulos, V.; Petrusheva, D.; Bogeva, E.; Hann, S. Fingerprinting of traditionally produced red wines using liquid chromatography combined with drift tube ion mobility-mass spectrometry. *Anal. Chim. Acta* **2019**, *1052*, 179–189. [CrossRef] [PubMed]
79. Garrido-Delgado, R.; Mercader-Trejo, F.; Sielemann, S.; de Bruyn, W.; Arce, L.; Valcárcel, M. Direct classification of olive oils by using two types of ion mobility spectrometers. *Anal. Chim. Acta* **2011**, *696*, 108–115. [CrossRef]
80. Gerhardt, N.; Birkenmeier, M.; Sanders, D.; Rohn, S.; Weller, P. Resolution-optimized headspace gas chromatography-ion mobility spectrometry (HS-GC-IMS) for non-targeted olive oil profiling. *Anal. Bianal. Chem.* **2017**, *409*, 3933–3942. [CrossRef] [PubMed]
81. Arroyo-Manzanares, N.; Martín-Gómez, A.; Jurado-Campos, N.; Garrido-Delgado, R.; Arce, C.; Arce, L. Target vs spectral fingerprint data analysis of Iberian ham samples for avoiding labelling fraud using headspace– gas chromatography—ion mobility spectrometry. *Food Chem.* **2018**, *246*, 65–73. [CrossRef] [PubMed]
82. Martín-Gómez, A.; Arroyo-Manzanares, N.; Rodríguez-Estévez, V.; Arce, L. Use of a non-destructive sampling method for characterization of Iberian cured ham breed and feeding regime using GC-IMS. *Meat Sci.* **2019**, 146–154. [CrossRef]
83. Garrido-Delgado, R.; Arce, L.; Guamán, A.V.; Pardo, A.; Marco, S.; Valcárcel, M. Direct coupling of a gas-liquid separator to an ion mobility spectrometer for the classification of different white wines using

chemometrics tools. *Talanta* **2011**, *84*, 471–479. [CrossRef] [PubMed]

84. Gerhardt, N.; Birkenmeier, M.; Schwolow, S.; Rohn, S.; Weller, P. Volatile-compound fingerprinting by headspace-gas-chromatography ion-mobility spectrometry (HS-GC-IMS) as a benchtop alternative to ^{1}H NMR profiling for assessment of the authenticity of honey. *Anal. Chem.* **2018**, *90*, 1777–1785. [CrossRef] [PubMed]
85. Wang, X.; Yang, S.; He, J.; Chen, L.; Zhang, J.; Jin, Y.; Zhou, J.; Zhang, Y. A green triple-locked strategy based on volatile-compound imaging, chemometrics, and markers to discriminate winter honey and *sapium* honey using headspace gas chromatography-ion mobility spectrometry. *Food Res. Int.* **2019**, *119*, 960–967. [CrossRef] [PubMed]
86. Chen, T.; Chen, X.; Lu, D.; Chen, B. Detection of adulteration in canola oil by using GC-IMS and chemometric analysis. *Int. J. Anal. Chem.* **2018**, *2018*, 3160265. [CrossRef] [PubMed]
87. Othman, A.; Goggin, K.A.; Tahir, N.I.; Brodrick, E.; Singh, R.; Sambanthamurthi, R.; Parveez, K.A.; Davies, A.N.; Murad, A.J.; Muhammad, N.H.; et al. Use of headspace-gas chromatography-ion mobility spectrometry to detect volatile fingerprints of palm fibre oil and sludge palm oil in samples of crude palm oil. *BMC Res. Notes* **2019**, *12*, 229. [CrossRef] [PubMed]
88. Garrido-Delgado, R.; Muñoz-Pérez, E.; Arce, L. Detection of adulteration in extra virgin olive oils by using UV-IMS and chemometric analysis. *Food Control.* **2018**, *85*, 292–299. [CrossRef]
89. Zhang, L.; Shuai, Q.; Li, P.; Zhang, Q.; Ma, F.; Zhang, W.; Ding, X. Ion mobility spectrometry fingerprints: A rapid detection technology for adulteration of sesame oil. *Food Chem.* **2016**, *192*, 60–66. [CrossRef]
90. Shuai, Q.; Zhang, L.; Li, P.; Zhang, Q.; Wang, X.; Ding, X.; Zhang, W. Rapid adulteration detection for flaxseed oil using ion mobility spectrometry and chemometric methods. *Anal. Methods* **2014**, *6*, 9575–9580. [CrossRef]
91. Khademi, S.M.S.; Telgheder, U.; Valadbeigi, Y.; Ilbeigi, V.; Tabrizchi, M. Direct detection of glyphosate in drinking water using corona-discharge ion mobility spectrometry: A theoretical and experimental study. *Int. J. Mass Spectrom.* **2019**, *442*, 29–34. [CrossRef]
92. European Commission. *Guidance Document on Analytical Quality Control and Method Validation Procedures for Pesticides Residues Analysis in Food and Feed*; SANTE/11813/2017; European Commission: Brussels, Belgium, 2017.
93. European Commission. Commission Decision (EEC) 2002/657/EC. *Off. J. Eur. Commun.* **2002**, *L221*, 8.
94. European Commission. *Guidance Document on Identification of Mycotoxins in Food and Feed*; SANTE/12089 /2016; European Commission: Brussels, Belgium, 2016.
95. Wageningen University & Research. Available online: https://www.wur.nl/upload_mm/e/6/e/0d3c53a2-28b8-4e4d-a436-b84cf471f20e_20181015-9%20Revision%202002-657%20confirmation%20criteria.pdf (accessed on 24 July 2019).
96. Weickhardt, C.; Kaiser, N.; Borsdorf, H. Ion mobility spectrometry of laser desorbed pesticides from fruit surfaces. *Int. J. Ion. Mobil. Spectrom.* **2012**, *15*, 55–62. [CrossRef]
97. Armenta, S.; de la Guardia, M.; Abad-Fuentes, A.; Abad-Somovilla, A.; Esteve-Turrillas, F.A. Off-line coupling of multidimensional immunoaffinity chromatography and ion mobility spectrometry: A promising partnership. *J. Chromatogr. A* **2015**, *1426*, 110–117. [CrossRef] [PubMed]
98. Sheibani, A.; Tabrizchi, M.; Ghaziaskar, H.S. Determination of aflatoxins B1 and B2 using ion mobility spectrometry. *Talanta* **2008**, *75*, 233–238. [CrossRef] [PubMed]
99. Aria, A.A.; Sorribes-Soriano, A.; Jafari, M.T.; Nourbakhsh, F.; Esteve-Turrilas, F.A.; Armenta, S.; Herrero-Martínez, J.M.; de la Guardia, M. Uptake and translocation monitoring of imidacloprid to chili and tomato plants by molecularly imprinting extraction—Ion mobility spectrometry. *Microchem. J.* **2019**, *144*, 195–202. [CrossRef]
100. Saraji, M.; Jafari, M.T.; Mossaddegh, M. Carbon nanotubes@silicon dioxide nanohybrids coating for solid-phase microextraction of organophosphorus pesticides followed by gas chromatography-corona discharge ion mobility spectrometric detection. *J. Chromatogr. A* **2016**, 30–39. [CrossRef] [PubMed]
101. Kermani, M.; Jafari, M.; Saraji, M. Porous magnetized carbon sheet nanocomposites for dispersive solid-phase microextraction of organophosphorus pesticides prior to analysis by gas chromatography-ion mobility spectrometry. *Microchim. Acta* **2019**, *186*, 88. [CrossRef] [PubMed]
102. McCooeye, M.; Kolakowski, B.; Boison, J.; Mester, Z. Evaluation of high-field asymmetric waveform ion mobility spectrometry mass spectrometry for the analysis of the mycotoxin zearalenone. *Anal. Chim. Acta* **2008**, *627*, 112–116. [CrossRef] [PubMed]

103. Xu, Z.; Li, J.; Chen, A.; Ma, X.; Yang, S. A new retrospective, multi-evidence veterinary drug screening method using drift tube ion mobility mass spectrometry. *Rapid Commun. Mass Spectrom.* **2018**, *32*, 1141–1148. [CrossRef]
104. Ahmed, E.; Kabir, K.M.M.; Wang, H.; Xiao, D.; Fletcher, J.; Donald, W.A. Rapid separation of isomeric perfluoroalkyl substances by high-resolution differential ion mobility mass spectrometry. *Anal. Chim. Acta* **2019**, *1058*, 127–135. [CrossRef]
105. Beach, D.G.; Melanson, J.E.; Purves, R.W. Analyis of paralytic shellfish toxins using high-field asymmetric waveform ion mobility spectrometry with liquid chromatography-mass spectrometry. *Anal. Bioanal. Chem.* **2015**, *407*, 2473–2484. [CrossRef]
106. Poyer, S.; Loutelier-Bourhis, C.; Coadou, G.; Mondeguer, F.; Enche, J.; Bossée, A.; Hess, P.; Afonso, C. Identification and separation of saxitoxins using hydrophilic interaction liquid chromatography coupled to travelling wave ion mobility-mass spectrometry. *J. Mass Spectrom.* **2015**, *50*, 175–181. [CrossRef] [PubMed]
107. Beucher, L.; Dervilly-Pinel, G.; Prévost, S.; Monteau, F.; Le Bizec, B. Determination of a large set of β-adrenergic agonists in animal matrices based on ion mobility and mass separations. *Anal. Chem.* **2015**, *87*, 9234–9242. [CrossRef] [PubMed]
108. Zheng, X.; Dupuis, K.T.; Aly, N.A.; Zhou, Y.; Smith, F.B.; Tang, K.; Smith, R.D.; Baker, E.S. Utilizing ion mobility spectrometry and mass spectrometry for the analysis of polyciyclic aromatic hydrocarbons, polychlorinated biphenyls, polybrominated diphenyl ethers and their metabolites. *Anal. Chim. Acta* **2018**, *1037*, 265–273. [CrossRef] [PubMed]
109. Varesio, E.; Le Blanc, J.C.Y.; Hopfgartner, G. Real-time 2D separation by LC x differential ion mobility hyphenated to mass spectrometry. *Anal. Bioanal. Chem.* **2012**, *402*, 2555–2564. [CrossRef] [PubMed]
110. Beach, D.G.; Kerrin, E.S.; Quilliam, M.A. Selective quantitation of the neurotoxin BMAA by use of hydrophilic-interaction liquid chromatography-differential mobility spectrometry-tandem mass spectrometry (HILIC-DMS-MS/MS). *Anal. Bioanal. Chem.* **2015**, *407*, 8397–8409. [CrossRef] [PubMed]
111. Goscinny, S.; Joly, L.; De Pauw, E.; Hanot, V.; Eppe, G. Travelling-wave ion mobility time-of-flight mass spectrometry as an alternative strategy for screening of multi-class pesticides in fruits and vegetables. *J. Chromatogr. A* **2015**, *1405*, 85–93. [CrossRef] [PubMed]
112. Goscinny, S.; McCullagh, M.; Far, J.; De Pauw, E.; Eppe, G. Towards the use of ion mobility mass spectrometry derived collision cross section as a screening approach for unambiguous identification of targeted pesticides in food. *Rapid Commun. Mass Spectrom.* **2019**, *33*, 34–48. [CrossRef]
113. Regueiro, J.; Negreira, N.; Berntssen, M.H.G. Ion-mobility-derived collision cross section as an additional identification point for multiresidue screening of pesticides in fish feed. *Anal. Chem.* **2016**, *88*, 11169–11177. [CrossRef]
114. Hines, K.M.; Ross, D.H.; Davidson, K.L.; Bush, M.F.; Xu, L. Large-scale structural characterization of drug and drug-like compounds by high-throughput ion mobility-mass spectrometry. *Anal. Chem.* **2017**, *89*, 9023–9030. [CrossRef]
115. Hernández-Mesa, M.; Le Bizec, B.; Monteau, F.; García-Campaña, A.M.; Dervilly-Pinel, G. Collision Cross Section (CCS) database: An additional measure to characterize steroids. *Anal. Chem.* **2018**, *90*, 4616–4625. [CrossRef]
116. Tejada-Casado, C.; Hernández-Mesa, M.; Monteau, F.; Lara, F.J.; del Olmo-Iruela, M.; García-Campaña, A.M.; Le Bizec, B.; Dervilly-Pinel, G. Collision cross section (CCS) as a complementary parameter to characterize human and veterinary drugs. *Anal. Chim. Acta* **2018**, *1043*, 52–63. [CrossRef] [PubMed]
117. Righetti, L.; Bergmann, A.; Galaverna, G.; Rolfsson, O.; Paglia, G.; Dall'Asta, C. Ion mobility-derived collision cross section database: Application to mycotoxin analysis. *Anal. Chim. Acta* **2018**, *1014*, 50–57. [CrossRef] [PubMed]
118. Righetti, L.; Fenclova, M.; Dellafiora, L.; Hajslova, J.; Stranska-Zachariasova, M.; Dall'Asta, C. High resolution-ion mobility mass spectrometry as an additional powerful tool for structural characterization of mycotoxin metabolites. *Food Chem.* **2018**, *245*, 768–771. [CrossRef] [PubMed]
119. Bauer, A.; Luetjohann, J.; Hanschen, F.S.; Schreiner, M.; Kuballa, J.; Jantzen, E.; Rohn, S. Identification and characterization of pesticide metabolites in *Brassica* species by liquid chromatography travelling wave ion mobility quadrupole time-of-flight mass spectrometry (UPLC-TWIMS-QTOF-MS). *Food Chem.* **2018**, *244*, 292–303. [CrossRef] [PubMed]
120. Dodds, J.N.; May, J.C.; McLean, J.A. Correlating resolving power, resolution, and collision cross section:

Unifying cross-platform assessment of separation efficiency in ion mobility spectrometry. *Anal. Chem.* **2019**, *89*, 12176–12184. [CrossRef] [PubMed]

121. Deng, L.; Webb, I.K.; Garimella, S.V.B.; Hamid, A.M.; Zheng, X.; Norheim, R.V.; Prost, S.A.; Anderson, G.A.; Sandoval, J.A.; Baker, E.S.; et al. Serpentine ultralong path with extended routing (SUPER) high resolution traveling wave ion mobility-MS using structures for lossless ion manipulations. *Anal. Chem.* **2017**, *89*, 4628–4634. [CrossRef] [PubMed]
122. Merenbloom, S.I.; Glaskin, R.S.; Henson, Z.B.; Clemmer, D.E. High-resolution ion cyclotron mobility spectrometry. *Anal. Chem.* **2009**, *81*, 1482–1487. [CrossRef] [PubMed]
123. Glaskin, R.S.; Ewing, M.A.; Clemmer, D.E. Ion trapping for ion mobility spectrometry measurements in a cyclical drift tube. *Anal. Chem.* **2013**, *85*, 7003–7008. [CrossRef]
124. Giles, K.; Wildgoose, J.; Pringle, S.; Garside, J.; Carney, P.; Nixon, P.; Langridge, D. Design and utility of a multi-pass cyclic ion mobility separator. In *Annual Conference Proceedings*; ASMS: Baltimore, MD, USA, 2014.
125. Ujma, J.; Ropartz, D.; Giles, K.; Richardson, K.; Langridge, D.; Wildgoose, J.; Green, M. Cyclic ion mobility mass spectrometry distinguishes anomers and open-ring forms of pentasaccharides. *J. Am. Soc. Mass Spectrom.* **2019**, *30*, 1028–1037. [CrossRef]
126. McCullagh, M.; Giles, K.; Richardson, K.; Stead, S.; Palmer, M. Investigations into the performance of travelling wave enabled conventional and cyclic ion mobility systems to characterize protomers of fluoroquinolone antibiotic residues. *Rapid Commun. Mass Spectrom.* **2019**, *33*, 11–21. [CrossRef]
127. Hamid, A.M.; Ibrahim, Y.M.; Garimella, S.V.B.; Webb, I.K.; Deng, L.; Chen, T.-C.; Anderson, G.A.; Prost, S.A.; Norheim, R.V.; Tolmachev, A.V.; et al. Characterization of traveling wave ion mobility separations in structures for lossless ion manipulations. *Anal. Chem.* **2015**, *87*, 11301–11308. [CrossRef] [PubMed]
128. Deng, L.; Ibrahim, Y.M.; Baker, E.S.; Aly, N.A.; Hamid, A.M.; Zhang, X.; Zheng, X.; Garimella, S.V.B.; Webb, I.K.; Prost, S.A.; et al. Ion mobility separations of isomers based upon long path length structures for lossless ion manipulations combined with mass spectrometry. *ChemistrySelect* **2016**, *1*, 2396–2399. [CrossRef] [PubMed]
129. Faleh, A.B.; Warnke, S.; Rizzo, T.R. Combining ultrahigh-resolution ion-mobility spectrometry with cryogenic infrared spectroscopy for the analysis of glycan mixtures. *Anal. Chem.* **2019**, *91*, 4876–4882. [CrossRef] [PubMed]
130. Zhou, Z.; Tu, J.; Zhu, Z.-J. Advancing the large-scale CCS database for metabolomics and lipidomics at the machine-learning era. *Curr. Opin. Chem. Biol.* **2018**, *42*, 34–41. [CrossRef] [PubMed]
131. Bijlsma, L.; Bade, R.; Celma, A.; Mullin, L.; Cleland, G.; Stead, S.; Hernandez, F.; Sancho, J.V. Prediction of collision cross-section values for small molecules: Application to pesticide residue analysis. *Anal. Chem.* **2017**, *89*, 6583–6589. [CrossRef] [PubMed]
132. Morris, C.B.; May, J.C.; Leaptrot, K.L.; McLean, J.A. Evaluating separation selectivity and collision cross section measurement reproducibility in helium, nitrogen, argon, and carbon dioxide drift gases for drift tube ion mobility-mass spectrometry. *J. Am. Soc. Mass Spectrom.* **2019**, *30*, 1059–1068. [CrossRef] [PubMed]
133. Kurulugama, R.T.; Darland, E.; Kuhlmann, F.; Stafford, G.; Fjeldsted, J. Evaluation of drift gas selection in complex sample analysis using a high performance drift tube ion mobility-QTOF mass spectrometer. *Analyst* **2015**, *140*, 6834–6844. [CrossRef]
134. Oranzi, N.R.; Kemperman, R.H.J.; Wei, M.S.; Petkovska, V.I.; Granato, S.W.; Rochon, B.; Kaszycki, J.; La Rotta, A.; Jeanne Dit Fouque, K.; Fernandez-Lima, F.; et al. Measuring the integrity of gas-phase conformers of sodiated 25-hydroxyvitamin d3 by drift tube, traveling wave, trapped, and high-field asymmetric ion mobility. *Anal. Chem.* **2019**, *91*, 4092–4099. [CrossRef]
135. Hinnenkamp, V.; Klein, J.; Mecklmann, S.W.; Balsaa, P.; Schmidt, T.C.; Schmitz, O.J. Comparison of CCS values determined by traveling wave ion mobility mass spectrometry and drift tube ion mobility mass spectrometry. *Anal. Chem.* **2018**, *90*, 12042–12050. [CrossRef]
136. Paglia, G.; Williams, J.P.; Menikarachchi, L.; Thompson, J.W.; Tyldesley-Worster, R.; Halldórson, S.; Rolfsson, O.; Moseley, A.; Grant, D.; Langridge, J.; et al. Ion mobility derived collision cross sections to support metabolomics applications. *Anal. Chem.* **2014**, *86*, 3985–3993. [CrossRef]
137. Paglia, G.; Astarita, G. Metabolomics and lipidomics using traveling-wave ion mobility mass spectrometry. *Nat. Protoc.* **2017** *12*, 797–813. [CrossRef] [PubMed]
138. Hofmann, J.; Struwe, W.B.; Scarff, C.A.; Scrivens, J.H.; Harvey, D.J.; Pagel, K. Estimating collision cross sections of negatively charged n-glycans using traveling wave ion mobility-mass spectrometry. *Anal. Chem.*

2014, *86*, 10789–10795. [CrossRef] [PubMed]
139. Hines, K.M.; May, J.C.; McLean, J.A.; Xu, L. Evaluation of collision cross section calibrants for structural analysis of lipids by travelling wave ion mobility-mass spectrometry. *Anal. Chem.* **2016**, *88*, 7329–7336. [CrossRef] [PubMed]

Identification of the *Pol* Gene as a Species-Specific Diagnostic Marker for Qualitative and Quantitative PCR Detection of *Tricholoma matsutake*

Luying Shan [1,†], Dazhou Wang [1,†], Yinjiao Li [1], Shi Zheng [1], Wentao Xu [2] and Ying Shang [1,2,*]

1 Yunnan Institute of Food Safety, Kunming University of Science and Technology, Yunnan 650500, China; shanluying1994@163.com (L.S.); wdz101545171@163.com (D.W.); 18487156903@163.com (Y.L.); zhengshi199201@163.com (S.Z.)

2 Beijing Laboratory of Food Quality and Safety, College of Food Science and Nutritional Engineering, China Agricultural University, Beijing 100083, China; xuwentao@cau.edu.cn

* Correspondence: shangying1986@163.com

† These authors contributed equally to this work.

Abstract: *Tricholoma matsutake* is a rare, precious, and wild edible fungus that could not be cultivated artificially until now. This situation has given way to the introduction of fake *T. matsutake* commodities to the mushroom market. Among the methods used to detect food adulteration, amplification of species-specific diagnostic marker is particularly important and accurate. In this study, the *Pol* gene is reported as a species-specific diagnostic marker to identify three *T. matsutake* varieties and 10 other types of edible mushrooms through qualitative and quantitative PCR. The PCR results did not reveal variations in the amplified region, and the detection limits of qualitative and quantitative PCR were found to be 8 ng and 32 pg, respectively. Southern blot showed that the Pol gene exists as a single copy in the *T. matsutake* genome. The method that produced the purest DNA of *T. matsutake* in this study was also determined, and the high-concentration salt precipitation method was confirmed to be the most suitable among the methods tested. The assay proposed in this work is applicable not only to the detection of raw materials but also to the examination of processed products containing *T. matsutake*.

Keywords: *Tricholoma matsutake*; *Pol* gene; qualitative and quantitative PCR; DNA extraction

1. Introduction

Tricholoma matsutake is an ectomycorrhizal agaricomycete predominantly associated with pines and oaks. It is a commercially valuable edible mushroom [1] with great significance, not only because of its delicate flavor but also because of its diverse biological properties [2], which include multiple immunostimulatory, hematopoietic, antineoplastic, antimutation, and antioxidation activities [3].

The growing number of *T. matsutake* consumers has steadily expanded market demands for this mushroom. On account of the continued deterioration of ecological systems and the environment and excessive picking, the natural productivity of this gourmet mushroom has gradually declined. To date, no artificial cultivation method has yet been developed for *T. matsutake* due to the lack of information concerning precise soil requirements and cues for sporophore formation [4]. Therefore, wild *T. matsutake* and its products are in a short supply in the market, and, as such, its economic value has risen sharply. To address demands for the mushroom and reap higher profits, merchants frequently pass off counterfeit or adulterated *T. matsutake* products as genuine items, for example, sliced dried *Agaricus blazei* [5]. Establishing an effective and convenient method for identifying authentic *T. matsutake* is an urgent necessity.

Traditional methods of identifying *T. matsutake* are generally divided into two categories, namely, morphological [6] and physicochemical methods [7], and each category presents some limitations [8]. Morphological methods are effective in identifying fresh and integrated sporophores. Unfortunately, identification becomes much more difficult when applied to processed products on account of the destruction of the morphological characteristics of the mushroom. Physicochemical methods, such as gas chromatography, high-performance liquid chromatography, and mass spectrometry, require selection of a unique compound or fingerprint to represent *T. matsutake*, and the detection result is easily influenced by variety, habitat, and processing method [9,10]. Thus, two or more physicochemical means are commonly combined to obtain a valid result. If an isomeride exists in the analyte, detection becomes even more complicated. When the unique compound is added artificially, the physicochemical method may fail.

Rapid developments in modern biotechnology have enabled the wide use of molecular biological methods for authentication due to their accuracy, convenience, and speed. PCR-based methods, in particular, have been broadly adopted by many laboratories and fields [11,12]. Compared with conventional PCR, quantitative PCR technology has realized the leap of PCR from qualitative to quantitative, and it has higher specificity, effective resolution and higher degree of automation, and widely used in many fields such as gene expression research, transgenic research, drug efficacy assessment, pathogen detection and food composition analysis, especially. Therefore, real-time quantitative PCR detection is considered to be an easy-to-use, accurate, specific, sensitive, and quantitative method [13–15]. Furthermore, the species-specific diagnostic marker (also denoted as endogenous reference gene in some literatures) is a significant parameter during PCR amplification, which can evaluate the quality of the extracted DNA and provide the means to quantify the amount of the tested DNA substance in the processed food samples [15].

Identification of species-specific diagnostic marker, which requires species specificity, a consistent and low copy number, and low heterogeneity in the same species [13,14], has principally focused on crops. Numerous species-specific diagnostic marker of have been developed and reported [13], *vicK* for *Staphylococcus aureus* [15], Ribosomal Protein *L21* [16] for genetically modified (GM) wheat; *CruA* [17], *PEP* [18], *HMG-I/Y* [19], *FatA* [20], and *BnAccg8* [21] for canola; *lectin* [17,22], *β-actin* [17], and *hsp* [23] for soybean; *Cotton-ppi-PPF* [24], *ACP*1 [25], *Sad1* [26], and *SAH7* [27] for cotton; *hmga* [28–30], *10kDa zein* [30–33], *Ivr1* [23,30,34,35], *zSSIIb* [22,36,37], and *Adh*1 [30] for maize; and *gos9* [38], *RBE4* [39], and *SPS* [40] for rice. Appropriate reference genes under abiotic stress of annual ryegrass also was selected [41]. However, species-specific diagnostic markers for *T. matsutake* have yet to be established or reported.

Extracting high-quality DNA from *T. matsutake* fruiting bodies is crucial for downstream molecular experiments. At present, the cetyltrimethyl ammonium bromide (CTAB) method, which is especially suitable for plant DNA extraction, is commonly used to extract mushroom DNA [42]. However, high-quality DNA from *T. matsutake* fruiting bodies is difficult to acquire using this method because the mushroom contains thick cytoderms or capsules [43] and is rich in polyphenols and viscous polysaccharides. Thus, exploring a suitable DNA isolation method for *T. matsutake* fruiting bodies is an important endeavor.

In this work, the *Pol* gene, which encodes RNase H and integrase, was selected as a candidate species-specific diagnostic DNA marker for *T. matsutake* through sequence alignment and BLAST. The gene was verified to be a valid species-specific diagnostic marker given its features of appearance as a single copy in genomic DNA, good species specificity, and real-time quantitative detection limit as low as 32 pg. A suitable method for extracting the DNA of *T. matsutake*, i.e., high-concentration salt precipitation method, was also determined. The benefits of the present study are multifold: it provides a convenient and accurate approach for detecting *T. matsutake* and its related processed products, it aids efforts to improve food safety and detect food adulteration, and it develops an effective technique for detecting adulterations in the wild-mushroom market.

2. Results and Discussion

2.1. Comparison of DNA Extraction Methods

The DNA of *T. matsutake* fruiting bodies from Yunnan was obtained using the CTAB, SDS-CTAB, high-concentration salt precipitation, and kit methods. The results of 1% agarose gel for genomic DNA (Figure 1A) indicated that the four DNA extraction methods could successfully extract genomic DNA with good integrity. All of the OD260/280 and DNA concentrations obtained are listed in Table 1, there are significant differences in the DNA concentrations, and the OD260/280 ratio of the High-concentration salt precipitation method compared with other three extraction methods reveals that the purity of genomic DNA extracted by this method is the highest.

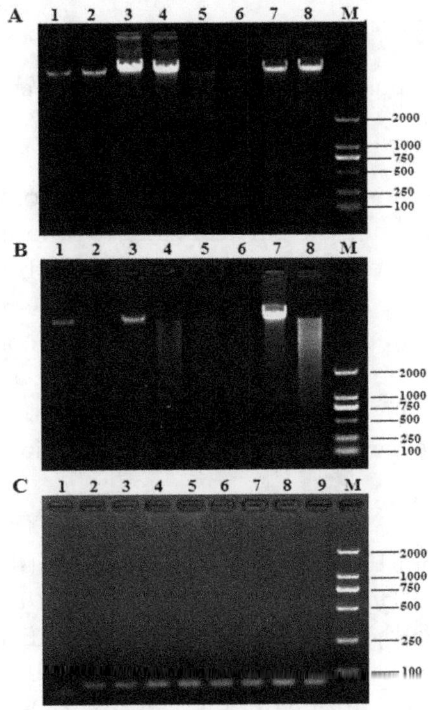

Figure 1. The effect of different DNA extraction methods. (**A**) The electrophoresis profile of *T. matsutake* genomic DNA; (**B**) the digestion result of *T. matsutake* genomic DNA. 1–2: cetyltrimethyl ammonium bromide (CTAB); 3–4: SDS-CTAB; 5–6: high-salt enzymolysis; 7–8: fungal genomic DNA isolation kit; M: DNA marker DL 2000; (**C**) the electrophoresis profile of PCR products of Tricholoma matsutake genomic DNA amplified by 18S universal primers. 1: negative control; 2–3: CTAB; 4–5: SDS-CTAB; 6–7: high-salt enzymolysis; 8-9: fungal genomic DNA isolation kit; M: DNA marker DL 2000.

Table 1. Comparison of purity and concentration of DNA extracted by four methods ($\bar{a} \pm$ SD, $n = 3$).

Method	OD260/280	Concentration (ng/μL)
CTAB	2.01 ± 0.02 ^{a}	983.33 ± 2.89 ^{a}
SDS-CTAB	1.97 ± 0.01 ^{a}	1138.33 ± 2.89 ^{b}
High-concentration salt precipitation	1.88 ± 0.02 ^{b}	676.67 ± 0.01 ^{c}
Kit method	1.66 ± 0.11 ^{c}	163.33 ± 2.89 ^{d}

$^{a\text{–}d}$ the upper letter of the OD260/280 and DNA concentrations indicates a significant difference (Duncan test, $p < 0.05$).

The restriction enzyme digestion result in Figure 1B showed that the genomic DNA extracted by the four methods could be digested by *EcoR* I, and the products were in a dispersive state. However, the digestibility of genomic DNA extracted by the CTAB, SDS-CTAB, and kit methods was lower than that of the high-concentration salt precipitation method, and there were a large number of large fragments, which showed brighter digestion products in the upstream part of the lanes, while the high-concentration salt precipitation method was more uniform. It revealed the high-concentration salt precipitation method was superior in quality and has less enzyme inhibitors.

The PCR products of the 18S rRNA region amplified with the fungal universal 18S primers were staining to identify whether the extracted *T. matsutake* genomic DNA can be used for subsequent PCR amplification. As shown in Figure 1C, the 18S rRNA fragments were efficiently amplified using the DNA extracted by the four methods.

In summary, by integrating the above four indicators shown in Table 1 and Figure 1, all genomic DNA extracted by the four methods could meet with the requirements of the PCR experiment, however, the high-concentration salt precipitation method was confirmed to be the most suitable DNA extraction way among the methods tested.

2.2. Species-Specific Diagnostic Marker of T. Matsutake

Among the genes selected, the *Pol* gene (*Tricholoma matsutake pol* gene for polyprotein encoding RNase H and integrase, Genbank No. AB016926) showed no homology with other genes from different varieties of the mushroom.

It has been reported that retrotransposons have been incorporated into the genome of their hosts and inherited to the host progenies since the earliest establishment of their parasitism. In a previous study, this *Pol* gene was a retroelement from *T. matsutake*, which related to RNase H and integrase of retrotransposons [44]. The reverse transcriptase domain was found in *T. matsutake* worldwide, this finding suggested that retroelements associate with ectomycorrhizal basidiomycetes and might be useful as genetic markers for identification, phylogenetic analysis, and mutagenesis of this fungal group [45]. After the BLAST analysis of *Pol* gene, it just confirmed the statement above. Hence, we chose the *Pol* gene as the candidate species-specific diagnostic marker of *T. matsutake*.

2.3. Species Specificity of Qualitative PCR aSSAYS

The genomic DNA isolated from 10 non-*T. matsutake* species (*R. virescens, A. deliciosus, B. speciosus, T. albuminosus, A. blazei, L. edodes, P. eryngii, F. velutipes*, and *P. ostreatus*) and three *T. matsutake* strains (Yunnan, Sichuan, and Jilin) was subjected to 18s rDNA amplification (Figure 2A). The 18s rDNA primer pair was used to confirm all the extracted DNA can be effectively amplified, and the quality of the extracted DNA clearly met the conditions of PCR (Figure 2A).

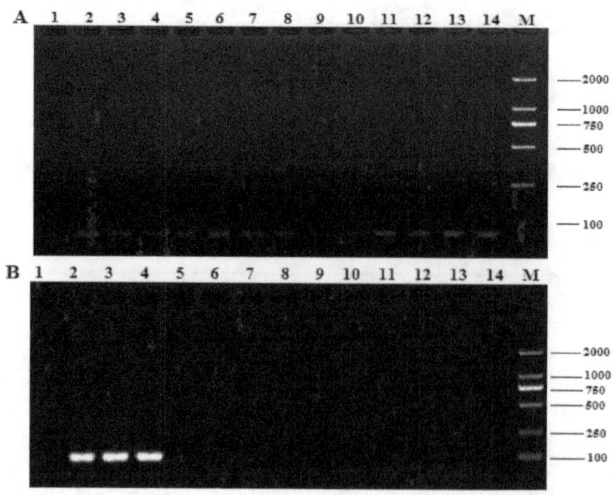

Figure 2. Specificity of the *Pol* gene detection in qualitative PCR. (**A**) The electrophoresis profiles of the DNA products amplified by the 18S rDNA; (**B**) the species specific identification of the *Pol* gene from *T. matsutake* by conventional PCR. 1: negative control, 2: Yunnan Shangri-La *T. matsutake*, 3: Sichuan Ganzi *T. matsutake*, 4: Jilin Yanbian *T. matsutake*, 5: *Russula virescens*, 6: *Agaricus deliciosus*, 7: *Boletus speciosus Forst*, 8: *Termitornyces albuminosus*, 9: *Agaricus blazei*, 10: *Agrocybe aegerita*, 11: *Lentinula edodes*, 12: *Pleurotus eryngii*, 13: *Flammulina velutipes*, 14: *Pleurotus ostreatus*, M: DNA marker DL 2000.

The primer pair *Pol*-F/R was applied to the qualitative PCR of the *Pol* gene; PCRs were also conducted using the genomic DNA of the 13 mushroom samples indicated above. Electrophoretic analysis of all qualitative PCR products (Figure 2B) revealed no objective product, except in the PCR products of the three *T. matsutake* samples. The results confirm that the qualitative PCR applied in this work are highly specific for *T. matsutake*.

2.4. Homology Analysis of the Pol Gene among Different T. Matsutake *Varieties*

An excellent species-specific diagnostic marker should have low heterogeneity and a consistent copy number in the same species. We carried out qualitative PCR using identical amounts of DNA from the three *T. matsutake* strains to determine whether the *Pol* gene undergoes any sort of variation. Amplification products with identical sizes and relative intensities were obtained for all varieties after qualitative PCR (Figure 2B, lines 2–4), and the slight differences, which were attributed to the quality of the isolated DNA [13], were considered negligible. As shown in Figure 3, the homologous similarity sequence identity between the PCR products and the reference *Pol* gene was 94.51%. These results indicate that the *Pol* gene did not show the sequence variation among the *T. matsutake* varieties studied.

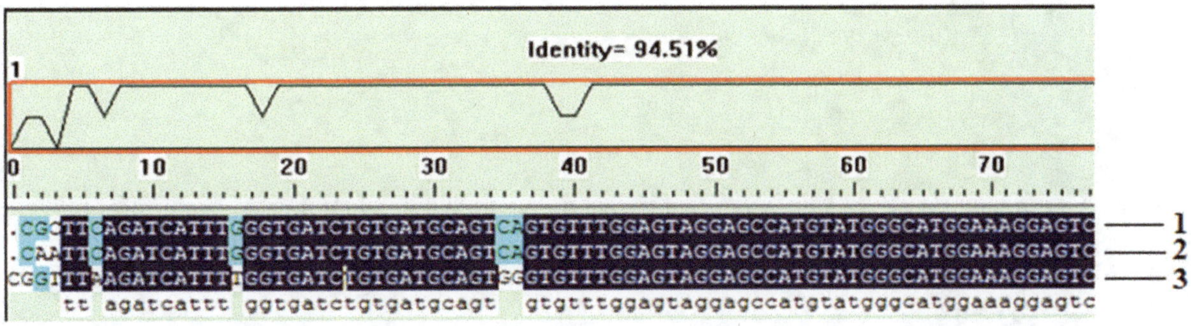

Figure 3. The sequencing result of *Pol* gene in different samples. 1, Yunnan Shangri-La *T. matsutake*; 2, Sichuan Ganzi *T. matsutake*; 3, Jilin Yanbian *T. matsutake*.

2.5. Confirmation of the Pol Gene Copy Number by Southern Blot

Besides species specificity, low sequence variation, and a consistent copy number, an excellent species-specific diagnostic marker is expected to possess a low copy number. Therefore, using Southern blot, we analyzed the copy number of the *Pol* gene in two *T. matsutake* varieties gathered from Yunnan and Jilin Provinces. Whether *Hind* III or *Eco*R I was used to digest the genomic DNA of *T. matsutake*, only one hybridization band was found in the nylon membrane (Figure 4), which means the *Pol* gene is only present as a single copy in the *T. matsutake* genome.

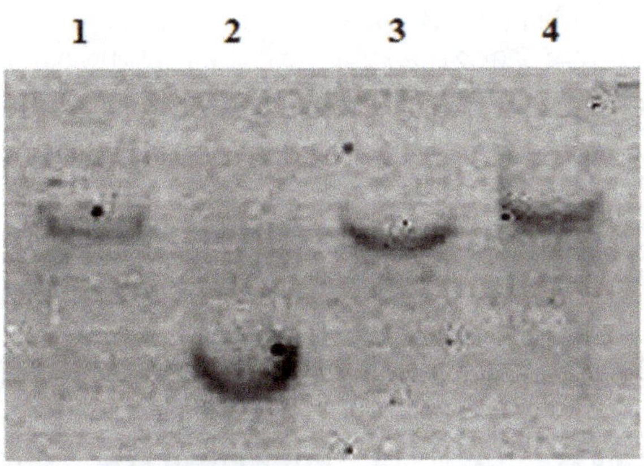

Figure 4. Southern blot result. 1, Jilin Yanbian *T. matsutake* digested by EcoR I; 2, Jilin Yanbian *T. matsutake* digested by Not I; 3, Yunnan Shangri-La *T. matsutake* digested by EcoR I; 4, Yunnan Shangri-La *T. matsutake* digested by Not I.

2.6. *Sensitivity of the Qualitative and Taqman-Based Real-Time Quantitative PCR Assays*

Genomic DNA from *T. matsutake* was diluted five times from 200 ng/μL to 12.8 pg/μL over a gradient, and the results (Figure 5A) showed a detection limit of 8 ng for qualitative PCR. With the same way, in Figure 5B, the sensitivity of Taqman quantitative PCR was found to be 32 pg. A standard curve of the *Pol* gene was then generated by using the proposed quantitative PCR system, and a linear relationship ($R^2 = 0.993$) with a slope of −3.081 was determined between the DNA quantities and Ct values (Figure 5C).

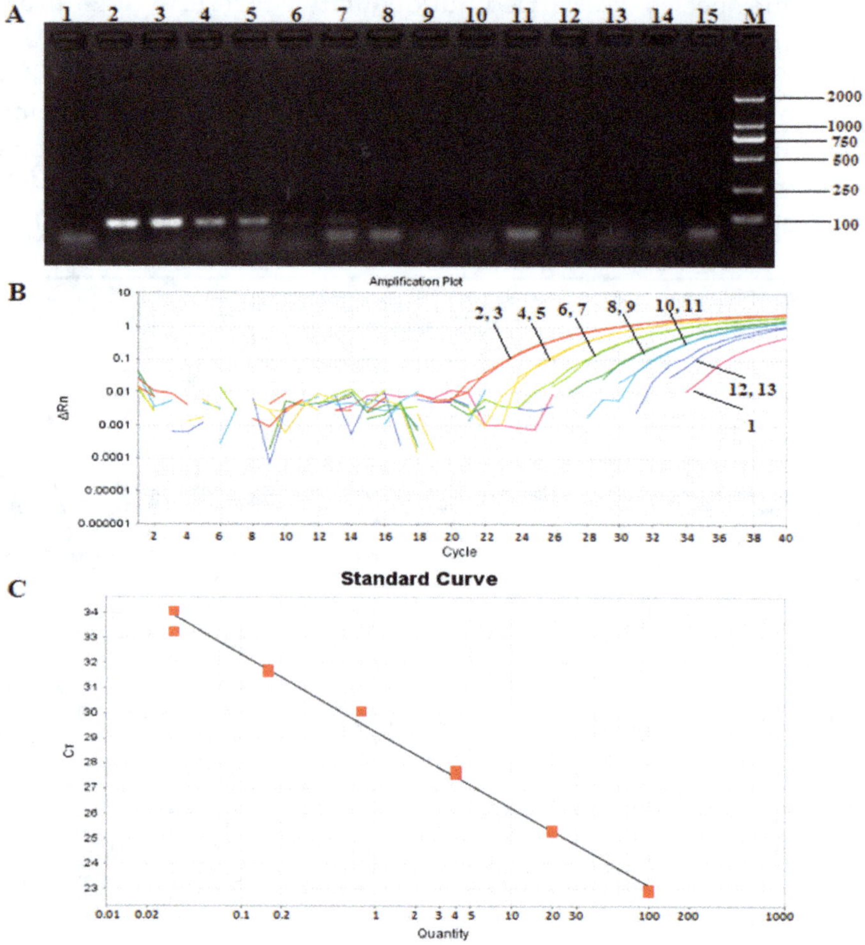

Figure 5. Sensitivity of the *Pol* gene detection in qualitative and Taqman quantitative PCR. (**A**) Sensitivity detection of *Pol* gene by conventional PCR. 1: negative control; 2–3: 200 ng; 4–5: 40 ng; 6–7: 8 ng; 8–9: 1.6 ng; 10–11: 0.32 ng; 12–13: 0.064 ng; 14–15: 0.0128 ng; M: DNA marker DL 2000; (**B**) the amplification curves of *Pol* gene by quantitative PCR. 1: negative control; 2–3: 100 ng; 4–5: 20 ng; 6–7: 4 ng; 8–9: 0.8 ng; 10–11: 160 pg; 12–13: 32 pg; (**C**) the standard curve of the sensitivity detection of *Pol* gene by qPCR.

2.7. *Application of the Pol gene to Detect Processed* T. Matsutake *Products*

We used the established Taqman-based PCR system to detect the source of *T. matsutake* in processed products, including mushroom biscuit, oil, and sauce. The target products could be amplified by employing Taqman-based quantitative PCR with the primer *Pol*-F/R and the probe *Pol*-P. The PCR results (Figure 6A) were consistent with the ingredients listed on the packaging of each products. The results of real-time quantitative PCR (Figure 6B) were in keeping with those of qualitative PCR. So, the results of both assays were prove that *T. matsutake* biscuits tested contained *T. matsutake* and the *Pol* gene was a practical and precise species-specific diagnostic marker for *T. matsutake* in highly processed foods, and affirmed its absence, as necessary.

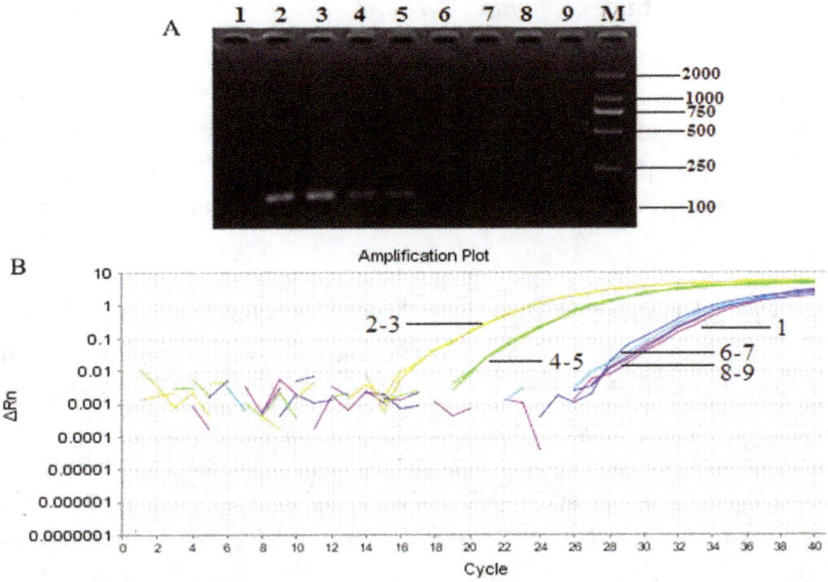

Figure 6. The detection of *T. matsutake* component in different kinds of food samples by qualitative (**A**) and real-time quantitative (**B**) PCR. 1: negative control; 2–3: positive control (Yunnan Shangri-La *T. matsutake*); 4–5: *T. matsutake* biscuits; 6–7: *T. matsutake* sauce; 8–9: Oil field *T. matsutake*; M: DNA marker DL 2000.

3. Materials and Methods

3.1. Materials

The following mushroom species were purchased from local farmers' markets: *Russula virescens*, *Agaricus deliciosus*, *Boletus speciosus*, *Termitornyces albuminosus*, *Agaricus blazei*, *Agrocybe cylindracea*, *Lentinula edodes*, *Pleurotus eryngii*, *Flammulina velutipes*, and *Pleurotus ostreatus*. Three *T. matsutake* varieties strains collected from the provinces of Yunnan, Sichuan, and Jilin, China, were kindly supplied by all China Federation of Supply and Marketing Cooperative's Kunming Institute of Edible Fungi. All of the samples were collected in the quantity of 200 g.

Three processed products labeled containing *T. matsutake* component, including mushroom biscuit, oil, and sauce, were gathered from Internet and local markets in Kunming, Yunnan province. They were used to verify the application of this selected species-specific diagnostic marker.

3.2. Genomic DNA Extraction

Four methods, including an improved CTAB method [46], the SDS-CTAB method [47], high-concentration salt precipitation method [48], and the kit method (Rapid Fungi Genomic DNA Isolation Kit, B518229, Sangon, Shanghai, China), were compared to determine the most suitable protocol for extracting the genomic DNA of *T. matsutake*. The *T. matsutake* fruiting bodies from Yunnan was used as the sample, each method was performed in triplicate. First, added the equal silica into 5 g mushroom sample, and then grounded them with liquid nitrogen. The target genomic DNA was isolated from the 0.2 g sample powder. All DNA extraction steps were performed in accordance with the references and manufacturer protocol. At last, the DNA was eluted in 50 μL EB buffer.

The subsequent DNA extraction was followed the selected optimal method. All of the OD260/280 and DNA concentrations were measured with a NanoDrop2000 spectrophotometer (Thermo Scientific, Waltham, MA, USA), the quality of genomic DNA was further analyzed on 1% agarose gel with ethidium bromide (0.1 μg /mL), and run with 1 × TAE buffer. Finally, the DNA solutions were stored at −20 °C.

3.3. *Enzyme Digestion of* T. Matsutake *Genomic DNA*

The DNA digestion was performed in a 20 μL reaction system containing 2 μL of 10× H Buffer, 1 μL *EcoR* I (15 U/μL) (TaKaRa Biotechnology Co. Ltd., Dalian, China), 3 μg genomic DNA, and added ultrapure water to 20 μL, and put the tube in a water bath (37 °C) for 2 h, after that added 2 μL 10× loading Buffer (Tiangen, Beijing, China) to stop the reaction. The digestion products was further analyzed on 1% agarose gel with ethidium bromide (0.1 μg /mL), and run with 1× TAE buffer.

3.4. *Southern Blot*

Complete enzyme cleavage of DNA from *T. matsutake* fruiting bodies obtained from Yunnan and Jilin was performed using *Hind* III and *Eco*R I, respectively, according to the manufacturer's instructions (TaKaRa) to ascertain the copy number of the *Pol* gene in the same species. The cleaved DNA was separated by 0.8% garose gel in 1×TAE at a constant voltage of 20 V overnight. Thereafter, the DNA fragments were transferred to a nylon membrane (Amersham Biosciences Shanghai Ltd., Darmstadt, Germany) from the 0.8% agarose gel. A 775 bp DNA fragment of the *Pol* gene was amplified with *Southern blot*-F/R and used as the hybridized probe, which was labeled with DIG-dUTP (DIG Hybridization Detection Kit, Mylab Co., Beijing, China). Pre-hybridization was carried out at 42 °C for 2 h, after which the pre-hybridization solution was poured off. The probe was denatured into single-strand DNA for 10 min at 100 °C and then cooled for 5 min. Exactly 4 μL of the denatured probe and 8 mL of the hybridization solution were added to a hybridization bag, and hybridization was performed at 42 °C overnight. Finally, the nylon membrane was washed twice with 3× SSC/0.5% SDS, and autoradiography was performed for 2–3 days.

3.5. *Species-Specific Diagnostic Marker Selection of* T. Matsutake

Genes belonging to *T. matsutake* were sought in Genbank, and several genes were preliminarily selected as detection targets according to their detailed gene information. Afterword, BLAST analysis was performed to determine the homology of the targets. The gene that with the lowest homology compared with all the DNA sequences in Genbank was selected as the candidate.

3.6. *Primers and Probe*

All primers and TaqMan fluorescent dye-labeled probes were designed using ABI Prism Primer Express version 3.0 software (Applied Biosystems, Foster City, CA, USA) and synthesized by Sangon Co. Ltd. (Shanghai, China).

The universal primer 18S rDNA-F/R was used to evaluate DNA quality. The *Pol* gene was assessed through qualitative PCR with the primer *pol*-F/R and through quantitative PCR with the primer *pol*-F/R, as well as the probe *pol*-P. Southern blot-F/R was used for Southern blot assay. The detailed sequences of these primers and probe are listed in Table 2.

Table 2. Primers used in qualitative and quantitative PCR.

Primer Name	Primer Sequence (5′→3′)	Length	Product Size (bp)	Reference
18S-F	CCTGAGAAACGGCTACCAT	19	80	This study
18S-R	ATCTTCACTACCTCCCCATTCTG	23		
pol-F	GACTCCCATACTGAAGCCAAT	21	107 (from 69 to 175 bp)	
pol-R	ACTCCTTTCCATGCCCATAC	20		
pol-probe	(FAM)-TGGCTCCTACTCCAAACACTGACAC-(TAMRA)			
Southern blot-F	CGTGATGGATGGAATACCTGT	21	775 (from 563 to 1157 bp)	
Southern blot-R	GTGTACCCCCCCTTAGACTGA	21		

3.7. *PCR Conditions*

Qualitative PCR was performed using an ABI SimpliAmp thermal cycler (Applied Biosystems) in a 25 μL reaction system containing 2.5 μL of 10×buffer, 0.2 mM dNTP, 0.4 μM of each primer, 2.5 units of Taq DNA Polymerase (TaKaRa Biotechnology Co. Ltd.), 100 ng DNA template, and 16.3 μL of ultrapure water. The 18S rDNA amplification was carried out to test DNA quality via the following program: 5 min at 95 °C; 40 cycles of 30 s at 95 °C, 30 s at 58 °C, and 30 s at 72 °C; and 10 min at 72 °C. During *Pol* gene amplification, the procedures applied for 18S rDNA amplification were followed, but annealing was optimized at 56 °C, 58 °C, 60 °C, 62 °C, and 64 °C separately. After comparison, the annealing temperature 60 °C was chosen.

The PCR amplified products were analyzed on 2% agarose gel with ethidium bromide (0.1 μg/mL), and run with 1× TAE buffer. The PCR products, that conducted the homology analysis, were sequenced using *pol*-F/R separately by Sangon (Shanghai, China) company.

Real-time quantitative PCR of the *Pol* gene was also conducted in an ABI StepOne Plus Real-Time System (Applied Biosystems, Thermo Fisher Scientific, Waltham, MA, USA) using the TaqMan probe methods. TaqMan probe-based qPCR was performed using a 20 μL reaction system containing 1×TaqMan Gene Expression Master Mix (TaKaRa, Dalian, China), 200 nM primers, 200 nM probe, and 50 ng of DNA. Each sample was quantified twice for each biological replicate.

3.8. Sensitivity of the Qualitative and Taqman-Based Real-Time Quantitative PCR Assays

Genomic DNA from *T. matsutake* was diluted seven times from 200 ng/μL to 12.8 pg/μL (five-fold serial dilutions) over a gradient using the non- *T. matsutake* DNA, and a series of PCRs with 1 μL DNA sample were conducted to determine the detection limit of qualitative PCR, and the products were analyzed using 2% agarose.

To evaluate the sensitivity of Taqman-based quantitative PCR, genomic DNA was serially diluted six times to final concentrations ranging from 100 ng/μL to 32 pg/μL (five-fold serial dilutions) using the non-*T. matsutake* DNA, and with 1 μL DNA sample, the detection limit of Taqman quantitative PCR was determined.

4. Conclusions

In conclusion, this work demonstrated that High-quality DNA from *T. matsutake* fruiting bodies was obtained following the high-concentration salt precipitation method, and the *Pol* gene was selected and validated as an ideal species-specific diagnostic marker for the PCR-based detection of *T. matsutake* sources after assay of its species specificity, copy number, high homology in different varieties, and sensitivity. The detection limit of Taqman-based quantitative PCR analysis was 32 pg, which means this method could be used to detect processed *T. matsutake* products containing low amounts of the target genomic DNA.

Author Contributions: Methodology, L.S.; formal analysis, D.W.; validation, Y.L. and S.Z.; writing—original draft preparation, L.S.; writing—review and editing, W.X. and Y.S.; supervision, Y.S.; funding acquisition Y.S.

References

1. Xu, J.P.; Sha, T.; Li, Y.C.; Zhao, Z.W.; Yang, Z.L. Recombination and genetic differentiation among natural populations of the ectomycorrhizal mushroom Tricholoma matsutake from southwestern China. *Mol. Ecol.* **2008**, *17*, 1238–1247. [CrossRef] [PubMed]
2. Geng, X.; Tian, G.; Zhao, Y.; Zhao, L.; Wang, H.; Ng, B.T. A Fungal α-Galactosidase from Tricholoma matsutake with Broad Substrate Specificity and Good Hydrolytic Activity on Raffinose Family Oligosaccharides. *Molecules* **2015**, *20*, 13550–13562. [CrossRef] [PubMed]
3. Yin, X.; You, Q.; Su, X. A comparison study on extraction of polysaccharides from Tricholoma matsutake by response surface methodology. *Carbohydr. Polym.* **2014**, *102*, 419–422. [CrossRef] [PubMed]

4. Vaario, L.-M.; Pennanen, T.; Lu, J.; Palmén, J.; Stenman, J.; Leveinen, J.; Kilpeläinen, P.; Kitunen, V. Tricholoma matsutake can absorb and accumulate trace elements directly from rock fragments in the shiro. *Mycorrhiza* **2015**, *25*, 325–334. [CrossRef] [PubMed]
5. Sun, Y.; Shi, Y. FT-TRS Analysis of Tricholoma matsutake Sing and Agaricus Blazei Murrill. *Phys. Test. Chem. Anal. (Part B Chem. Anal.)* **2013**, *49*, 1076–1078.
6. Zhu, Y.; Zhang, H.; Zhu, Y.; Pu, X. An Appraisal of Past and Present Methods for Identifying Poisonous Mushrooms. *J. Gansu Sci.* **2008**, *20*, 44–48.
7. Liu, Y.; Liu, N.; Zhang, Y. Research progress of food authentication technology. *Sci. Technol. Food Ind.* **2016**, *37*, 374–383.
8. Chen, Y.; Dong, W.; Wu, Y.; Yuan, F.; Huang, W.; Ge, Y. Research and application of food authentication system. *Sci. Technol. Food Ind.* **2008**, *29*, 216–218.
9. Qiang, L.; Li, Z.; Li, W.; Li, X.; Huang, W.; Hua, Y.; Zheng, L. Chemical compositions and volatile compounds of Tricholoma matsutake from different geographical areas at different stages of maturity. *Food Sci. Biotechnol.* **2016**, *25*, 71–77.
10. Ding, X.; Hou, Y.L. Identification of genetic characterization and volatile compounds of Tricholoma matsutake from different geographical origins. *Biochem. Syst. Ecol.* **2012**, *44*, 233–239. [CrossRef]
11. Yang, L.; Yang, Y.; Jin, W.; Zhang, X.; Li, X.; Wu, Y.; Li, J.; Li, L. Development and Interlaboratories Validation of Event-Specific Quantitative Real-Time PCR Method for Genetically Modified Rice G6H1 Event. *J. Agric. Food Chem.* **2018**, *66*, 8179. [CrossRef]
12. Liu, Y.; Jiang, D.; Lu, X.; Wang, W.; Xu, Y.; He, Q. Phage Medicated Immuno-PCR for Ultrasensitive Detection of Cry1Ac protein based on Nanobody. *J. Agric. Food Chem.* **2016**, *64*, 7882–7889. [CrossRef]
13. Xu, W.; Bai, W.; Guo, F.; Luo, Y.; Yuan, Y.; Huang, K. A papaya-specific gene, papain, used as an endogenous reference gene in qualitative and real-time quantitative PCR detection of transgenic papayas. *Eur. Food Res. Technol.* **2008**, *228*, 301–309. [CrossRef]
14. Shang, Y.; Zhu, P.; Huang, K.; Liu, W.; Tian, W.; Luo, Y.; Xu, W. A peach (Prunus persica)-specific gene, Lhcb2, used as an endogenous reference gene for qualitative and real-time quantitative PCR to detect fruit products. *LWT-Food Sci. Technol.* **2014**, *55*, 218–223. [CrossRef]
15. Liu, Z.-M.; Shi, X.-M.; Pan, F. Species-specific diagnostic marker for rapid identification of Staphylococcus aureus. *Diagn. Microbiol. Infect. Dis.* **2007**, *59*, 379–382. [CrossRef]
16. Liu, Y.K.; Li, H.P.; Huang, T.; Cheng, W.; Gao, C.S.; Zuo, D.Y.; Zhao, Z.X.; Liao, Y.C. Wheat-Specific Gene, Ribosomal Protein L21, Used as the Endogenous Reference Gene for Qualitative and Real-Time Quantitative Polymerase Chain Reaction Detection of Transgenes. *J. Agric. Food Chem.* **2014**, *62*, 10405–10413. [CrossRef]
17. James, D.; Schmidt, A.-M.; Wall, E.; Green, M.; Masri, S. Reliable Detection and Identification of Genetically Modified Maize, Soybean, and Canola by Multiplex PCR Analysis. *J. Agric. Food Chem.* **2003**, *51*, 5829–5834. [CrossRef]
18. Zeitler, R.; Pietsch, K.; Waiblinger, H.-U. Validation of real-time PCR methods for the quantification of transgenic contaminations in rape seed. *Eur. Food Res. Technol.* **2002**, *214*, 346–351. [CrossRef]
19. Weng, H.; Yang, L.; Liu, Z.; Ding, J.; Pan, A.; Zhang, D. Novel Reference Gene, High-mobility-group protein I/Y, Used in Qualitative and Real-Time Quantitative Polymerase Chain Reaction Detection of Transgenic Rapeseed Cultivars. *J. AOAC Int.* **2005**, *88*, 577–584.
20. Demeke, T.; Ratnayaka, I. Multiplex qualitative PCR assay for identification of genetically modified canola events and real-time event-specific PCR assay for quantification of the GT73 canola event. *Food Control* **2008**, *19*, 893–897. [CrossRef]
21. Hernández, M.; Río, A.; Esteve, T.; Prat, S.; Pla, M. A Rapeseed-Specific Gene, Acetyl-CoA Carboxylase, Can Be Used as a Reference for Qualitative and Real-Time Quantitative PCR Detection of Transgenes from Mixed Food Samples. *J. Agric. Food Chem.* **2001**, *49*, 3622–3627. [CrossRef]
22. Yoshimura, T.; Kuribara, H.; Matsuoka, T.; Kodama, T.; Iida, M.; Watanabe, T.; Akiyama, H.; Maitani, T.; Furui, S.; Hino, A. Applicability of the Quantification of Genetically Modified Organisms to Foods Processed from Maize and Soy. *J. Agric. Food Chem.* **2005**, *53*, 2052–2059. [CrossRef]
23. Van Duijn, G.; Biert, R.V.; Bleeker-Marcelis, H.; Peppelman, H.; Hessing, M. Detection methods for genetically modified crops. *Food Control* **1999**, *10*, 375–378. [CrossRef]

24. Chaouachi, M.; Giancola, S.; Romaniuk, M.; Laval, V.; Bertheau, Y.; Brunel, D. A Strategy for Designing Multi-Taxa Specific Reference Gene Systems. Example of Application—ppi Phosphofructokinase (ppi-PPF) Used for the Detection and Quantification of Three Taxa: Maize (*Zea mays*), Cotton (*Gossypium hirsutum*) and Rice (*Oryza sativa*). *J. Agric. Food Chem.* **2007**, *55*, 8003–8010. [CrossRef]
25. Lee, S.-H.; Kim, J.-K.; Yi, B.-Y. Detection Methods for Biotech Cotton MON 15985 and MON 88913 by PCR. *J. Agric. Food Chem.* **2007**, *55*, 3351–3357. [CrossRef]
26. Yang, L.; Chen, J.; Huang, C.; Liu, Y.; Jia, S.; Pan, L.; Zhang, D. Validation of a cotton-specific gene, Sad1, used as an endogenous reference gene in qualitative and real-time quantitative PCR detection of transgenic cottons. *Plant Cell Rep.* **2005**, *24*, 237–245. [CrossRef]
27. Baeumler, S.; Wulff, D.; Tagliani, L.; Song, P. A Real-Time Quantitative PCR Detection Method Specific to Widestrike Transgenic Cotton (Event 281-24-236/3006-210-23). *J. Agric. Food Chem.* **2006**, *54*, 6527–6534. [CrossRef]
28. Zimmermann, A.; Hemmer, W.; Liniger, M.; Lüthy, J.; Pauli, U. A Sensitive Detection Method For Genetically Modified MaisGardTMCorn Using a Nested PCR-system. *LWT-Food Sci. Technol.* **1998**, *31*, 664–667. [CrossRef]
29. Pardigol, A.; Guillet, S.; Pöpping, B. A simple procedure for quantification of genetically modified organisms using hybrid amplicon standards. *Eur. Food Res. Technol.* **2003**, *216*, 412–420. [CrossRef]
30. Hernández, M.; Duplan, M.-N.; Berthier, G.; Vaïtilingom, M.; Hauser, W.; Freyer, R.; Pla, M.; Bertheau, Y. Development and Comparison of Four Real-Time Polymerase Chain Reaction Systems for Specific Detection and Quantification of *Zea mays* L. *J. Agric. Food Chem.* **2004**, *52*, 4632–4637. [CrossRef]
31. Vaïtilingom, M.; Pijnenburg, H.; Gendre, F.; Brignon, P. Real-Time Quantitative PCR Detection of Genetically Modified Maximizer Maize and Roundup Ready Soybean in Some Representative Foods. *J. Agric. Food Chem.* **1999**, *47*, 5261–5266. [CrossRef] [PubMed]
32. Höhne, M.; Santisi, C.; Meyer, R. Real-time multiplex PCR: An accurate method for the detection and quantification of 35S-CaMV promoter in genetically modified maize-containing food. *Eur. Food Res. Technol.* **2002**, *215*, 59–64. [CrossRef]
33. Matsuoka, T.; Kuribara, H.; Takubo, K.; Akiyama, H.; Miura, H.; Goda, Y.; Kusakabe, Y.; Isshiki, K.; Toyoda, M.; Hino, A. Detection of Recombinant DNA Segments Introduced to Genetically Modified Maize (*Zea mays*). *J. Agric. Food Chem.* **2002**, *50*, 2100–2109. [CrossRef] [PubMed]
34. Brodmann, P.D.; Ilg, E.C.; Berthoud, H.; Herrmann, A. Real-Time Quantitative Polymerase Chain Reaction Methods for Four Genetically Modified Maize Varieties and Maize DNA Content in Food. *J. AOAC Int.* **2002**, *85*, 646–653. [PubMed]
35. Rønning, S.B.; Vaïtilingom, M.; Berdal, K.G.; Holst-Jensen, A. Event specific real-time quantitative PCR for genetically modified Bt11 maize (*Zea mays*). *Eur. Food Res. Technol.* **2003**, *216*, 347–354. [CrossRef]
36. Kuribara, H.; Shindo, Y.; Matsuoka, T.; Takubo, K.; Futo, S.; Aoki, N.; Hirao, T.; Akiyama, H.; Goda, Y.; Toyada, M.; et al. Novel Reference Molecules for Quantitation of Genetically Modified Maize and Soybean. *J. AOAC Int.* **2002**, *85*, 1077–1089.
37. Lee, S.-H.; Min, D.-M.; Kim, J.-K. Qualitative and Quantitative Polymerase Chain Reaction Analysis for Genetically Modified Maize MON863. *J. Agric. Food Chem.* **2006**, *54*, 1124–1129. [CrossRef]
38. Hernândez, M.; Esteve, T.; Pla, M. Real-Time Polymerase Chain Reaction Based Assays for Quantitative Detection of Barley, Rice, Sunflower, and Wheat. *J. Agric. Food Chem.* **2005**, *53*, 7003–7009. [CrossRef]
39. Jeong, S.-C.; Pack, I.S.; Cho, E.-Y.; Youk, E.S.; Park, S.; Yoon, W.K.; Kim, C.-G.; Choi, Y.D.; Kim, J.-K.; Kim, H.M. Molecular analysis and quantitative detection of a transgenic rice line expressing a bifunctional fusion TPSP. *Food Control* **2007**, *18*, 1434–1442. [CrossRef]
40. Ding, J.; Jia, J.; Yang, L.; Wen, H.; Zhang, C.; Liu, W.; Zhang, D. Validation of a Rice Specific Gene, Sucrose Phosphate Synthase, Used as the Endogenous Reference Gene for Qualitative and Real-Time Quantitative PCR Detection of Transgenes. *J. Agric. Food Chem.* **2004**, *52*, 3372–3377. [CrossRef]
41. Liu, Q.; Qi, X.; Yan, H.; Huang, L.; Nie, G.; Zhang, X. Reference Gene Selection for Quantitative Real-Time Reverse-Transcriptase PCR in Annual Ryegrass (*Lolium multiflorum*) Subjected to Various Abiotic Stresses. *Molecules* **2018**, *23*, 172. [CrossRef] [PubMed]
42. Jiang, S.; Shao, B.; Chen, W.; Li, S.; Wang, Z.; Zhun, X.; Liao, J. Study on the Comparison of three Genomic DNA Extraction Methods for 15 Familiar Edulis and Medicincal Fungi. *Food Sci.* **2004**, *25*, 36–40.

43. González-Mendoza, D.; Argumedo-Delira, R.; Morales-Trejo, A.; Pulido-Herrera, A.; Cervantes-Díaz, L.; Grimaldo-Juarez, O.; Alarcón, A. A rapid method for isolation of total DNA from pathogenic filamentous plant fungi. *Genet. Mol. Res.* **2010**, *9*, 162–166. [CrossRef] [PubMed]
44. Murata, H.; Yamada, A.; Babasaki, K. Identification of Repetitive Sequences Containing Motifs of Retrotransposons in the Ectomycorrhizal Basidiomycete Tricholoma matsutake. *Mycologia* **1999**, *91*, 766–775. [CrossRef]
45. Murata, H.; Yamada, A. marY1, a Member of the gypsy Group of Long Terminal Repeat Retroelements from the Ectomycorrhizal Basidiomycete Tricholoma matsutake. *Appl. Environ. Microbiol.* **2000**, *66*, 3642–3645. [CrossRef]
46. Nan, Z.; Xu, W.; Bai, W.; Zhai, Z.; Luo, Y.; Yan, X.; Jing, H.; Huang, K. Event-specific qualitative and quantitative PCR detection of LY038 maize in mixed samples. *Food Control* **2011**, *22*, 1287–1295.
47. Shang, Y.; Yan, Y.; Xu, W.; Tian, W.; Huang, K. Research on Gene Mobility and Gene Flow Between Genetically Modified Mon 15985 Cotton and Pleurotus Ostreatus. *J. Food Saf.* **2016**, *36*, 423–432. [CrossRef]
48. Yan, Y.; Xu, W.; Su, C.; Luo, Y.; Wang, Y.; Gu, X.; Dai, Y.; Tian, H. Study on the methods of extraction of Pleurotus ostreatus genomic DNA. *Sci. Technol. Food Ind.* **2011**, 190–193.

6

Phytochemical and Biological Characteristics of Mexican Chia Seed Oil

Yingbin Shen [1,†], Liyou Zheng [2,†], Jun Jin [2], Xiaojing Li [2], Junning Fu [1], Mingzhong Wang [3], Yifu Guan [4,*] and Xun Song [5,*]

1 Department of Food Science and Engineering, Jinan University, Guangzhou 510632, China; shenybin412@gmail.com (Y.S.); juningf0313@163.com (J.F.)
2 State Key Laboratory of Food Science and Technology, Synergetic Innovation Center of Food Safety and Nutrition, School of Food Science and Technology, Jiangnan University, 1800 Lihu Road, Wuxi 214122, China; liyou890513@sina.com (L.Z.); zgzjjin@126.com (J.J.); Lixiaojing19900810@163.com (X.L.)
3 Shenzhen Kivita Innovative Drug Discovery Institute, Shenzhen 518110, China; mzwang2000@163.com
4 Research Institute for Marine Drugs, Guangxi University of Chinese Medicine, Nanning 530200, China
5 School of Pharmaceutical Science, Shenzhen University, Shenzhen 518060, China
* Correspondence: 000917@gxtcmu.edu.cn (Y.G.); 13480243@life.hkbu.edu.hk (X.S.)

† These authors contributed equally to this work and share first authorship.

Academic Editors: Alessandra Gentili and Chiara Fanali

Abstract: The purpose of this research was to investigate the chemical profile, nutritional quality, antioxidant and hypolipidemic effects of Mexican chia seed oil (CSO) in vitro. Chemical characterization of CSO indicated the content of α-linolenic acid (63.64% of total fatty acids) to be the highest, followed by linoleic acid (19.84%), and saturated fatty acid (less than 11%). Trilinolenin content (53.44% of total triacylglycerols (TAGs)) was found to be the highest among seven TAGs in CSO. The antioxidant capacity of CSO, evaluated with $ABTS^{\bullet+}$ and $DPPH^{\bullet}$ methods, showed mild antioxidant capacity when compared with Tocopherol and Catechin. In addition, CSO was found to lower triglyceride (TG) and low-density lipoprotein-cholesterol (LDL-C) levels by 25.8% and 72.9%respectively in a HepG2 lipid accumulation model. As CSO exhibits these chemical and biological characteristics, it is a potential resource of essential fatty acids for human use.

Keywords: Chia seed oil; polyunsaturated fatty acid; antioxidant; lipid-lowering effect

1. Introduction

Vegetable oils such as palm, canola, soybean, rapeseed, corn, olive, and sunflower oils are commonly used in cooking and industrial food manufacturing. The chemical constituents and content of cooking oils, particularly the type of fatty acids, are considered important criteria for quality of oils and their health benefits. Nutrition studies indicate that the type and content of dietary fat intake is closely related to various diseases, such as cardiovascular disease, diabetes, and depression [1]. Thus, consumption of healthier polyunsaturated fatty acids (PUFAs), which are considered 'good fat', is a vitally important prevention strategy for fat related health complications. However, as PUFAs cannot be synthesized in the human body, it is essential to supplement them through diet to meet daily needs [2]. Due to the huge market of PUFA rich oils, a number of industrial food manufacturing plants tend to seek alternative, plant-based sources of PUFAs.

Chia, botanically known as *Salvia hispanica* L., native to southern Mexico and northern Guatemala, is a 2 m tall herbaceous plant that belongs to the Lamiaceae family [3]. Currently, chia is grown on a commercial scale in the South America, Australia, Europe, and Southeast Asia, due to the high edible value in whole seeds, flour, and oil in food industry. In general, the major constituents in chia seeds include 25–40% oil, 17–24% protein, and 18–30% dietary fiber [4]. It also contains significant quantities of minerals and phytonutrients, including tocopherols, carotenoids and phytosterols, and phenolic compounds [5]. Chia seed oil is probably one of the healthiest oils on the market, containing predominantly essential fatty acids such as α-linolenic acid and linoleic acid. Numerous findings have confirmed that chia seed oil is a healthy oil for lowering the risk for cardiovascular disease, hepatoprotective effect, inflammation, and prevention of obesity-related disorders [6]. At present, the in vitro cancer cytotoxic properties of CSO have been reported by Ramzi et al., who showed that the CSO significantly inhibited the proliferation of human lymphoblastic leukemic cell lines, HeLa, and MCF-7 cells [7]. Up to now, there has been no evidence of adverse effects of chia seeds, and toxicological data on CSO from animal and controlled human trials on the safety and efficacy are still limited. However, experience gained from previous and current use of chia seeds for food purposes in developing countries can be regarded as supportive evidence to allow a positive conclusion on the safety of CSO [8].

Research findings by Ayerza et al. indicated that the content of the bioactive nutrients of chia seed can be affected by the geographical location and climate condition [9]. To date, research has been carried out on basic chemical and physical characteristics, including quantification of fatty acids, tocopherol, and polyphenols. However, to the best of our knowledge, sn-2 fatty acid composition and minor components of Mexican chia seed oil have not been extensively studied. The purpose of this research was to elaborate on most chemical compositions (fatty acids, triacylglycerol, tocopherols, sterols, polyphenols, metal elements, and PAHs) and functional values (antioxidant activity and lipid-lowering effects) of oil extracted from chia seeds grown in Mexico, to further evaluate its nutritional value and thus contribute towards identification of a potential food with medicinal and industrial applications.

2. Results and Discussion

2.1. Physical and Chemical Profiles

The physical and chemical properties analysis of CSO are presented in Table 1. The oil content of chia seed ranged from 31.39 to 32.39 g/100 g, much higher than that in soybean (17.6 to 25.4 g/100 g) reported by Dornbos et al. [10]. The result was consistent with the published data (31.2 g/100 g, on average, of Brazil chia seed) [11] and 28.5–32.70 g/100 g in chia seeds from Colombia, Argentina, Peru and Bolivia [12]. The results obtained in this study suggested that chia seed was a good source of crude oil based on the high yield of oil content.

Table 1. Physical and chemical characters of chia seed oil.

Oil Parameters	Content	
Oil content (g/100 g)	31.89 ± 0.50	
Oil stability (induction period)/h	0.68 ± 0.03	
Oil color/(units)		
	R	Y
	1.65 ± 0.07	13.00 ± 0.00

Table 1. *Cont.*

Oil Parameters	Content
[a] **Fatty Acid Composition**	
Palmitic acid C16:0	7.07 ± 0.01
Palmitoleic acid C16:1 (n-9)	0.06 ± 0.00
Trianoic acid C17:0	0.16 ± 0.01
Stearic acid C18:0	2.81 ± 0.04
Oleic acid C18:1 (n-9)	5.50 ± 0.01
Vaccenic acid C18:1 (n-7)	0.80 ± 0.01
Linoleic acid C18:2 (n-6)	19.84 ± 0.01
α-Linolenic C18:3 (n-3)	63.64 ± 0.06
Arachidic acid C20:0	0.12 ± 0.01
SFA	10.16 ± 0.06
PUFA	89.84 ± 0.07
PUFA/SFA	8.85 ± 0.06
n-3/n-6 FA ratio	3.21 ± 0.00
sn-2 Fatty Acid Composition	
C16:0	1.10 ± 0.05
C18:0	0.88 ± 0.09
C18:1	6.38 ± 0.12
C18:2	25.07 ± 0.06
C18:3	63.76 ± 0.66

[a] Values reported as means ± SD of three replicate analyses (n = 3). SFA = total saturated fatty acids, PUFA = total polyunsaturated fatty acids, n-6 = total omega-6, n-3 = total omega-3 fatty acids.

2.2. Oxidative Stability

The induction period, widely used to determine stability of edible oils, is an indicator of oxidative processes. The oxidative stability of oils was measured as the induction period in response to forced oxidation. The results in Table 1 indicate that the induction period of CSO was found to be 0.68 h, which is similar to the reported value for Camelina oil (0.63 h) [13], but much lower than that of soybean oil [14]. The possible reason for lower induction period is the high content of unsaturated fatty acids in CSO (89.84% of total lipids). However, as there is limited data available on the stability of CSO, there is rising need for investigation on the stability of CSO during processing and storage.

At present there aren't any color standards for CSO, and thus the Lovibond colorimeter was used firstly for color measurement. The color of CSO, represented by R and Y Lovibond scale, displayed more yellow (13 units) than red (1.65 units), which was consistent with that reported by Maira et al. [15]. The color ratio of CSO was also similar to that of linseed oil with yellow (70 units) and red (8.6 units) [16]. Nevertheless, CSO has different color than oils such as palm (yellow, 3.2 units; red, 27.4 units) and soy oil (yellow, 4.6 units; red, 10.6 units), which show more red than yellow [17]. Total pigment and carotenoid content are responsible for the natural color of vegetable oil, which could be an indicator of metamorphic oils. Significant changes of the R and Y values were found on the CSO during the metamorphic process, which could be used as a mark in distinguishing the degree of metamorphism.

2.3. Fat and Triacylglycerol Composition

As shown in Table 2, the most abundant fatty acid was α-linolenic acid (63.64% of total lipids), followed by linoleic acid (19.84% of total lipids), palmitic acid (7.07% of total lipids), and oleic acid (5.5% of total lipids). Among the fatty acids in CSO, the order of abundance of the identified components: α-linolenic acid > linoleic acid > palmitic acid > oleic acid > stearic acid > vaccenic acid > trianoic acid > arachidic acid > palmitoleic acid (Table 2 and Figure S1). This study showed that the major type of fatty acid in CSO was polyunsaturated fatty acid (PUFA, 89.84%). Interestingly, some research has reported that animal feed with chia seed of high PUFA content can increase the level of PUFA in meat fats, as well as aroma and flavor [18].

Table 2. Fat, TAGs and minor components of chia seed oil.

Oil Parameters	Percentages
Fat Compositions/%	
TAG	82.60 ± 0.15
1,3-DAG	0.82 ± 0.03
1,2(2,3)-DAG	0.74 ± 0.02
Total DAG	1.56 ± 0.02
FFA	15.18 ± 0.11
TAG Compositions/%	
aLnaLnaLn	53.44 ± 0.47
aLnaLnL	23.76 ± 0.22
aLnLL	8.22 ± 0.24
aLnaLnP	6.25 ± 0.05
aLnLO	1.80 ± 0.24
aLnOP	4.43 ± 0.28
aLnOO	2.10 ± 0.35
Di-UTAG	10.69 ± 0.33
Tri-UTAG	89.31 ± 0.34
Minor components	[a] Content (mg/kg)
Tocopherols	
α-tocopherol	5.10 ± 0.42
γ-tocopherol	70.38 ± 7.99
δ-tocopherol	1.48 ± 0.06
Total tocopherols	76.96 ± 8.47
Squalene	226.43 ± 38.19
Phytosterols	
Campesterol	387.77 ± 59.05
Stigmasterol	177.47 ± 31.57
β-Sitosterol	2433.56 ± 71.69
Total phytosterols	2998.80 ± 162.30
Mineral contents	
Boron	0.193 ± 0.012
Magnesium	3.566 ± 0.185
Aluminum	4.104 ± 0.644
Calcium	1.226 ± 0.082
Manganese	0.098 ± 0.010
Zinc	0.153 ± 0.017
Arsenic	0.014 ± 0.004
Strontium	0.071 ± 0.014

Tri-UTAG, triunsaturated triacylglycerols; Di-UTAG, diunsaturated triacylglycerols; P, palmitic acid; S, stearic; O, oleic; L, linoleic acid; Ln, linolenic acid; aLn, α- linolenic acid. [a] All the measurements are in terms of mean ± standard deviation.

CSO has very high contents of α-linolenic (63.64%) and linoleic acids (19.84%), together accounting for more than 83%. In addition, the level of palmitic acid was 7.07%. We found that the ratio n-3/n-6 in CSO was 3.21, in agreement with the range of 3.18–4.18 reported by Ixtaina et al. [19], which is much higher than that of most vegetable oils, such as canola oil (0.45), olive oil (0.13), soybean oil (0.15), and walnut oil (0.20) [20]. Furthermore, the sn-2 fatty acid composition of CSO is indicated in Table 1 and Figure S2. Obviously, α-linolenic, linoleic, and oleic acids in CSO were the major sn-2 fatty acids, which were responsible for more than 95%. This result is in line with the recognized principle that unsaturated fatty acids mainly occupy the sn-2 position.

Based on a daily nutrient criterion for linoleic acid and α-linolenic acid proposed by National Institutes of Health [21], CSO could be applied as a good dietary supplement. In addition, CSO was characterized by a high polyunsaturated fatty acid/saturated fatty acid (PUFA/SFA) ratio, which is highly favorable for the reduction of serum cholesterol and atherosclerosis, and the prevention of cardiovascular disease [22]. Many studies have shown that CSO plays a role in lowering serum triglyceride, and also raises the level of HDL-C in rats [23]. As shown in Table 1, the PUFA/SFA ratio of CSO was 8.85. Thus, the incorporation of CSO into the diet could bring great beneficial effects to the cardiovascular system due to the high content of PUFAs.

The compositions of acylglycerol and free fatty acids are indicated in Table 2. The main composition of CSO was triacylglycerol, with its content up to 82.60%. The content of total diacylglycerol and FFA were 1.56% (1,3-DAG = 0.82% and 1,2 or 2,3-DAG = 0.74%) and 15.18%, respectively. As shown in Table 2, seven different TAGs were found in CSO, including aLnaLnaLn, aLnaLnL, aLnLL, aLnaLnP, aLnLO, aLnOP and aLnOO (Figure S3). The main triacylglycerols in CSO were aLnaLnaLn, aLnaLnL, aLnLL and aLnaLnP, and their levels were 53.44%, 23.76%, 8.22% and 6.25%, respectively. We found that α-linolenic acid was present in all the measured TAGs, which was also discovered by Ixtaina and coworkers [19].

2.4. Tocopherol and Phytosterol Levels

Tocopherols are natural antioxidants which can stabilize oils. The α-, γ- and δ-tocopherols, and total tocopherol content, in CSO are shown in Table 2 and Figure S4. CSO contained 76.96 mg/kg of tocopherols, mainly γ-tocopherol (>91%) and α-tocopherol (6.6%). δ-Tocopherol was present in low concentration (1.48 mg/kg). However, β-tocopherol was not detected in CSO. The content of total tocopherols (76.96 $\pm$ 8.47 mg/kg) in CSO was much lower than that of research (238–427 mg/kg) reported by Matthäus [24]. This difference may be explained by different varieties, processing, and storage conditions.

Table 2 and Figure 1A also show the content and the composition of sterols in CSO. Three phytosterols including campesterol, stigmasterol, and β-sitosterol were identified, with β-sitosterol as the most abundant sterol, accounting for more than 81% of the total amount of sterols. β-Sitosterol was predominant (2433.56 mg/kg), followed by campesterol (387.77 mg/kg) and stigmasterol (177.47 mg/kg). The total phytosterols of the CSO amounted to 3 g/kg, lower than the previous data ranged from 7 to 17 g/kg [25], which is probably linked to the loss of sterols during oil refining. Compared with other common edible oils, CSO contained a higher total tocopherol content than olive, soybean, peanut, corn, sunflower, and canola oils (260–1000 mg/kg) [26], indicating that CSO probably has relatively better oxidative stability.

It is well known that squalene is an important contributor to reduction of cholesterol levels. In this study, squalene was found in CSO at 226.43 mg/kg, much higher than the concentrations in walnuts, almonds, peanuts, hazelnuts, and macadamia nuts revealed by Maguire et al. [27]. The range of squalene content in CSO was 50–500 mg/kg, dependent on the extraction method used [27]. The high content of squalene in CSO is probably an important contribution to the beneficial effects of cancer prevention and health in human diet.

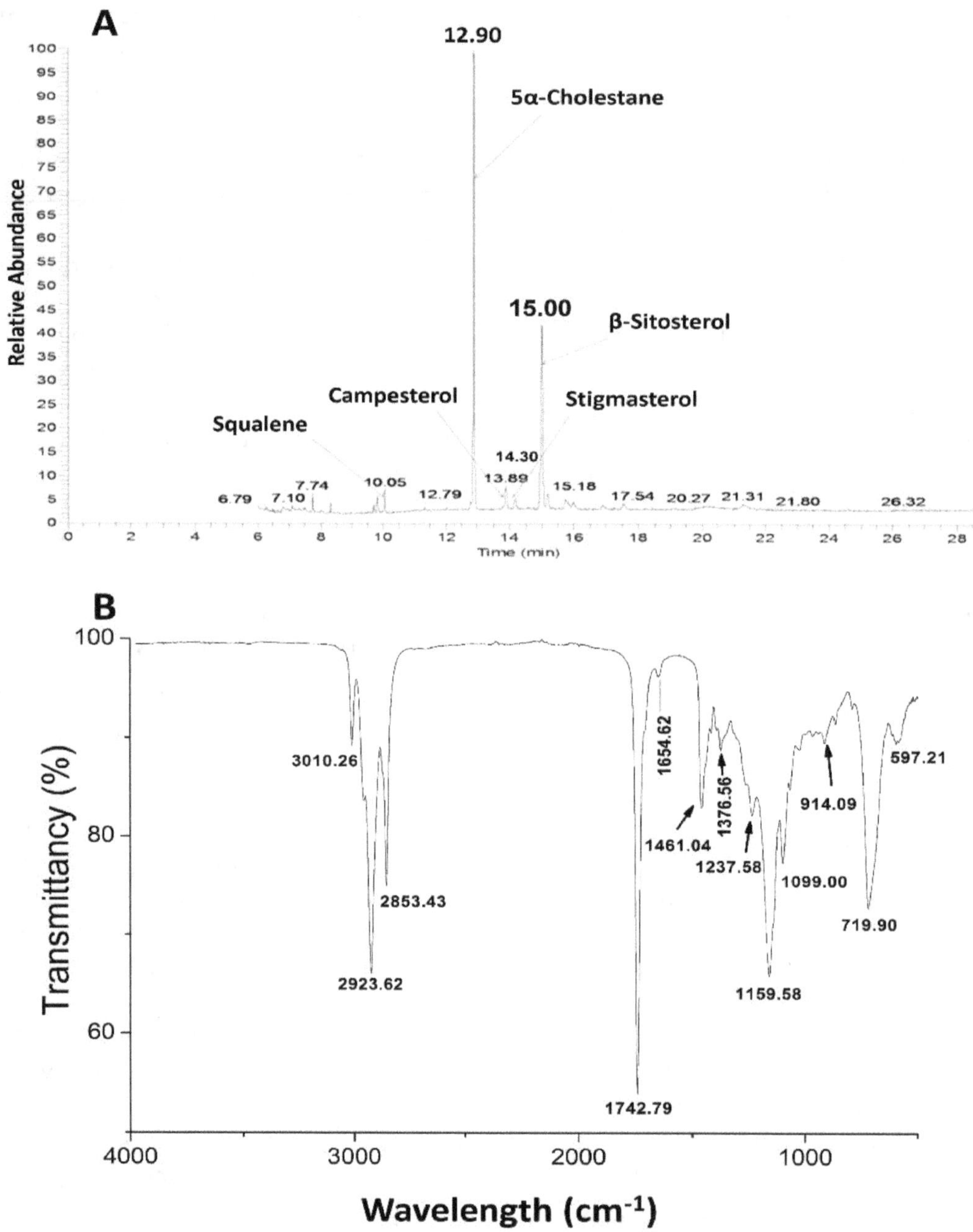

Figure 1. (**A**) Sterols in Mexican chia seed oil were analyzed using a gas chromatograph-mass spectrometer system by standards. (tR = 10.05 min), squalene; (tR = 12.90 min), 5-α-cholestan; (tR = 13.89 min), campesterol; (tR = 14.30 min), stigmasterol; (tR = 15.00 min), β-sitosterol. (**B**) FTIR spectra for chia seed oil extracted.

2.5. Determination of Trace Element Levels

Trace elements like manganese (Mn) increase the rate of oxidation of oil by the formation of free radicals of fatty acids and hydroperoxides [28], so they are undesirable in oils. The content of trace elements in CSO are shown in Table 2. Aluminum (Al), magnesium (Mg), calcium (Ca), boron (B), zinc (Zn), manganese (Mn), strontium (Sr) and arsenic (As) were detected in CSO. As indicated in Table 2, mineral composition of chia seed oil is low in elements, like all other oils (Ca = 1.226 mg/kg; Mg = 3.566 mg/kg; Zn = 0.153 mg/kg; Al = 4.104 mg/kg; Mn = 0.098 mg/kg; B = 0.193 mg/kg; Sr = 0.071 mg/kg). The most abundant trace element was aluminum (4.104 mg/kg). CSO is a rich source of minerals such as Ca, Mg and Zn. There were no toxic elements (Ni, Pb, Cd, Tl and Hg) detected except arsenic. However, mean levels of arsenic (0.014 mg/kg) do not raise concern. It is possible that concentrations of some elements were influenced by the growth, manufacturing process, and equipment. As arsenic was found in CSO, further investigation would be needed to determine its valence state, since different valence states vary in toxicity. All metals measured in CSO were lower than the maximum level accepted for virgin vegetable oils [29], and also lower than those in crude and degummed sunflower oils [30]. Thus, it could be used in food, medical and cosmetics industry.

2.6. Determination of Polycyclic Aromatic Hydrocarbons (PAHs)

PAHs constitute the critical hazards which are widespread in vegetable oil such as sunflower, olive, peanut, soybean, corn, canola, and palm oils. PAHs might occur from raw materials or the refining process [31]. Table 3 indicates the levels of the PAHs detected in CSO. Ten PAHs were revealed in CSO, including pyrene, fluoranthene, BaA, BbFlu, DBahA, Chr, BaP, BkFlu, IP and BghiP. Pyrene had the highest level (180.24 μg/100g) of the PAHs found in this oil. The concentrations of fluoranthene, BaA, and BbFlu were 84.72, 66.92, and 31.49 μg/100g, respectively. The total concentration of the other 6 PAHs was 45.18 μg/100g. The average sum of PAHs in CSO (41.06 μg/kg) was found to be higher than that in sunflower oil (17.36 μg/kg), and lower than that in soybean oil (65.33 μg/kg) [32]. One reason for PAHs introduced in vegetable oils is because of the heating-drying and extraction processes on oilseed or raw material [33]. Maximum concentration for total PAHs in edible oil in Regulation (European Union) is 10 μg/kg [34]. Thus, using uncontaminated raw materials, improving the refining process, and using activated carbon to remove PAHs during the oil refining process, are necessary requirements to decrease PAHs before market use [35].

Table 3. Polycyclic aromatic hydrocarbons (PAH) in chia seed oil.

PAHs	Concentration (μg/100 g)
Fluoranthene	84.72 ± 9.85
Pyrene	180.24 ± 18.84
Benzo (*a*) anthracene (BaA)	66.92 ± 10.05
Chrysene (Chr)	5.69 ± 0.98
Benzo (*b*) fluoranthene (BbFlu)	31.49 ± 8.84
Benzo (*k*) fluoranthene (BkFlu)	2.21 ± 0.95
Benzo (*a*) pyrene(BaP)	3.68 ± 1.02
Dibenzo (*a.h*) anthracene (DBahA)	30.04 ± 6.62
Indeno(1,2,3-*cd*) pyrene and Benzo (*g,h,i*) perylene (IP and BghiP)	3.56 ± 0.88

2.7. FT-IR Spectrum Analysis

The FT-IR spectrum of CSO in Figure 1B shows an absorption band at about 3000 cm^{-1} corresponding with C–H stretching, and we observed the =C–H, asymmetric and symmetric methyl groups at 3010.26, 2923.62 and 2853.43 cm^{-1}, respectively [36]. The frequencies at 1461.04 and 1742.79 cm^{-1} were assigned to bending vibration of lipid CH_2 groups and the ester carbonyl stretching (C=O) fatty acids, respectively. According to the study by Guillen and Cabo, the distinct weak peak at 1654 cm^{-1} refers to C=C stretching absorptions of disubstituted *cis*-olefins, which could be used to evaluate the level of total lipids and identify total unsaturation in the oil [37].

Absorption bands associated with (C–C(=O)–O) and (OC–C) coupled bonds stretching are usually very strong at 1159.58 and 1099 cm^{-1}. The peak at 1099 cm^{-1} is related to C–O–C stretch of triglyceride ester linkage. The bands at 719.9 cm^{-1} are due to a combined vibration of *cis*-disubstituted olefins having seven or more carbon atoms, and CH_2 out-of-plane rocking [36].

The FT-IR spectrum clearly indicates that CSO contains C=C, representing the unsaturated fatty acids, and ester functional groups. The high level of α-linolenic acid (~64% in CSO), with three double bonds, led to a high degree of unsaturation in the spectrum. It was interesting to note that characteristic absorption of OH at 3470 cm^{-1} is not found on the FT-IR spectrum, which suggests that the moisture content could be neglected in CSO.

FTIR is one of the most practical, non-destructive and relatively cost-efficient techniques to evaluate relative levels of unsaturated fatty acid in vegetable oils by their characteristic peaks. The integration area of absorption peaks at 3010 cm^{-1} and 1743 cm^{-1} were representative of the content of PUFAs and total lipid respectively. Furthermore, the percentage of unsaturated fatty acid could be calculated by the area ratio of 3010/1743 cm^{-1}.The percentage of unsaturated fatty acid found in this study indicates that CSO contains a high concentration of UFAs, which provided the unsaturation observed using individual feature of the peak at 3010 and 1654 cm^{-1} [36].

2.8. Antioxidant and Cytotoxicity

The antioxidant capability of CSO was evaluated by the inactivation of DPPH and ABTS assays. The radical inhibition decreased after treatment with CSO, as shown in Table 4. The IC_{50} value (IC_{50} value is the concentration of the sample required to inhibit 50% of radicals) of DPPH of the chia seed oil was 33.94 mg/mL, while the IC_{25} value (IC_{25} value is the concentration of the sample required to inhibit 25% of radicals) of ABTS was 28.51 mg/mL. Overall, catechin and tocopherol demonstrated superior scavenging activity than CSO. Both the ABTS and DPPH methods as investigated in our study gave similar results as the study by Xuan et al. [38] and by Scapin et al. for ethanol extract of Brazil chia seeds (3.84 mg/mL) [38], and by Scapin et al. for a Brazil chia seed ethanol extraction (3.84 mg/mL) obtained by solvent extraction [39]. Compared with sunflower, safflower, canola, and soybean oil, CSO presented the lowest DPPH and ABTS radical scavenging capability [38].

Table 4. Antioxidant activity and cytotoxicity of chia seed oil.

Samples	Antioxidant Activity		Cytotoxicity [c] IC_{25} (μg/mL)	
	DPPH [a] IC_{50} (mg/mL)	ABTS [b] IC_{25} (mg/mL)	LNcap	HepG2
Chia seed oil	33.94	28.51	580.12	889.68
Catechin	0.005	-	-	-
Tocopherol	-	0.004	-	-

[a] IC_{50} value is the concentration of the sample required to inhibit 50% of radical of DPPH, [b] IC_{25} value is the concentration of the sample required to inhibit 25% of radical of ABTS, [c] IC_{25} value is the concentration of the sample required to inhibit 25% of cell growth. IC_{25} value was much more reliable in ABTS and cytotoxicity assay because it was in the range of tested concentration.

Further, the cytotoxic property of chia seed oil was assessed in LNcap and HepG 2 cells. Our data indicated the IC_{25} values (IC_{25} value is the concentration of the sample required to inhibit 25% of cell growth) of CSO were up to 580 and 889.68 μg/mL in LNcap and HepG 2 cell lines for 48 h, separately, which means cell toxic effects can be excluded by this extract method on this cell line (Table 4).

2.9. The Effect of Chia Seed Oil on the Hepg2 Lipid Accumulation Model

Hepatic steatosis can be induced in HepG2 cells by exposing cells to pathophysiological levels of oleic-acid (OA) to mimic the influx of excess lipo-toxicity of free fatty acids in hepatocytes [40]. To date there have been no reports published indicating anti-hepatic lipogenic effect in OA-induced HepG2 Cells. Thus, this study was designed to evaluate the anti-hepatic lipogenic effect of chia seed oil in the selected in vitro model. In the present study, upon establishing this successful hepatic steatosis model, intracellular TG, TC, HDL-C and LDL-C levels were significantly increased (Figure 2) after 0.2 mM OA stimulation. CSO (500 μg/mL) exhibited no cytotoxicity on HepG2 cells (Table 4). As showed in Figure 2, significant decreases of TG and LDL-C levels by 25.8% and 72.9% were seen in the CSO treatment group, separately, compared with OA-group, similar to the tocopherol group. Moreover, the group treated with CSO (500 μg/mL) demonstrated a higher reduction of LDL-C than the tocopherol treated group. We have demonstrated that chia seed oil does not reduce the TC secreted in HepG2 cells, but increases the levels of cellular HDL-C. These results could be supplementary in vitro evidence to the in vivo studies by Ayerza et al., which indicated that chia seed diets for 4 weeks decreased TG levels and increased HDL-C cholesterol content in rat serum [41]. In terms of lipid lowering efficacy, this study indicates that chia seed oil is more effective at reducing triglycerides than total cholesterol, similar to tocopherol. Thus, chia seed oil has great potential as nutraceutical in the food industry with its TG and LDL-C reducing effects, and increasing effect on the secretion of HDL-C.

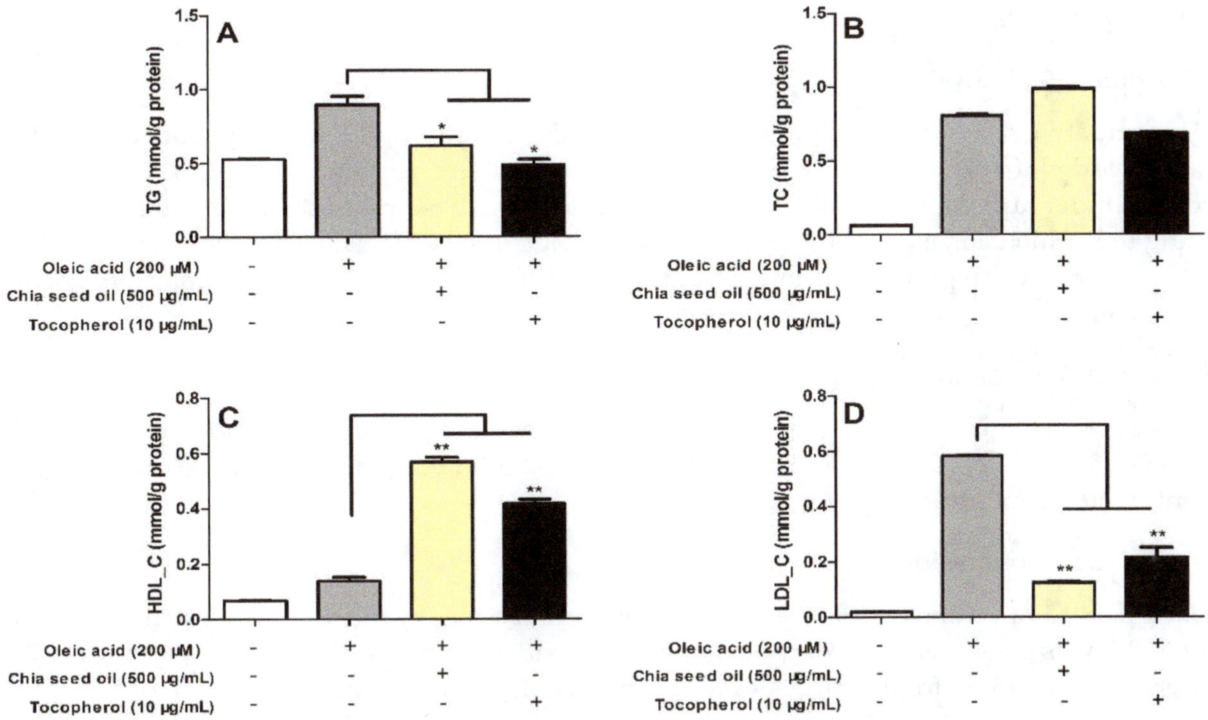

Figure 2. Effects of chia seed oil on accumulation of the intracellular (**A**) TG, (**B**) TC, (**C**) HDL-C, and (**D**) LDL-C in oleic acid (OA)-induced HepG2 cells. OA-induced accumulation of TG and LDL-C was significantly reduced by chia seed oil co-treatment in HepG2 cells. Vertical bars represent the mean ± SE of three independent experiments. * $p < 0.05$, ** $p < 0.01$ versus OA group.

3. Materials and Methods

3.1. Chemicals

Standards of tocopherols, phytosterols and squalene were purchased from Sigma Aldrich (St. Louis, MO, USA). Triacylglycerols were obtained from J&K Scientific (Beijing, China). Target standard compounds of polycyclic aromatic hydrocarbons (PAHs, dissolved in acetonitrile, 200 μg/mL), including benzo(*a*)anthracene (BaA), chrysene (Chr), benzo(*b*)fluoranthene BbFlu, benzo(*k*)fluoranthene (BkFlu), benzo(*a*)pyrene (BaP), dibenzo(*a.h*)anthracene (DBahA), benzo(*g.h.i*)perylene (BghiP), and indeno(1,2,3-*cd*)pyrene (IP), were purchased from Sigma-Aldrich (St. Louis, MO, USA). Chemical solvents were of HPLC grade. Commercial detection kits of triglyceride (TG), total cholesterol (TC), low-density lipoprotein-cholesterol (LDL-C) and high-density lipoprotein-cholesterol (HDL-C) were obtained from Nanjing Jiancheng Bioengineering Institute (Nanjing, China).

3.2. Extraction of Chia Seed Oil

Chia seeds (*S. hispanica* L.) were purchased from Guangzhou Yongheng Biotechnology Company Co., Ltd (Guangzhou, China). The chia seeds (100 g) were ground, and extracted with 0.5 L of *n*-hexane at 60 °C for 4 h in three repetitions. The obtained extracts were then combined and concentrated using a rotavapor apparatus. The resulting oil was stored at −20 °C for further analysis.

3.3. Oil Content of Chia Seeds

Determination of chia seed oil (CSO) content was adopted by the China National GB Standard for Oilseeds (GB/T 14488.1-2008) [42]. The oil content was reported as g/100 g seed.

3.4. Chemical Properties

3.4.1. Oxidative Stability

The oxidative stability of CSO was evaluated by the widely used Rancimat analysis [43]. Briefly, 3 g oil was added to the reaction vessel in triplicate, followed by heating to 120 °C. The samples were then dried under airflow at the rate of 20 L/h. A receiver with 60 mL of distilled water was used for collecting the effluent air, containing volatile organic acids from CSO. The conductivity of the water as the process of oxidation proceeded, and the induction period in hours (h) was automatically measured by the machine.

3.4.2. Color Determination

Color measurement was performed via official AOCS method (Cc 13b-45) [44]. In brief, CSO was loaded to a 25.4 mm cell, and the R/Y value was recorded following the instructions of Lovibond PFX 995 (Tintometer, Amesbury, UK).

3.4.3. Fatty Acid Composition

Fatty acid composition of CSO was detected as methyl esters by the following method [45]. Briefly, 100 mg CSO was dissolved in 2 mL potassium hydroxide solution (KOH, 10% *w/v* in MeOH) and placed in water at 85 °C for 45 min. As the solution cooled, 5 mL of H_2O and 5 mL of hexane were added and mixed thoroughly. Then mixed solution was subjected to liquid–liquid partitioning 3 times, and hexane fractions were combined. The hexane fractions with unsaponifiable substances were then washed with 10% alcohol until neutral pH was obtained, and finally dried by rotary evaporator. The hexane extract containing unsaponifiable substances was kept at −20 °C until further analysis for determination of sterols and squalene.

Saponified fatty acids were extracted from the aqueous layer (previously acidified with 2 M HCl) three times using hexane. The hexane layer containing saponified fatty acids was dried, and dissolved in 2 mL Boron trifluoride-methanol solution (14% methanol). The mixture reaction was then incubated at 60 °C for 45 min to obtain methylation of fatty acids. Hexane was used to extract fatty acid methyl esters (FAMEs) from the cooled mixture. Analysis of FAMEs was further carried out by gas chromatography (7820A, Agilent, Santa Clara, CA, USA), coupled with a FID and a Trace TR-FAME capillary column (i.d. 0.25 μm, 60 m × 0.25 mm, Thermo Fisher, Grand Island, NY, USA). The analysis condition was programmed as follows: the injector and FID temperatures were kept at 250 °C, and initial oven temperature was held at 80 °C for 3 min. It was then increased to 215 °C (15 °C/min), and finally up to 215 °C (20 min hold time). Nitrogen was used as carrier gas at 1 mL/min, with a split ratio of 1:20. Retention times (RT) of fatty acids were compared with those of authentic standards, and contents of individual components were expressed in relative percentages.

3.4.4. Sn-2 Fatty Acid Composition

Analysis of sn-2 fatty acids of CSO was conducted by the assay used in reference [46]. Briefly, 10 mg CSO was added to a reaction solution containing 1 mL of 1 M Tris buffer (pH 8.0), 0.25 mL of bile salts (0.05%), 0.1 mL of calcium chloride (2.2%), and 10 mg of pancreatic lipase. The reaction mixture was kept incubated at 40 °C, with shaking for 3 min. The reaction was stopped by adding 1 mL of 6 M HCl. Separation of above reaction product was performed by preparative TLC on silica gel, with hexane/diethyl ether/acetic acid (1/1/0.02, *v/v/v*) as the developing solvent system. The band responding to the 2-monoacylglycerol (2-MAG) on the TLC plate was scraped into a glass tube and extracted 3 times with diethyl ether. The 2-MAG was dried and methylated to esters following the same saponification process as above. The resultant product was finally analyzed by the same analytical method.

3.4.5. Triacylglycerol Composition

Analysis of triacylglycerol components of CSO was conducted by the official assay (AOCS Ce 5c-93), with HPLC-ELSD (Agilent 1200, Agilent Technologies, Santa Clara, CA, USA) equipped with a C18 column (150 mm × 4.6 mm, 5 μm, Varian, Palo Alto, CA, USA) [47]. CSO was prepared in hexane at 2 mg/mL and analyzed using the following conditions. Briefly, the mobile phase consisted of acetonitrile and isopropanol with a flow rate of 0.8 mL/min. The initial mobile phase was held at 60% acetonitrile for 40 min, changed linearly to 55% (40–80 min), returned to 60% (80–85 min) and maintained at 60% for 5 min (85–90 min). High sensitivity was obtained by using ELSD temperature set at 55 °C with a gain value of 8, and gas flow rate of 1.5 mL/min. Peaks were identified by comparing their retention times with standard reference compounds. Triacylglycerols were confirmed by comparison of RT with their standards. Relative content was reported as a percentage.

3.4.6. FT-IR Spectrum Analysis

The infrared spectrum of CSO was measured by Fourier transform infrared spectroscopy (FTIR) with a Vector 70 model FT-IR instrument (Bruker, Billerica, MA, USA) in the infrared region of 4000−400 cm^{-1}. This enabled investigation of structural information with a resolution of 4 cm^{-1}. The assignment of functional groups corresponding to IR absorption bands was performed by comparison with distinctive bands of functional groups of the chemicals in edible oils.

3.5. Minor Components

3.5.1. Tocopherol and Phytosterol Levels

Analysis of the composition of tocopherols was performed by HPLC using Waters 2475 multi λ fluorescence detector (λ_{Ex} 293 nm; λ_{Em} 325 nm) (Waters, Milford, MA, USA) based on reported methods [48]. Briefly, the column used in this analysis was a DIOL column (100 mm × 3 mm, 7 μm, Varian, Palo Alto, CA, USA). CSO was dissolved in hexane at 0.25 mg/mL and injected directly into the HPLC column. The optimized conditions for the run were as follows: in a linear gradient elution with the flow rate of 0.5 mL/min, and solvents hexane/tetrahydrofuran (A, 98/2, *v*/*v*) and isopropanol (B) with the following gradient timetable: 0–40 min, 100% A; 40–45 min, 100–95% A; 50–51 min, 95–100% A; 51–60 min, 100% A. The identification and quantification of α-, β-, γ- and δ-tocopherols in CSO was determined by comparison with their retention times and the area of the standards of tocopherols, and the unit was expressed as mg/kg of CSO.

The analysis of sterols and squalene was carried out by GC-MS equipped with a flame ionization detector (FID) and using a DB-5MS capillary column (60 m × 0.25 mm, 0.25 μm, Agilent, Santa Clara, CA, USA). Gas chromatograph conditions were as follows: helium was used as carrier gas at the flow rate of 1 mL/min, then the split ratio was 1:100. Oven temperature program: initial temperature 200 °C for 1 min, then 200 °C to 300 °C at 1 °C/min, and maintained at 300 °C for 18 min; detector temperature set at 290 °C, injector temperature 290 °C, injection volume, 1.0 μL. Sterol and squalene were determined and quantified by the internal standard (5α-cholestane), and the contents were expressed as mg/kg.

3.5.2. Determination of Polycyclic Aromatic Hydrocarbons (PAH)

PAH levels were analyzed by LC/MS/MS. Briefly, CSO (0.5 g) was added a 10 ml Pyrex tube, and mixed with 5 μL of internal standard (2 mg/mL) and 3 mL of *n*-hexane by shaking. SupelMIP solid-phase extraction (SPE, Anpel, Shanghai, China) was applied to extract PAHs from the oil. The mixture contents were fully loaded to the SPE column, to allow adhesion of PAHs to the membrane. The column was then washed with hexane (5 mL), and PAHs were eluted with methylene dichloride (10 mL) at a low speed. Eluent was then dried and dissolved in acetonitrile (200 μL) for further experiments.

LC/MS/MS with APCI Source (Agilent, Santa Clara, CA, USA) was applied to detect PAHs in CSO. Separation was achieved using a C18 column (100 × 2.1 mm, 1.8 μm; Agilent, Santa Clara, CA, USA) with a mobile phase comprising of solvent A (0.1% formic acid in acetonitrile) and solvent B (0.1% formic acid in water). The column oven temperature was set at 25 °C, and flow rate was 0.5 mL/min. The optimized conditions for separation were as follow: linear gradient elution was started from 45% A and then increased from 45 to 100% A (1–15 min), held 100% A from 15 to 21 min, next decreased 100% to 45% A from 21–22 min, lastly maintaining 45% A for 3 min. The PAH concentrations in CSO were quantified with internal standard.

3.5.3. Determination of Trace Element Levels

The content of trace elements of CSO was determined by inductively coupled plasma-mass spectrometry (ICP-MS), based on previous methods with minor modifications [49]. Briefly, the oven temperature was programmed from 60 °C (3 min hold) to 175 °C at 5 °C/min, maintained at 175 °C for 15 min, then raised to 220 °C at 2 °C/min, and finally held at 220 °C for 10 min.

3.5.4. Antioxidant Activity In Vitro

The in vitro antioxidant effects of CSO were evaluated by DPPH and ABTS assays [50]. Briefly, 50 μL volumes of different concentrations of CSO were added to 150 μL of DPPH solution (200 μM) with shaking. After incubation of 40 min, optical density was recorded at 517 nm (OD_{517}). Tocopherol was used as a positive control. In the same manner, ABTS radical scavenging capability of CSO was measured at a wavelength of 734 nm. Catechin was used as the positive control in ABTS assay. The formula for calculating % scavenging effects (SE) by DPPH and ABTS was as follows:

$$SE(\%) = \frac{\text{ODcontrol} - \text{ODsample}}{\text{ODcontrol}} \times 100\%$$

3.6. Cell Cytotoxicity

The cytotoxic effects of the chia seed oil against LNcap cells and HepG2 were tested using a 3-(4,5-dimethyl-2-yl)-2, 5 diphenyltetrazolium bromide (MTT) assay [51]. Cells were obtained from Stem Cell Bank, Chinese Academy of Sciences, Shanghai, China. Cells were cultured in Dulbecco's modified Eagle's medium (DMEM; Thermo, Grand Island, NY, USA) containing 10% fetal bovine serum (FBS; Gibco, USA) in an incubator (37 °C, 5% CO2). The dried CSO was dissolved in DMSO to 20 mg/mL and sterilized by filtration through a 0.22 μm membrane filter (Millipore, Burlington, MA, USA) as a stock solution. To each well of a 96-microwell plate, 190 μL of LNcap cells or HepG2 (5×10^4 cell/mL) was added with or without 10 μL of different concentrations of CSO. After 48 h incubation, 20 μL of MTT (5 mg/mL) was added and then incubated for an additional 4 h. The supernatant was then removed completely, and 150 μL of DMSO was added to each well. Lastly, the OD value at 570 nm was measured using a microplate reader (Bio Tek, San Diego, CA, USA). Cell viability was measured and calculated by the formula: cell viability (%) = [OD_{570} (sample)/OD_{570} (control)] × 100%.

3.7. CSO Inhibits OA-Induced Intracellular Lipid Level in HepG2 Cells

3.7.1. Cell Treatment

HepG2 was cultured in DMEM with high glucose (Gibco, USA) supplement with 10% FBS (Gibco, USA) at 37 °C with 5% CO_2 in a cell incubator. The oleic acid (OA) was dissolved in 10% bovine serum albumin (BSA, fatty acid free, Sigma, St. Louis, MO, USA), and diluted in a culture medium with 1%

BSA as the final concentration of 200 μM. CSO was prepared in DMEM as stock solution. The cells were seeded in 6-well plates and incubated for 24 h, before being cultured as follows: DMEM plus 1% BSA as a control group; DMEM with OA-BSA at 200 μM as a model group; CSO at final concentration of 500 μg/mL was applied with OA-BSA simultaneously as a co-treatment.

3.7.2. Measurement of Cellular Lipid Levels in HepG2 Cells

Contents of TG, TC, LDL-C and HDL-C in HepG2 cell model were measured by commercial kits (Nanjing Jiancheng Bioengineering Institute, Nanjing, China). Briefly, cells were collected and washed twice with pre-chilled PBS to remove culture medium after treatment for 24 h. Then, 200 μL of 2% Triton X-100 was added for lysing cells by sonication. The supernatant was collected for analysis of TG, TC, LDL-C and HDL-C. Results were normalized to protein concentration using the Bradford protein assay kit (Bio-Rad, Hercules, CA, USA).

3.8. Statistical Analysis

Statistical analysis was carried out using GraphPad Prism 7.0 (GraphPad, San Diego, CA, USA). Differences were considered to be statistically significant at $p < 0.05$, or very significant at $p < 0.01$, by the Duncan test.

4. Conclusions

In conclusion, PUFA analysis in chia seed oil from Mexico was greater than 80% of fatty acids. This research indicates that α-linolenic acid was the most abundant fatty acid, followed by linoleic acid with an n-3/n-6 ratio of 3.21:1. Due to the large amount of easily-oxidized PUFAs in chia seed oil (induction period = 0.68 h), raw chia seeds might need special treatment during sample collection, processing, storage, and transportation, to prevent oxidation and maintain the quality of the oil. We also found that CSO presented some antioxidant effects in vitro, which are beneficial for human health. This is the first study to report the ability of CSO to reduce the levels of hepatic TG and LDL-C markedly, and increase the content of HDL-C in HepG2 cell model. Thus, this indicates that chia seed oil may help in reducing the risk for cardiovascular disease, and it can be widely used as cooking oil and in manufacture of healthy food supplements.

Supplementary Materials. Figure S1: HPLC chromatogram of fatty acids or chia seed oils, Figure S2: HPLC chromatography of sn-2 fatty acids from chia seed oil, Figure. S3: Analysis of triacylglycerols from chia seed oil by HPLC, Figure S4: HPLC chromatogram of tocopherols in chia seed oil.(A) standards of tocopherols, and (B) tocoperols in chia seed oil.

Author Contributions: Extraction, IR, and GC/MS analysis, Y.S. and L.Z.; Antioxidant and cytotoxicity assay, J.J. and M.W.; PAHs analysis, X.L. and J.F.; Writing improvement, M.W.; Project administration, supervision and writing, Y.G. and X.S. All authors read and approved the final manuscript.

References

1. Simopoulos, A.P. The importance of the ratio of omega-6/omega-3 essential fatty acids. *Biomed. Pharmacother.* **2002**, *56*, 365–379. [CrossRef]
2. Singh, M. Essential fatty acids, DHA and human brain. *Indian J. Pediatr.* **2005**, *72*, 239–242. [CrossRef] [PubMed]
3. Watchareewan, J.T.; Phillips, R.; Geneve, J.; Cahill, D.H. Extending the range of an ancient crop, *Salvia hispanica L.*—A new x3 source. *Genet. Resour. Crop Evol.* **2012**, *59*, 171–178.
4. Timilsena, Y.P.; Wang, B.; Adhikari, R.; Adhikari, B. Preparation and characterization of chia seed protein isolate—Chia seed gum complex coacervates. *Food Hydrocoll.* **2016**, *52*, 554–563. [CrossRef]

5. Martinez-Cruz, O.; Paredes-Lopez, O. Phytochemical profile and nutraceutical potential of chia seeds (*Salvia hispanica L.*) by ultra high performance liquid chromatography. *J. Chromatogr. A* **2014**, *1346*, 43–48. [CrossRef] [PubMed]
6. Wu, H.; Sung, A.; Burns-Whitmore, B.; Jo, E.; Wien, M. Effect of chia seed (*Salvia hispanica, L.*) supplementation on body composition, weight, post-prandial glucose and satiety. *Faseb J.* **2016**, *30*, lb221.
7. Gazem, R.A.A.; Ramesh Puneeth, H.; Shivmadhu Madhu, C.; Chandrashekaraiah Sharada, A. In vitro anticancer and anti-lipoxygenase activities of chia seed oil and its blends with selected vegetable oils. *Asian J. Pharm. Clin. Res.* **2017**, *10*, 124–128. [CrossRef]
8. Authority, E.F.S. Opinion on the safety of 'Chia seeds (*Salvia hispanica L.*) and ground whole Chia seeds' as a food ingredient. *EFSA J.* **2009**, *996*, 1–26.
9. Ayerza, R. The Seed's protein and oil content, fatty acid aomposition, and growing cycle length of a single genotype of chia (*Salvia hispanica L.*) as affected by environmental factors. *J. Oleo Sci.* **2009**, *58*, 347–354. [CrossRef] [PubMed]
10. Dornbos, D.L.; Mullen, R.E. Soybean seed protein and oil contents and fatty-acid composition adjustments by drought and temperature. *J. Am. Oil Chem. Soc.* **1992**, *69*, 228–231. [CrossRef]
11. Silva, B.P.; Anunciacao, P.C.; Matyelka, J.; Della Lucia, C.M.; Martino, H.S.D.; Pinheiro-Sant'Ana, H.M. Chemical composition of Brazilian chia seeds grown in different places. *Food Chem.* **2017**, *221*, 1709–1716. [CrossRef] [PubMed]
12. Ayerza, R.; Coates, W. Composition of chia (*Salvia hispanica*) grown in six tropical and subtropical ecosystems of South America. *Trop. Sci.* **2004**, *44*, 131–135. [CrossRef]
13. Ratusz, K.; Popis, E.; Ciemniewska-Zytkiewicz, H.; Wroniak, M. Oxidative stability of camelina (*Camelina sativa L.*) oil using pressure differential scanning calorimetry and Rancimat method. *J. Therm. Anal. Calorim.* **2016**, *126*, 343–351. [CrossRef]
14. Farhoosh, R. The effect of operational parameters of the rancimat method on the determination of the oxidative stability measures and shelf-life prediction of soybean oil. *J. Am. Oil Chem. Soc.* **2007**, *84*, 205–209. [CrossRef]
15. Uzunova, G.; Nikolova, K.; Perifanova, M.; Gentscheva, G.; Marudova, M.; Antova, G. Physicochemical characterization of chia (*Salvia hispanica*) seed oil from Argentina. *Bulg. Chem. Commun.* **2016**, *48*, 131–135.
16. Hosseinian, F.S.; Rowland, G.G.; Bhirud, P.R.; Dyck, J.H.; Tyler, R.T. Chemical composition and physicochemical and hydrogenation characteristics of high-palmitic acid solin (low-linolenic acid flaxseed) oil. *J. Am. Oil Chem. Soc.* **2004**, *81*, 185–188. [CrossRef]
17. Patil, R.T.; Ali, N. Effect of pre-treatments on mechanical oil expression of soybean using a commercial oil expeller. *Int. J. Food Prop.* **2006**, *9*, 227–236. [CrossRef]
18. Coates, W.; Ayerza, R. Chia (*Salvia hispanica L.*) seed as an n-3 fatty acid source for finishing pigs: Effects on fatty acid composition and fat stability of the meat and internal fat, growth performance, and meat sensory characteristics. *J. Anim. Sci.* **2009**, *87*, 3798–3804. [CrossRef]
19. Ixtaina, V.Y.; Martínez, M.L.; Spotorno, V.; Mateo, C.M.; Maestri, D.M.; Diehl, B.W.K.; Nolasco, S.M.; Tomás, M.C. Characterization of chia seed oils obtained by pressing and solvent extraction. *J. Food Compos. Anal.* **2011**, *24*, 166–174. [CrossRef]
20. Boschin, G.; D'Agostina, A.; Annicchiarico, P.; Arnoldi, A. The fatty acid composition of the oil from *Lupinus albus cv. Luxe* as affected by environmental and agricultural factors. *Eur. Food Res. Technol.* **2007**, *225*, 769–776. [CrossRef]
21. Food and Nutrition Board. *Nutrient Recommendations: Dietary Reference Intakes (Dri), Dri Table: Recommended Dietary Allowances and Adequate Intakes, Total Water and Macronutrients*; National Academy of Sciences: Washington, DC, USA, 2017.
22. Reena, M.B.; Lokesh, B.R. Hypolipidemic effect of oils with balanced amounts of fatty acids obtained by blending and interesterification of coconut oil with rice bran oil or sesame oil. *J. Agric. Food Chem.* **2007**, *55*, 10461–10469. [CrossRef] [PubMed]
23. Mohd Ali, N.; Yeap, S.K.; Ho, W.Y.; Beh, B.K.; Tan, S.W.; Tan, S.G. The promising future of chia, *Salvia hispanica* L. *J. Biomed. Biotechnol.* **2012**, *2012*, 1–9. [CrossRef] [PubMed]
24. Matthaus, B.; Ozcan, M.M. Quantitation of fatty acids, sterols, and tocopherols in turpentine (*Pistacia terebinthus Chia*) growing wild in Turkey. *J. Agric. Food Chem.* **2006**, *54*, 7667–7671. [CrossRef] [PubMed]
25. Alvarez-Chavez, L.M.; Valdivia-Lopez, M.D.; Aburto-Juarez, M.D.; Tecante, A. Chemical characterization of

the lipid fraction of Mexican chia seed (*Salvia hispanica L.*). *Int. J. Food Prop.* **2008**, *11*, 687–697. [CrossRef]
26. Bakowska-Barczak, A.M.; Schieber, A.; Kolodziejczyk, P. Characterization of saskatoon berry (*Amelanchier alnifolia Nutt.*) seed oil. *J. Agric. Food Chem.* **2009**, *57*, 5401–5406. [CrossRef]
27. Maguire, L.S.; O'Sullivan, S.M.; Galvin, K.; O'Connor, T.P.; O'Brien, N.M. Fatty acid profile, tocopherol, squalene and phytosterol content of walnuts, almonds, peanuts, hazelnuts and the macadamia nut. *Int. J. Food Sci. Nutr.* **2004**, *55*, 171–178. [CrossRef]
28. Karadjova, I.; Zachariadis, G.; Boskou, G.; Stratis, J. Electrothermal atomic absorption spectrometric determination of aluminium, cadmium, chromium, copper, iron, manganese, nickel and lead in olive oil. *J. Anal. Atom Spectrom.* **1998**, *13*, 201–204. [CrossRef]
29. Commission, C.A. Codex Standard for Edible Fats and Oils not Covered by Individual Standards (Codex stan 19–1981, Rev. 2-1999). Available online: http://www.fao.org/docrep/004/y2774e/y2774e03.htm (accessed on 30 October 2018).
30. Brevedan, M.I.V.; Carelli, A.A.; Crapiste, G.H. Changes in composition and quality of sunflower oils during extraction and degumming. *Grasas Aceites* **2000**, *51*, 417–423.
31. Ciecierska, M.; Obiedzinski, M.W. Polycyclic aromatic hydrocarbons in vegetable oils from unconventional sources. *Food Control* **2013**, *30*, 556–562. [CrossRef]
32. Mafra, I.; Amaral, J.S.; Oliveira, M.B.P.P. Polycyclic aromatic hydrocarbons (PAH) in olive oils and other vegetable oils; Potential for carcinogenesis. *Olives Olive Oil Health Dis. Prev.* **2010**, 489–498. [CrossRef]
33. Moreda, W.; Perez-Camino, M.C.; Cert, A. Gas and liquid chromatography of hydrocarbons in edible vegetable oils. *J. Chromatogr. A* **2001**, *936*, 159–171. [CrossRef]
34. Zelinkova, Z.; Wenzl, T. The occurrence of 16 EPA PAHs in food—A review. *Polycycl. Aromat. Comp.* **2015**, *35*, 248–284. [CrossRef] [PubMed]
35. Arrebola, F.J.; Frenich, A.G.; Rodriguez, M.J.G.; Bolanos, P.P.; Vidal, J.L.M. Determination of polycyclic aromatic hydrocarbons in olive oil by a completely automated headspace technique coupled to gas chromatography-mass spectrometry. *J. Mass Spectrom.* **2006**, *41*, 822–829. [CrossRef] [PubMed]
36. Guillen, M.D.; Cabo, N. Infrared spectroscopy in the study of edible oils and fats. *J. Sci. Food Agr.* **1997**, *75*, 1–11. [CrossRef]
37. Timilsena, Y.P.; Vongsvivut, J.; Adhikari, R.; Adhikari, B. Physicochemical and thermal characteristics of Australian chia seed oil. *Food Chem.* **2017**, *228*, 394–402. [CrossRef] [PubMed]
38. Xuan, T.D.; Gu, G.Q.; Minh, T.N.; Quy, T.N.; Khanh, T.D. An overview of chemical profiles, antioxidant and antimicrobial activities of aommercial vegetable edible oils marketed in Japan. *Foods* **2018**, *7*, 21. [CrossRef] [PubMed]
39. Scapin, G.; Schmidt, M.M.; Prestes, R.C.; Rosa, C.S. Phenolics compounds, flavonoids and antioxidant activity of chia seed extracts (*Salvia hispanica*) obtained by different extraction conditions. *Int. Food Res. J.* **2016**, *23*, 2341–2346.
40. Li, X.; Zhao, M.; Fan, L.; Cao, X.; Chen, L.; Chen, J.; Lo, Y.M.; Zhao, L. Chitobiose alleviates oleic acid-induced lipid accumulation by decreasing fatty acid uptake and triglyceride synthesis in HepG2 cells. *J. Funct. Foods* **2018**, *46*, 202–211. [CrossRef]
41. Ayerza, R.; Coates, W. Ground chia seed and chia oil effects on plasma lipids and fatty acids in the rat. *Nutr. Res.* **2005**, *25*, 995–1003. [CrossRef]
42. China National Standardization Management Committee. *Determination of Oil Content Valid GB/T 14488.2-2008*; China National Standardization Management Committee: Beijing, China, 2008.
43. Azadmard-Damirchi, S.; Habibi-Nodeh, F.; Hesari, J.; Nemati, M.; Achachlouei, B.F. Effect of pretreatment with microwaves on oxidative stability and nutraceuticals content of oil from rapeseed. *Food Chem.* **2010**, *121*, 1211–1215. [CrossRef]
44. Committee, A.C. Color of Fats and Oils, Lovibond (Wesson) Wesson Method Using Color Glasses Calibrated in Accordance with the Aocs-Tintometer Color Scale. AOCS Official Method Cc 13b-45. Available online: https://aocs.personifycloud.com/PersonifyEBusiness/Default.aspx?TabID=251&productId=111497 (accessed on 30 October 2018).
45. Jin, J.; Wang, Y.; Su, H.; Warda, P.; Xie, D.; Liu, Y.J.; Wang, X.S.; Huang, J.H.; Jin, Q.Z.; Wang, X.G. Oxidative stabilities of mango kernel fat fractions produced by three-stage fractionation. *Int. J. Food Prop.* **2017**, *20*, 2817–2829. [CrossRef]
46. Jin, J.; Warda, P.; Mu, H.Y.; Zhang, Y.F.; Jie, L.; Mao, J.H.; Xie, D.; Huang, J.H.; Jin, Q.Z.; Wang, X.G.

Characteristics of mango kernel fats extracted from 11 China-specific varieties and their typically fractionated fractions. *J. Am. Oil Chem. Soc.* **2016**, *93*, 1115–1125. [CrossRef]

47. AOCS. Individual Triglycerides in Oils and Fats by HPLC. AOCS Official Method Ce 5c-93. Available online: https://www.aocs.org/attain-lab-services/methods/methods/method-detail?productId=114619 (accessed on 30 October 2018).
48. Moreau, R.A.; Hicks, K.B. Reinvestigation of the effect of heat pretreatment of corn fiber and corn germ on the levels of extractable tocopherols and tocotrienols. *J. Agric. Food Chem.* **2006**, *54*, 8093–8102. [CrossRef] [PubMed]
49. Llorent-Martinez, E.J.; Ortega-Barrales, P.; Fernandez-de Cordova, M.L.; Dominguez-Vidal, A.; Ruiz-Medina, A. Investigation by ICP-MS of trace element levels in vegetable edible oils produced in Spain. *Food Chem.* **2011**, *127*, 1257–1262. [CrossRef] [PubMed]
50. Shen, Y.; Zhang, H.; Cheng, L.; Wang, L.; Qian, H.; Qi, X. In vitro and in vivo antioxidant activity of polyphenols extracted from black highland barley. *Food Chem.* **2016**, *194*, 1003–1012. [CrossRef]
51. Srivastava, M.; Nambiar, M.; Sharma, S.; Karki, S.S.; Goldsmith, G.; Hegde, M.; Kumar, S.; Pandey, M.; Singh, R.K.; Ray, P.; et al. An inhibitor of nonhomologous end-joining abrogates double-strand break repair and impedes cancer progression. *Cell* **2012**, *151*, 1474–1487. [CrossRef] [PubMed]

7

Chemical Composition, Antioxidant and Antihyperglycemic Activities of the Wild *Lactarius deliciosus* from China

Zhou Xu [1,†], Liang Fu [2,†], Shiling Feng [1], Ming Yuan [1], Yan Huang [1], Jinqiu Liao [1], Lijun Zhou [1], Hongyu Yang [1] and Chunbang Ding [1,*]

1 College of Life Sciences, Sichuan Agricultural University, Yaan 625014, China; xzhsicau@163.com (Z.X.); fengshilin@outlook.com (S.F.); yuanming@sicau.edu.cn (M.Y.); shirley11hy@163.com (Y.H.); liaojinqiu630@sicau.edu.cn (J.L.); zhouzhou124@126.com (L.Z.); yhy4868135@163.com (H.Y.)
2 Dazhou Institute of Agricultural Sciences, Dazhou 635000, China; fuliangrain@126.com
* Correspondence: dcb@sicau.edu.cn
† These authors contributed equally to this paper.

Academic Editors: Alessandra Gentili and Chiara Fanali

Abstract: The wild mushroom *Lactarius deliciosus* from China was studied for the first time to obtain information about its chemical composition, antioxidant, and antihyperglycemic activities. Nutritional value, dietary fiber, fatty acids, metal elements, free sugars, free amino acids, organic acids, flavor 5′-nucleotides, and volatile aroma compounds were determined. Potential antioxidant and antihyperglycemic activities were also tested by investigating 1,1-diphenyl-2-picrylhydrazyl (DPPH) and 2,2′-Azino-bis(3-ethylbenzothiazoline-6-sulfonic acid) (ABTS) radicals scavenging activities, ferric ion reducing activity, as well as α-amylase and α-glucosidase inhibitory activities using ethanol and aqueous extracts. The results showed that *L. deliciosus* was a good wild mushroom with high protein, carbohydrate, and dietary fiber contents, while low in fat and calorie, extensive unsaturated fatty acids contents, with negligible health risks about harmful metal elements. Twenty kinds of free amino acids were detected with a total content 3389.45 mg per 100 g dw. Flavor 5′-nucleotides including 5′-CMP, 5′-UMP, 5′-IMP, and 5′-AMP were 929.85, 45.21, 311.75, and 14.49 mg per 100 g dw, respectively. Mannitol (7825.00 mg per 100 g dw) was the main free sugar, and quininic acid (729.84 mg per 100 g dw) was the main organic acid. Twenty-five kinds of volatile aroma compounds were identified, acids (84.23%) were the most abundant compounds based on content, while aldehydes (15 of 25) were the most abundant compounds based on variety. In addition, both ethanol and aqueous extracts from *L. deliciosus* exhibited excellent antioxidant activity. While in antihyperglycemic activity tests, only ethanol extracts showed inhibitory effects on α-amylase and α-glucosidase.

Keywords: *Lactarius deliciosus*; chemical composition; antioxidant; antihyperglycemic

---

1. Introduction

Edible mushrooms consist of Basidiomycota and Ascomycota members [1]. Abundant protein, essential amino acids, mineral elements, dietary fiber, flavor 5′-nucleotides, and volatile aroma components endowed mushrooms with great nutritional value and unique flavor [2,3]. Thus, mushrooms have been consumed as popular food stuff and flavoring for centuries, and consumer demand has continued to increase in recent years. Nowadays, dozens of cultivated mushrooms such as *Agaricus bisporus*, *Lentinula edodes*, *Pleurotus ostreatus*, *Flammulina velutipes*, and *Auricularia auricula* occupy the main consumer market, but there are far more mushrooms that cannot be artificially cultivated called wild mushrooms that only come from field acquisition [4,5]. Although difficult to collect, wild mushrooms are still pursued by innumerable gluttons due to their unique flavors, and *Tuber melanosporum*, *Tricholoma matsutake* and *Collybia albuminosa*, etc., are regarded as gems [6,7]. Moreover, in the east, traditional medicine theory holds the concept that "drug homologous food", hence, mushrooms also serve as indispensable raw materials for pharmaceuticals [8].

Lactarius Persoon is an ectomycorrhizal group belonging to the family of Russulaceae and widespread in temperate, subtropical and tropical forests. To date, ~500 species of *Lactarius* Per. have been discovered around the world and *Lactarius deliciosus* is one of them [9]. Owing to its excellent taste, texture, and flavor, *L. deliciosus* is popular in China, with some chemical substances and bioactivities of *L. deliciosus* being reported. Research has shown that sesquiterpenoids contribute to the unique color of *L. deliciosus*, some of which possess potent biological activities such as antimicrobial and anticancer [10–12]. Ding et al. and Hou et al. reported that *L. deliciosus* polysaccharides exhibited significant anti-tumor activity in mice in vivo and immunomodulatory effects through proliferative growth of both B cell and macrophages cell in vitro [13,14]. Meanwhile, other *L. deliciosus* extracts also have various biological activities such as enzymes inhibition, antioxidant, antimicrobial, and anti-inflammatory [15–17].

As an essential class of food raw material, the nutritional value of mushrooms is comprehensively decided by proteins, carbohydrates, fats, minerals, etc., and this is crucial consumer focus point [18]. Other factors such as free sugars, free amino acids, organic acids, flavor 5′-nucleotides, and volatile aroma components could dramatically affect taste and flavor, and should not be ignored [19,20]. In China, although *L. deliciosus* is a popular wild edible mushroom, comprehensive studies of this fungi are relatively lacking. Based on the above, we collected wild *L. deliciosus* from Sichuan Province (South-west of China), and chemical compositions including proteins, carbohydrates, fats, dietary fiber, minerals, free sugars, free amino acids, organic acids, flavor 5′-nucleotides, and volatile aroma components were analyzed. In addition, we evaluated bioactivities by assaying antioxidant and anti-hyperglycemic activities of ethanol and aqueous extracts.

2. Results and Discussion

2.1. General Nutritional Value

Generally, mushrooms are considered as a valuable health food since they have perfect proportions of protein, fat, and carbohydrate. As summarized in Table 1, fresh *L. deliciosus* was preponderantly moist, and dry matter content was relatively low (8.00%). This result was in accordance with previous studies demonstrating that dry weight content of fresh mushrooms was generally 5–15% [21,22]. In the dried fruiting body, carbohydrate was the most abundant substance with 66.61 g per 100 g dw, followed by protein (17.19 g per 100 g dw), ash (8.62 g per 100 g dw), and fat (4.82 per 100 g dw). Moreover, 31.81 g per 100 g dw total dietary fiber was detected in *L. deliciosus*, indicating that consuming this mushroom is a great way for dietary fiber intake. On the whole, *L. deliciosus* is a good food that can meet the low-calorie requirements.

Table 1. Proximate composition, energetic value, dietary fiber, and fat composition of wild *L. deliciosus*.

Component	*L. deliciosus*
Moisture (g per 100 g)	92.00 ± 0.64
Dry matter (g per 100 g)	8.00 ± 0.64
Total carbohydrate (g per 100 g dw)	66.61 ± 1.02
Crude fat (g per 100 g dw)	4.82 ± 0.15
Crude Protein (g per 100 g dw)	17.19 ± 0.06
Ash (g per 100 g dw)	8.62 ± 0.25
Energy (kcal per 100 g dw)	378.60 ± 2.74
Total dietary fiber (g per 100 g dw)	31.81 ± 1.51
Insoluble dietary fiber (g per 100 g dw)	26.51 ± 1.54
Soluble dietary fiber (g per 100 g dw)	5.30 ± 0.36
C16:0 (% of total fatty acids)	5.17 ± 0.30
C18:0 (% of total fatty acids)	16.96 ± 0.19
C18:1 (% of total fatty acids)	48.37 ± 0.62
C18:2 (% of total fatty acids)	29.49 ± 0.55

Four fatty acids were identified in the crude fat and their constituents were as follows: palmitic acid (C16:0, 5.17%), stearic acid (C18:0, 16.96%), oleic acid (C18:1, 48.37%), and linoleic acid (C18:2, 29.49%) (Table 1 and Figure S1). Unsaturated fatty acids (C18:1 and C18:2) were dominant in *L. deliciosus* fat, which was in agreement with previous research on wild edible mushrooms [23,24]. Moreover, recent research has documented that unsaturated fatty acids may lower the risk of cardiovascular disease, type two diabetes, and cancer [25,26]. Nevertheless, considering the low level of fat in *L. deliciosus*, the health effects of various fatty acids are very limited.

Compared to green plants, the metal content in mushrooms is higher due to their effective mechanism of easily accumulating metals from the ecosystem [27]. Thus, wild edible mushrooms are regarded as an excellent choice for dietary mineral requirement. Concurrently, a hidden danger is arising with excess heavy metal ingestion. As summarized in Table 2, magnesium and calcium contents of *L. deliciosus* were 1244.29 and 247.07 mg per kg dw, this result was in agreement with a previous conclusion that calcium content in mushrooms is 100–500 mg/kg dw, and magnesium is 800–1800 mg per kg dw, based on data collected from over 1000 samples of 400 mushroom species [28]. Among trace elements, iron content (197.01 mg per kg dw) was notably the highest. Of note, compared with numerous studies, copper content (1.28 mg per kg dw) in this sample was relatively low, which might be related to the physiological property of this species due to *L. deliciosus* always possessing lower copper content in other comparative studies [29,30]. Usually, daily intake is 300 g of fresh mushroom, which contains ~30 g of dry matter [31]. Compared with recommended dietary allowance (RDA) and adequate intake (AI) for females and males (aged from 19 to 30) recommended by the Institute of Medicine, consumption of *L. deliciosus* does not provide significant contribution to calcium and copper supplementation, while good contribution of magnesium, zinc, manganese, iron, and chromium [32,33]. Notably, although a large daily intake value of chromium was observed (up to 344.57% for male and 482.40% for female of AI%), this poses no risk to the human body, with reference to the tolerable upper intake level recommended by the Institute of Medicine [33]. Moreover, toxic arsenic, cadmium, and plumbum were also detected in *L. deliciosus*. EU scientific committee standards stipulated provisional tolerable daily intake values for arsenic, cadmium, and plumbum for adults (of 60 kg body weight) were 0.13, 0.06, and 0.21 mg, respectively. Therefore, the intake of heavy metals (arsenic, cadmium, and plumbum) via consuming *L. deliciosus* is risk-free for the consumers.

Table 2. Contents and daily intake estimations of metal elements of wild *L. deliciosus*.

Element	Content (mg per kg dw)	Daily Intake (mg/day) [a]	RDA or AI (mg/d) [d]		RDA or AI % [d]	
			Male	Female	Male	Female
Magnesium	1244.29 ± 42.16	37.33	400 [b]	310 [b]	9.33	12.04
Calcium	247.07 ± 4.23	7.41	1000 [c]	1000 [c]	0.74	0.74
Zinc	52.34 ± 2.68	1.57	11 [b]	8 [b]	14.28	19.63
Manganese	23.12 ± 0.75	0.69	2.3 [c]	1.8 [c]	30.16	38.53
Iron	197.01 ± 13.14	5.91	8 [b]	18 [b]	73.88	32.84
Chromium	4.02 ± 0.69	0.12	0.035 [c]	0.025 [c]	344.57	482.40
Copper	1.28 ± 0.02	0.04	0.9 [b]	0.9 [b]	4.27	4.27
Arsenic	0.75 ± 0.04	0.02	-	-	-	-
Cadmium	1.91 ± 0.05	0.06	-	-	-	-
Plumbum	0.85 ± 0.04	0.03	-	-	-	-

[a] Daily element intake values calculated using 30 g *L. deliciosus* dry matter; [b] RDA recommended by the Institute of Medicine; [c] AI recommended by the Institute of Medicine; [d] RDA or AI for both males and females aged 19–30; - Toxic elements without RDA and AI.

2.2. Non-Volatile Compounds Relating to Special Flavour

Edible mushrooms are considered a valuable food for their abundance of nutrients and desirable complex delicious taste. The taste of mushrooms is primarily ascribed to abundant soluble non-volatile taste components, such as free amino acids, flavor 5′-nucleotides, free sugars, and organic acids [34,35]. As shown in Table 3 and Figure S2, twenty-two kinds of amino acids have been measured, of which, twenty kinds were found in the free amino acids of *L. deliciosus*, and the content of total free amino acids (TAA) was 3389.45 mg per 100 g dw. Among these amino acids, glutamic acid, glutamine, histidine, and alanine were found in relatively high concentrations. Moreover, the content of nine kinds of essential amino acids (EAA) in *L. deliciosus* was 1026.29 mg per 100 g dw, accounting for 30.28% of total free amino acids. This proportion was approximate to a previous report by Sun et al., in which they found the ratios of EAA/TAA in *Tricholomopsis lividipileata, Boletinus pinetorus*, and *Amanita hemibapha* were 26.82, 27.94, and 34.75%, respectively [36]. Free amino acids can be classified into four groups based on their taste characteristics. Aspartic acid and glutamic acid were monosodium glutamate-like (MSG-like) components responsible for umami taste. The level of MSG-like free amino acids in *L. deliciosus* was 415.71 mg per 100 g dw, which was at low-level (< 500 mg per 100 g dw) of MSG-like components according to the standard defined by Yang et al. [37]. Threonine, serine, glycine, and alanine were sweet taste amino acids, amounting to 734.82 mg per 100 g dw. Valine, methionine, isoleucine, leucine, phenylalanine, histidine, arginine, tryptophan, and tyrosine were classified as bitter amino acids. In *L. deliciosus*, the amount of bitter amino acids was 1033.29 mg per 100 g dw. Although bitter amino acids might bring bitterness, while sweet taste amino acids and soluble sugars could mask this unpleasant taste. Asparagine, glutamine, citrulline, proline, and lysine did not have effects on taste.

The chromatograms concerning flavor 5′-nucleotides, free sugars, and organic acids are presented in Figures S3–S5, and the compositions of these components are summarized in Table 4. The amount of total flavor 5′-nucleotides in *L. deliciosus* was 1301.30 mg per 100 g dw, and the individual contents of 5′-CMP, 5′-UMP, 5′-IMP, and 5′-AMP were 929.85, 45.21, 311.75, and 14.49 mg per 100 g dw, respectively, while 5′-GMP and 5′-XMP were not detected. Based on the amount of total flavor 5′-nucleotides, Yang et al. divided flavor 5′-nucleotides into three ranges, low (<100 mg per 100 g dw), medium (100–500 mg per 100 g dw), and high (>500 mg per 100 g dw) [37]. Accordingly, the amount of flavor 5′-nucleotides in *L. deliciosus* fell within the high category. Regarding free sugars, trehalose and mannitol have been detected, and their contents were 4990.09 and 7825.00 mg per 100 g dw, respectively. Compared with other wild mushrooms, the content of total free sugars in *L. deliciosus* was in the mid range, which would contribute a moderate sweet taste perception [38–40]. In the context of organic acids, quininic acid predominated in *L. deliciosus*, followed by L-malic acid and fumaric acid.

Oxalic acid was not found in *L. deliciosus* in this study, in which was however identified in another report of *L. deliciosus* from Portugal [41].

Table 3. Free amino acids composition of wild *L. deliciosus*.

Free Amino Acid	Content (mg per 100 g dw)
Aspartic acid [b]	39.43 ± 0.87
Glutamic acid [b]	376.29 ± 4.12
Asparagine	79.41 ± 0.84
Serine [c]	136.35 ± 0.66
Glutamine	794.06 ± 54.85
Histidine [a,d]	278.57 ± 23.62
Glycine [c]	91.32 ± 0.59
Threonine [a,c]	131.76 ± 3.45
Citrulline	16.26 ± 0.59
Arginine [d]	155.70 ± 11.34
Alanine [c]	375.38 ± 2.59
Tyrosine [d]	91.53 ± 1.38
Cystine	-
Valine [a,d]	122.17 ± 4.93
Methionine [a, d]	25.41 ± 0.62
Tryptophan [a, d]	68.95 ± 0.43
Phenylalanine [a, d]	133.61 ± 0.45
Isoleucine [a, d]	47.82 ± 0.95
Leucine [a, d]	109.53 ± 1.61
Lysine [a]	208.46 ± 16.92
Hydroxyproline	-
Proline	107.43 ± 12.44
Total free amino acids (TAA)	3389.45 ± 38.13

[a] Essential amino acids; [b] Monosodium glutamate-like (MSG-like) amino acids; [c] sweet amino acids; [d] bitter amino acids; - Not detected.

Table 4. Flavor 5′-nucleotides, free sugars and organic acids composition of wild *L. deliciosus*.

Component	Content (mg per 100 g dw)
5′-CMP	929.85 ± 42.33
5′-UMP	45.21 ± 6.72
5′-IMP	311.75 ± 13.43
5′-AMP	14.49 ± 3.37
Trehalose	4990.09 ± 307.95
Mannitol	7825.00 ± 466.72
Quininic acid	729.84 ± 71.80
L-Malic acid	415.63 ± 87.44
Fumaric acid	120.71 ± 11.45

Food flavor is a comprehensive concept that contains sweet, sour, bitter, spicy, astringent, umami, etc. Umami is especially important for edible mushrooms because they are usually used as natural freshness-enhancing materials. Equivalent umami concentration (EUC) value calculated using an equation from sensory evaluation, based on the report of Yamaguchi et al., has often been used to evaluate umami-like taste characteristics of mushrooms [42]. For *L. deliciosus*, the EUC value was 145.32 g per 100 g dw. Mau (2005) grouped mushrooms EUC values into four levels: the first level, >1000 g per 100 g dw, the second level, 100–1000 g per 100 g dw; the third level, 10–100 g per 100 g dw; and the fourth level, <10 g per 100 g dw [43]. Thus, the EUC value for *L. deliciosus* belonged to the second level.

2.3. Volatile Aroma Compounds Relating to Special Flavor

Although mushroom tastes depend on water-soluble non-volatile compounds, the function of volatile aroma compounds could not be ignored due to the fact that they directly influence consumer acceptability. Typical volatile aroma compounds in mushrooms originated mostly from chemical or enzymatic oxidation of unsaturated fatty acids and further interactions with proteins, peptides, and free amino acids [44]. In this research, volatile aroma compounds of *L. deliciosus* were estimated by headspace solid phase micro-extraction GS-MS combining library catalog. As shown in Table 5 and Figure S6, 25 compounds were identified, including 15 aldehydes, six acids, two alkanes, one alcohol, and one ester. Based on qualitative and quantitative analysis, acids were confirmed to be the most important aroma volatile compounds accounting for a 84.23% total chromatographic area, followed by aldehydes with 14.77%. Acids dominated in *L. deliciosus* aroma volatile compounds, which was in accordance with a previous study on *L. edodes* and *Pleurotus sajor-caju* conducted by Çağlarırmak [45]. However, other studies have shown that alcohols are important aromatic substances in mushrooms, which differed to our findings in the present research, this might be caused by the drying process, since Tian also found that drying leads to a sharp drop in alcohols while acids and aldehydes increase in *L. edodes* [44]. Moreover, volatile aroma compounds of mushrooms were also affected by growth conditions and genetic differences.

Table 5. Aroma volatile compounds of wild *L. deliciosus*.

Component	Composition (%)
Heptanal	2.03 ± 0.15
Benzaldehyde	2.23 ± 0.09
Hexanoic acid	0.67 ± 0.02
Octanal	1.19 ± 0.07
Benzyl alcohol	0.08 ± 0.00
2-Octenal	0.58 ± 0.11
Nonanal	1.58 ± 0.16
2-Nonenal	0.83 ± 0.05
Dodecane	0.29 ± 0.04
Decanal	0.79 ± 0.05
2-Decenal	1.62 ± 0.29
Undecanal	0.46 ± 0.02
2,4-Decadienal	0.29 ± 0.01
2-Undecenal	1.48 ± 0.18
2-Butyl-2-octenal	0.16 ± 0.02
n-Decanoic acid	0.76 ± 0.05
Decanoic acid, ethyl ester	0.24 ± 0.01
Tetradecane	0.38 ± 0.06
Dodecanal	0.47 ± 0.08
2-Dodecenal	0.65 ± 0.03
Tridecanal	0.41 ± 0.10
n-Hexadecanoic acid	12.82 ± 0.67
9,12-Octadecadienoic acid	3.11 ± 0.21
9-Octadecenoic acid	60.57 ± 2.89
Octadecanoic acid	6.30 ± 0.74

2.4. Antioxidant Activity

Free radicals induce cell-damage, which can cause DNA mutation, proteins damage, lipid peroxidation, and low-density lipoproteins modification. Free radicals can also cause several diseases including diabetes, cancer, neurodegenerative and cardiovascular diseases [46,47]. Usually, food or food extracts can act as antioxidant to counteract damage caused by free radicals. However, the antioxidant capacity of food is determined by complicated factors with various action mechanisms [48]. Thus, several evaluation methods are always used simultaneously to evaluate the antioxidant capacity of food. Due to this effect, we evaluated the antioxidant activity of *L. deliciosus* through DPPH and ABTS radical scavenging and ferric ion reducing (FRAP) assays using ethanol and aqueous extracts. The antioxidant capacity was expressed as trolox equivalent antioxidant capacity (TEAC) values. As shown in Table 6, the TEAC values of ethanol extracts with DPPH, ABT, S and FRAP assays were 18.38, 20.07, and 10.72 μmol Trolox/g dw, respectively. For aqueous extracts, these values were 45.63, 48.05, and 22.28 μmol Trolox/g dw, respectively. Aqueous extracts showed 2–3 fold higher antioxidant capacity compared to ethanol extracts, which might be related to the content of phenols in two extracts, since the total phenol content in aqueous extract was 3.01 fold higher than in ethanol extract.

Table 6. Antioxidant capacity and total phenols in ethanol and aqueous extracts obtained from wild *L. deliciosus*.

	TEAC $_{DPPH}$ (μmolTrolox/g dw) [a]	**TEAC $_{ABTS}$ (μmolTrolox/g dw) [b]**	**TEAC $_{FRAP}$ (μmolTrolox/g dw) [c]**	**Total Phenols Content (mg GAE/g dw) [d]**
Ethanol extract	18.38 ± 1.31	20.07 ± 1.75	10.72 ± 1.04	4.55 ± 0.24
Aqueous extract	45.63 ± 4.40	48.05 ± 3.37	22.28 ± 3.25	13.68 ± 0.26

[a] Trolox equivalent antioxidant capacity (TEAC), 1,1-diphenyl-2-picrylhydrazyl (DPPH); [b] 2,2′-Azino-bis(3-ethylbenzothiazoline-6-sulfonic acid) diammonium salt (ABTS); [c] Ferric ion reducing antioxidant power (FRAP); [d] Gallic acid equivalent weight (GAE).

2.5. Antihyperglycemic Activity

α-Amylase and α-glucosidase are key enzymes in the digestive system that catalyze carbohydrate hydrolysis to enhance blood glucose concentration. A proportion of the population, especially diabetics, suffer from hyperglycemia. A therapeutic approach to hyperglycaemia is to retard the absorption of glucose by inhibiting carbohydrate-hydrolyzing enzymes [49]. Thus, effective and nontoxic inhibitors of α-amylase and α-glucosidase are crucial for the treatment of hyperglycemia. As shown in Figure 1, ethanol extracts exhibited a dose-dependent increase in both α-amylase and α-glucosidase inhibitory assays, while compared with acarbose, the antienzyme activity of ethanol extract was weaker. At 5.0 mg/mL, ethanol extracts exhibited 29.53 and 52.36% on α-amylase and α-glucosidase inhibition, respectively. However, regarding aqueous extracts, no inhibitory effects on α-amylase and α-glucosidase were observed.

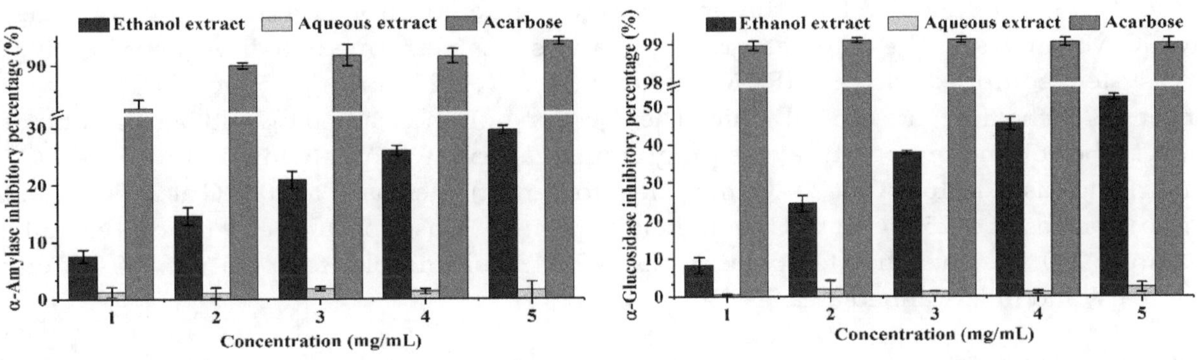

Figure 1. Antihyperglycemic activity of ethanol extract and aqueous extract obtained from wild *L. deliciosus*.

3. Materials and Methods

3.1. Mushroom Species

Wild growing *Lactarius deliciosus* fruiting bodies were collected from Dazhou (Southwest China) pine forests, in the autumn of 2016. All the samples were selected and authenticated at the Dazhou Institute of Agricultural Sciences based on their microscopic and macroscopic characteristics.

3.2. Standards and Reagents

1,1-diphenyl-2-picrylhydrazyl (DPPH), 2,2′-Azino-bis(3-ethylbenzothiazoline-6-sulfonic acid) diammonium salt (ABTS) were purchased from Sigma Chemical Co. (St. Louis, MO, USA). Ferric-tripyridyltriazine (Fe^{3+}-TPTZ) is purchased from Beyotime Biotechnology (Shanghai, China). Chromatographic grade analytical standards asparaginic acid, glutamic acid, asparagine, serine, glutamine, histidine, glycine, threonine, citrulline, arginine, alanine, tyrosine, cystine, valine, methionine, tryptophan, phenylalanine, isoleucine, leucine, lysine, hydroxyproline, proline, 5′-CMP, 5′-UMP, 5′-GMP, 5′-IMP, 5′-XMP, 5′-AMP, trehalose, mannitol, quininic acid, L-malic acid and fumaric acid purchased from Solarbio Science & Technology Co., Ltd. (Beijing, China). α-Amylase, α-glucosidase, 4-nitrophenyl α-*d*-glucopyranoside (PNPG), and soluble starch were purchased from Shanghai Yuanye Bio-Technology Co., Ltd. (Shanghai, China). All other reagents were analytical grade and obtained from Chengdu Kelong Chemical Factory (Chengdu, China).

3.3. Nutritional Value Assay

Moisture, ash, crude fat, crude proteins, and dietary fiber of the research sample were analyzed by the Association of Official Analytical Chemists methods [50]. Briefly, moisture content was measured by hot air heating at 105 °C until constant weight; ash content was measured by calcination at 600 ± 15 °C using a muffle furnace; crude fat was measured by Soxhlet extraction method with petroleum; crude protein content was estimated by the macro-Kjeldahl method using convert coefficient as 4.38; dietary fiber content was estimated by enzymatic hydrolysis method. Total carbohydrate content was measured by the phenol-sulfuric acid method after test sample completely hydrolyzed by hydrochloric acid. Total energy contribution was calculated according to the following equation [40]:

$$Energy\ (kcal) = 4 \times (g\ proteins + g\ carbohydrate) + 9 \times (g\ fat) \tag{1}$$

3.4. Fatty Acids Composition Assay

Fatty acids composition of crude fat was measured by GC-MS method. Briefly, 20 mg crude fat was methylated with 1 mL of sodium hydroxide (1 M) methanol solution. Then, 2 mL *n*-hexane was added to the mixture. Finally, the *n*-hexane phase was filtered through a 0.22-μm membrane filter before GC-MS analysis. The chromatographic analysis was performed on an Agilent 7890B-5977A GC-MS system equipped with an HP-5MS column (30 m × 0.25 mm × 0.25 μm). Helium was used as carrier gas at the flow rate of 1 mL/min. The injection volume was 1.0 μL with a split ratio of 1:50 at 250 °C. The column temperature was programmed as follow: initial temperature at 50 °C (held for 1 min), increased to 160 °C at 20 °C/min (held for 1 min), increased to 200 °C at 20 °C/min (held for 1 min), increased to 250 °C at 5 °C/min (held for 5 min). Mass spectrometry conditions: interface temperature 250 °C, ion source temperature 230 °C, MS quadrupoles temperature 150 °C, electron energy 70 eV, and m/z scanned area 35–550.

3.5. Metal Elements Assay

Metal elements of *L. deliciosus* were analyzed by ICP-MS method. Briefly, 0.2 g lyophilized sample powder was put into digestion tank, 5 mL HNO_3 was added and fully digested in a Multiwave Pro microwave digester (Anton Paar GmbH, Graz, Austria). Then, excess acid was expelled under 150 °C, and residual liquid was diluted with deionized water to 25 mL. Finally, metal elements analysis was performed on a PerkinElmer NexION350D system (PerkinElmer Co., Waltham, MA, USA). The working parameters were as following: Radio-frequency power 1500 W, nebulizer flow rate 0.8 L/min, coolant gas flow 15 L/min, scanning mode peak hopping, sampling depth 10, and isotopes of selected $^{24}Mg^+$, $^{43}Ca^+$, $^{53}Cr^+$, $^{55}Mn^+$, $^{57}Fe^+$, $^{63}Cu^+$, $^{66}Zn^+$, $^{75}As^+$, $^{111}Cd^+$, $^{208}Pb^+$.

3.6. Free Sugars and Free Amino Acids Assay

Free sugars and free amino acids were analyzed according to previous methods [23,36] with slight modification. Briefly, 0.2 g lyophilized sample powder was suspended in 20 mL aqueous ethanol (80%, *v*/*v*), and extracted 30 min at 80 °C. The clear supernatant was obtained through centrifuging at 10,000× *g* for 10 min. The above extraction process repeated again, and supernatant was collected together. Then, the supernatant was analyzed by an Agilent 1260 HPLC system equipped with a Hi-Plex Ca column (300 × 7.7 mm, 8.0 µm) and a Refractive Index Detector (RID) for free sugars assay. The analysis conditions of free sugars were as follows: injection volume: 5 µL; mobile phase: H_2O; flow rate of mobile phase: 0.5 mL/min; column temperature: 80 °C; detector temperature: 40 °C. Furthermore, free amino acids were analyzed by HPLC equipped with a Zorbax Eclipse AAA column (150 × 4.6 mm, 5.0 µm) and a Fluorescence Detector (FLD) (Agilent Technologies, Inc., Santa Clara, CA, USA) using above supernatant by online pre-column derivatization high performance liquid chromatography that has been detailed reported by Sun et al. [36].

3.7. Flavor 5′-Nucleotides Assay

Flavor 5′-nucleotides were analyzed according to the previous method [51] with slight modification. Briefly, 0.5 g lyophilized sample powder was mixed with 20 mL distilled water and extracted by boiling water bath for 1 min. Then, the clear supernatant was obtained through centrifuging at 10,000× *g* for 10 min. The above extraction process repeated again, and the supernatant was collected together. Finally, flavor 5′-nucleotides were analyzed by HPLC equipped with a Zorbax SB-C18 column (150 × 4.6 mm, 5.0 µm) and a Diode Array Detector (DAD) (Agilent Technologies, Inc., Santa Clara, CA, USA) using above supernatant. The analysis conditions were as follows: injection volume: 10 µL; mobile phase A: H_2O; mobile phase B: 0.01 M KH_2PO_4 (pH = 4.0; containing 1.45 M tetrabutylammonium hydrogen sulfate); mobile phase C: Acetonitrile; mobile phase gradient: 80-80-77-77% of mobile phase B and 3-3-6-6% of mobile phase C along with analysis time linear increased from 0-9.5-13-15 min; flow rate of mobile phase: 1.0 mL/min; column temperature: 25 °C; detector wavelength: 254 nm.

3.8. Organic Acids Assay

Organic acids were analyzed by SPE-HPLC method. Briefly, 0.15 g lyophilized sample powder was extracted twice by 20 mL methanol at 40 °C for 30 min. The supernatant was collected through centrifugation and dried under negative pressure condition. Then, the residue was re-dissolved by 1 mL distilled water and purified by SAX solid phase extraction column. Finally, the purified liquid was diluted to 100 mL and analyzed by HPLC equipped with a Zorbax SB-C18 column (150 × 4.6 mm, 5.0 µm) and a Diode Array Detector. The analysis conditions were as follows: injection volume: 10 µL; mobile phase A: 0.1% phosphoric acid solution; mobile phase B: methanol; mobile phase ratio: mobile phase A: mobile phase A = 97.5:2.5; flow rate of mobile phase: 1.0 mL/min; column temperature: 40 °C; detector wavelength: 210 nm.

3.9. Equivalent Umami Concentration

The equivalent umami concentration (EUC, g monosodium glutamate (MSG) per 100 g) was used to reflect the umami intensity of *L. deliciosus*, and is represented by the following addition equation according to Yamaguchi report [42].

$$Y = \sum a_i b_i + 1218\left(\sum a_i b_i\right)\left(\sum a_j b_j\right) \tag{2}$$

where Y is the EUC of the mixture in terms of g MSG/100 g; a_i is the concentration (g per 100 g dw) of each umami amino acid (aspartic acid or glutamic acid); a_j is the concentration (g per 100 g dw) of each umami 5′-nucleotide (5′-IMP, 5′-GMP, 5′-XMP or 5′-AMP); b_i is the relative umami concentration for each umami amino acid to MSG (aspartic acid, 0.077; glutamic acid, 1); b_j is the relative umami concentration for umami 5′-nucelotide to 5′-IMP (5′-IMP, 1; 5′-GMP, 2.3; 5′-XMP, 0.61; 5′-AMP, 0.18); and 1218 is a synergistic constant based on the concentration g per 100 g used.

3.10. Volatile Aroma Components Assay

Volatile aroma components were analyzed using the headspace SPME-GC-MS method. Briefly, 2.0 g lyophilized sample powder was put into a sample bottle, a polydimethylsiloxane SPME fiber (100 μm, Supelco, Bellefonte PA) was used to adsorb statically for 40 min at 50 °C. Then, the volatile aroma components were released at 250 °C and analyzed on an Agilent 7890B-5977C GC-MS system equipped with DB-5MS column (30 m × 0.25 mm × 0.25 μm). Helium was used as carrier gas at the flow rate of 1 mL/min. The column temperature was programmed as follow: initial temperature at 40 °C (held for 1 min), increased to 70 °C at 10 °C/min (held for 2 min), increased to 105 °C at 3 °C/min (held for 1 min), increased to 180 °C at 5 °C/min (held for 1 min), increased to 220 °C at 10 °C/min (held for 5 min). Mass spectrometry conditions: interface temperature 280 °C, ion source temperature 230 °C, MS quadrupoles temperature 150 °C, electron energy 70 eV, and m/z scanned area 35–550.

3.11. Ethanol and Aqueous Extracts Preparation

The lyophilized sample was extracted with ethanol at 50 °C for 1 h three times, the supernatant was collected, concentrated and dried to obtain ethanol extracts. Then, the residue was extracted with distilled water at 80 °C for 1 h three times to obtain aqueous extract.

3.12. Antioxidant Activity Assay

3.12.1. DPPH Radical Scavenging Activity Assay

DPPH radical scavenging activity was measured according to the previous method described in [52] with slight modification. Briefly, 70 μL sample solution was mixed with 140 μL DPPH-ethanol solution. Then, the mixture was incubated in the dark for 30 min at room temperature, and the absorbance at 517 nm was measured using a spectrophotometric micro-plate reader. Trolox solution was used as a positive control and the antioxidant properties of samples were expressed as μmol trolox per g dry weight extract (μmol Trolox/g dw).

3.12.2. ABTS Radical Scavenging Activity Assay

ABTS radical scavenging activity was measured according to previous method described in [53] with slight modification. Briefly, 100 μL sample solution was mixed with 100 μL ABTS solution. Then, the mixture was incubated in the dark for 6 min at room temperature, and the absorbance at 734 nm was measured. Trolox solution was used as a positive control and the antioxidant properties of samples were expressed as μmol trolox per g dry weight extract (μmol Trolox/g dw).

3.12.3. Ferric Ion Reducing Activity Assay

Ferric ion reducing activity was measured according to previous method described in [54] with slight modification. Briefly, 100 μL sample solution was mixed with 100 μL Ferric-tripyridyltriazine (Fe^{3+}-TPTZ) solution. Then, the mixture was incubated in the dark for 10 min at room temperature, and the absorbance at 593 nm was measured. Trolox solution was used as a positive control and the antioxidant properties of samples were expressed as μmol trolox per g dry weight extract (μmol Trolox/g dw).

3.12.4. Total Polyphenols Content Assay

Total polyphenols contents of ethanol and aqueous extracts were determined by the Folin-Ciocalteu assay [55] with some modifications. Briefly, 1 mL sample solution was mixed with 2 mL Folin-Ciocalteu reagent. After incubated for two minutes, 2 mL Na_2CO_3 (10%, w/v) was added and the resulting mixture was incubated for 15 min at 50 °C. Finally, absorbance at 775 nm was measured and polyphenol contents was expressed as mg gallic acid equivalents per g dry weight extract (mg GAE/g dw).

3.13. Antihypertensive Activity Assay

3.13.1. α-Amylase Inhibition Activity Assay

α-Amylase inhibition activity was carried out according to previous method [56] with slight modification. Briefly, α-amylase (1 U/mL), soluble starch (1%) and a series concentration of sample were dissolved by 0.02 M sodium phosphate buffer (pH = 6.9, containing 0.0067 M sodium chloride). Sample solution (100 μL) and 100 μL α-amylase solution were mixed and incubated at 37 °C for 10 min. After adding 500 μL soluble starch solution, the mixture was incubated at 37 °C for 10 min. Then, the reaction was terminated by 100 μL hydrochloric acid solution (5 M). After diluting above mixture six times with distilled water, 50 μL liquid was taken, added into 96-well plate, and 150 μL of iodine solution was added as color developing reagent. The absorbance at 660 nm was measured using a spectrophotometric micro-plate reader. Finally, the α-amylase inhibition activity was calculated as the following formula:

$$\alpha - Amylase\ inhibition\ activity\ (\%) = \frac{Abs_1}{Abs_2} \times 100 \tag{3}$$

where Abs_1 was the absorbance of sample solution mixed with α-amylase solution and soluble starch solution; and Abs_2 was the absorbance of sample solution mixed with soluble starch solution (α-amylase solution was replaced by sodium phosphate buffer).

3.13.2. α-Glucosidase Inhibition Activity Assay

α-Glucosidase inhibition activity was carried out according to previous method described in [57] with slight modification. Briefly, α-glucosidase (0.5 U/mL), 4-nitrophenyl α-d-glucopyranoside (PNPG, 3 mM) a series concentration of sample were dissolved by 0.1 M sodium phosphate buffer (pH = 6.8). Sample solution (100 μL) and 100 μL α-glucosidase solution were mixed and incubated at 37 °C for 10 min. After adding 100 μL PNPG solution, the mixture was incubated at 37 °C for 10 min. Then, the reaction was terminated by 300 μL sodium carbonate solution (0.2 M). Finally, the absorbance at 405 nm was measured using a spectrophotometric micro-plate reader, and the α-glucosidase inhibition activity was calculated as the following formula:

$$\alpha - Glu\cos idase\ inhibition\ activity\ (\%) = \left[1 - \frac{(Abs_1 - Abs_2)}{Abs_0}\right] \times 100 \tag{4}$$

where Abs_0 was the absorbance of the α-glucosidase solution mixed with PNPG solution; Abs_1 was the absorbance of sample solution mixed with α-glucosidase solution and PNPG solution; and Abs_2 was the absorbance of sample solution mixed with PNPG solution (α-glucosidase solution was replaced by sodium phosphate buffer).

3.14. Statistical Analysis

All assays were carried out in triplicate, and the results are expressed as mean ± standard deviation (SD).

4. Conclusions

In conclusion, chemical analysis, GC-MS, ICP-MS, HPLC, SPE-HPLC, HS-SPME-GC-MS, radicals scavenging assays, ferric ion reducing activity assay, and enzymes inhibitory assays were conducted to evaluate nutritional value, non-volatile flavor compounds, volatile aroma compounds, potential antioxidant and anti-hyperglycemic activities of wild mushroom *L. deliciosus* from China for the first time. Experimental data indicated that *L. deliciosus* is a good wild edible mushroom with high nutritional value, low calorie level, and extensive flavor compounds. Moreover, *L. deliciosus* also have potential as natural antioxidant and anti-hyperglycemic agents in food and pharmaceutical industry in the future.

Supplementary Materials: Figure S1: Gas chromatography mass spectrum of fatty acids of *L. deliciosus* fat; Figure S2: High performance liquid chromatography of amino acids analyzed by online OPA-FMOC derivation; Figure S3: High performance liquid chromatography of flavor 5′-nucleotides; Figure S4: High performance liquid chromatography of organic acids; Figure S5: High performance liquid chromatography of free sugars; and Figure S6: Headspace solid phase micro-extraction gas chromatography mass spectrum of aroma volatile compounds from *L. deliciosus*.

Author Contributions: Z.X. and C.D. designed the research; Z.X., L.F., S.F. and M.Y. performed the experiments; Z.X., L.F., Y.H., J.L., L.Z. and H.Y. analyzed the data and wrote the draft manuscript; and C.D. provided financial support. All authors revised the manuscript.

References

1. Okoro, I.O. Antioxidant activities and phenolic contents of three mushroom species, *Lentinus squarrosulus* Mont. *Volvariella esculenta* (Massee) Singer and *Pleurocybella porrigens* (Pers.) Singer. *Int. J. Nutr. Metab.* **2012**, *4*, 72–76.
2. Ouzouni, P.K.; Petridis, D.; Koller, W.D.; Riganakos, K.A. Nutritional value and metal content of wild edible mushrooms collected from West Macedonia and Epirus, Greece. *Food Chem.* **2009**, *115*, 1575–1580. [CrossRef]
3. Finimundy, T.C.; Dillon, A.J.P.; Henriques, J.A.P.; Ely, M.R. A review on general nutritional compounds and pharmacological properties of the Lentinula edodes mushroom. *Food Nutr. Sci.* **2014**, *5*, 1095–1105.
4. Reis, F.S.; Barros, L.; Martins, A.; Ferreira, I.C.F.R. Chemical composition and nutritional value of the most widely appreciated cultivated mushrooms: An inter-species comparative study. *Food Chem. Toxicol.* **2012**, *50*, 191–197. [CrossRef] [PubMed]
5. Miles, P.G.; Chang, S.T. *Mushrooms: Cultivation, Nutritional Value, Medicinal Effect, and Environmental Impact*, 2nd ed.; CRC Press: Boca Raton, FL, USA, 2004.
6. Culleré, L.; Ferreira, V.; Chevret, B.; Venturini, M.E.; Sánchez-Gimeno, A.C.; Blanco, D. Characterisation of aroma active compounds in black truffles (*Tuber melanosporum*) and summer truffles (*Tuber aestivum*) by gas chromatography–olfactometry. *Food Chem.* **2010**, *122*, 300–306. [CrossRef]
7. Cho, I.H.; Namgung, H.J.; Choi, H.K.; Kim, Y.S. Volatiles and key odorants in the pileus and stipe of pine-mushroom (*Tricholoma matsutake* Sing.). *Food Chem.* **2008**, *106*, 71–76. [CrossRef]
8. Mizuno, T. Bioactive biomolecules of mushrooms: Food function and medicinal effect of mushroom fungi. *Food Rev. Int.* **1995**, *11*, 5–21. [CrossRef]
9. Verbeken, A.; Nuytinck, J. Not every milkcap is a *Lactarius*. *Scr. Bot. Belg.* **2013**, *51*, 162–168.
10. Feussi Tala, M.; Qin, J.; Ndongo, J.T.; Laatsch, H. New azulene-type sesquiterpenoids from the fruiting bodies of *Lactarius deliciosus*. *Nat. Prod. Bioprospect.* **2017**, *7*, 269–273. [CrossRef]
11. Anke, H.; Bergendorff, O.; Sterner, O. Assays of the biological activities of guaiane sesquiterpenoids isolated from the fruit bodies of edible lactarius species. *Food Chem. Toxicol.* **1989**, *27*, 393–397. [CrossRef]
12. Bergendorff, O.; Sterner, O. The sesquiterpenes of *Lactarius deliciosus* and *Lactarius deterrimus*. *Phytochemistry* **1988**, *27*, 97–100. [CrossRef]
13. Ding, X.; Hou, Y.; Hou, W. Structure feature and antitumor activity of a novel polysaccharide isolated from *Lactarius deliciosus* Gray. *Carbohyd. Polym.* **2012**, *89*, 397–402. [CrossRef]
14. Hou, Y.; Liu, L.; Ding, X.; Zhao, D.; Hou, W. Structure elucidation, proliferation effect on macrophage and its mechanism of a new heteropolysaccharide from *Lactarius deliciosus* Gray. *Carbohyd. Polym.* **2016**, *152*, 648–657. [CrossRef]
15. Vetter, J. Trypsin inhibitor activity of basidiomycetous mushrooms. *Eur. Food Res. Technol.* **2000**, *211*, 346–348. [CrossRef]
16. Kosanić, M.; Ranković, B.; Rančić, A.; Stanojković, T. Evaluation of metal concentration and antioxidant, antimicrobial, and anticancer potentials of two edible mushrooms *Lactarius deliciosus* and *Macrolepiota procera*. *J. Food Drug Anal.* **2016**, *24*, 477–484. [CrossRef]

17. Moro, C.; Palacios, I.; Lozano, M.; D'Arrigo, M.; Guillamón, E.; Villares, A.; Martínez, J.A.; García-Lafuente, A. Anti-inflammatory activity of methanolic extracts from edible mushrooms in LPS activated RAW 264.7 macrophages. *Food Chem.* **2012**, *130*, 350–355. [CrossRef]
18. Rathore, H.; Prasad, S.; Sharma, S. Mushroom nutraceuticals for improved nutrition and better human health: A review. *Pharma Nutr.* **2017**, *5*, 35–46. [CrossRef]
19. Miyazawa, M.; Takahashi, T.; Horibe, I.; Ishikawa, R. Two new aromatic compounds and a new D-arabinitol ester from the mushroom *Hericium erinaceum*. *Tetrahedron* **2012**, *68*, 2007–2010. [CrossRef]
20. Myrdal Miller, A.; Mills, K.; Wong, T.; Drescher, G.; Lee, S.; Sirimuangmoon, C.; Schaefer, S.; Langstaff, S.; Minor, B.; Guinard, J.X. Flavor-enhancing properties of mushrooms in meat-based dishes in which sodium has been reduced and meat has been partially substituted with mushrooms. *J. Food Sci.* **2014**, *79*, S1795–S1804. [CrossRef]
21. Beluhan, S.; Ranogajec, A. Chemical composition and non-volatile components of Croatian wild edible mushrooms. *Food Chem.* **2011**, *124*, 1076–1082. [CrossRef]
22. Wang, X.M.; Zhang, J.; Wu, L.H.; Zhao, Y.L.; Li, T.; Li, J.Q.; Wang, Y.Z.; Liu, H.G. A mini-review of chemical composition and nutritional value of edible wild-grown mushroom from China. *Food Chem.* **2014**, *151*, 279–285. [CrossRef]
23. Barros, L.; Baptista, P.; Correia, D.M.; Casal, S.; Oliveira, B.; Ferreira, I.C.F.R. Fatty acid and sugar compositions, and nutritional value of five wild edible mushrooms from Northeast Portugal. *Food Chem.* **2007**, *105*, 140–145. [CrossRef]
24. Yilmaz, N.; Solmaz, M.; Türkekul, İ.; Elmastaş, M. Fatty acid composition in some wild edible mushrooms growing in the middle Black Sea region of Turkey. *Food Chem.* **2006**, *99*, 168–174. [CrossRef]
25. Orsavova, J.; Misurcova, L.; Ambrozova, J.; Vicha, R.; Mlcek, J. Fatty acids composition of vegetable oils and its contribution to dietary energy intake and dependence of cardiovascular mortality on dietary intake of fatty acids. *Int. J. Mol. Sci.* **2015**, *16*, 12871–12890. [CrossRef]
26. Calder, P.C. Functional roles of fatty acids and their effects on human health. *J. Parenter. Enter.* **2015**, *39*, 18S–32S. [CrossRef]
27. Ouzouni, P.K.; Veltsistas, P.G.; Paleologos, E.K.; Riganakos, K.A. Determination of metal content in wild edible mushroom species from regions of Greece. *J. Food Compos. Anal.* **2007**, *20*, 480–486. [CrossRef]
28. Kalač, P. Chemical composition and nutritional value of European species of wild growing mushrooms: A review. *Food Chem.* **2009**, *113*, 9–16. [CrossRef]
29. Çayır, A.; Çoşkun, M.; Coşkun, M. The heavy metal content of wild edible mushroom samples collected in Canakkale province, Turkey. *Biol. Trace Elem. Res.* **2010**, *134*, 212–219. [CrossRef]
30. Mendil, D.; Uluözlü, Ö. D.; Hasdemir, E.; Çağlar, A. Determination of trace elements on some wild edible mushroom samples from Kastamonu, Turkey. *Food Chem.* **2004**, *88*, 281–285. [CrossRef]
31. Kalač, P.; Svoboda, L. A review of trace element concentrations in edible mushrooms. *Food Chem.* **2000**, *69*, 273–281. [CrossRef]
32. Food and Nutrition Board; Institute of Medicine. *Dietary Reference Intakes for Calcium, Phosphorus, Magnesium, Vitamin D, and Fluoride*; National Academy Press: Washington, DC, USA, 1997.
33. Trumbo, P.; Yates, A.A.; Schlicker, S.; Poos, M. Dietary reference intakes: Vitamin A, vitamin K, arsenic, boron, chromium, copper, iodine, iron, manganese, molybdenum, nickel, silicon, vanadium, and zinc. *J. Acad. Nutr. Diet.* **2001**, *101*, 294–301.
34. Pei, F.; Shi, Y.; Gao, X.; Wu, F.; Mariga, A.M.; Yang, W.; Zhao, L.; An, X.; Xin, Z.; Yang, F.; Hu, Q. Changes in non-volatile taste components of button mushroom (*Agaricus bisporus*) during different stages of freeze drying and freeze drying combined with microwave vacuum drying. *Food Chem.* **2014**, *165*, 547–554. [CrossRef] [PubMed]
35. Yin, C.; Fan, X.; Fan, Z.; Shi, D.; Yao, F.; Gao, H. Comparison of non-volatile and volatile flavor compounds in six *Pleurotus* mushrooms. *J. Sci. Food Agric.* **2018**. [CrossRef] [PubMed]
36. Sun, L.; Liu, Q.; Bao, C.; Fan, J. Comparison of free total amino acid compositions and their functional classifications in 13 wild edible mushrooms. *Molecules* **2017**, *22*, 350. [CrossRef] [PubMed]
37. Yang, J.H.; Lin, H.C.; Mau, J.L. Non-volatile taste components of several commercial mushrooms. *Food Chem.* **2001**, *72*, 465–471. [CrossRef]
38. Vieira, V.; Barros, L.; Martins, A.; Ferreira, I. Nutritional and biochemical profiling of *Leucopaxillus candidus* (Bres.) Singer wild mushroom. *Molecules* **2016**, *21*, 99. [CrossRef]

39. Mocan, A.; Fernandes, A.; Barros, L.; Crisan, G.; Smiljkovic, M.; Sokovic, M.; Ferreira, I. Chemical composition and bioactive properties of the wild mushroom *Polyporus squamosus* (Huds.) Fr: A study with samples from Romania. *Food Funct.* **2018**, *9*, 160–170. [CrossRef]
40. Toledo, C.V.; Barroetaveña, C.; Fernandes, Â.; Barros, L.; Ferreira, I.C.F.R. Chemical and antioxidant properties of wild edible mushrooms from native Nothofagus spp. forest, Argentina. *Molecules* **2016**, *21*, 1201. [CrossRef]
41. Barros, L.; Pereira, C.; Ferreira, I.C.F.R. Optimized analysis of organic acids in edible mushrooms from Portugal by ultra fast liquid chromatography and photodiode array detection. *Food Anal. Method.* **2013**, *6*, 309–316. [CrossRef]
42. Yamaguchi, S.; Yoshikawa, T.; Ikeda, S.; Ninomiya, T. Measurement of the relative taste intensity of some L-α-amino acids and 5′-nucleotides. *J. Food Sci.* **1971**, *36*, 846–849. [CrossRef]
43. Mau, J. The umami taste of edible and medicinal mushrooms. *Int. J. Med. Mushrooms* **2005**, *7*, 119–126. [CrossRef]
44. Tian, Y.; Zhao, Y.; Huang, J.; Zeng, H.; Zheng, B. Effects of different drying methods on the product quality and volatile compounds of whole shiitake mushrooms. *Food Chem.* **2016**, *197*, 714–722. [CrossRef]
45. Çağlarırmak, N. The nutrients of exotic mushrooms (*Lentinula edodes* and *Pleurotus* species) and an estimated approach to the volatile compounds. *Food Chem.* **2007**, *105*, 1188–1194. [CrossRef]
46. Ziegler, D.V.; Wiley, C.D.; Velarde, M.C. Mitochondrial effectors of cellular senescence: Beyond the free radical theory of aging. *Aging cell* **2015**, *14*, 1–7. [CrossRef] [PubMed]
47. Liochev, S.I. Reactive oxygen species and the free radical theory of aging. *Free Radic. Biol. Med.* **2013**, *60*, 1–4. [CrossRef] [PubMed]
48. Moo-Huchin, V.M.; Moo-Huchin, M.I.; Estrada-León, R.J.; Cuevas-Glory, L.; Estrada-Mota, I.A.; Ortiz-Vázquez, E.; Betancur-Ancona, D.; Sauri-Duch, E. Antioxidant compounds, antioxidant activity and phenolic content in peel from three tropical fruits from Yucatan, Mexico. *Food Chem.* **2015**, *166*, 17–22. [CrossRef]
49. Bhandari, M.R.; Jong-Anurakkun, N.; Hong, G.; Kawabata, J. α-Glucosidase and α-amylase inhibitory activities of Nepalese medicinal herb Pakhanbhed (*Bergenia ciliata*, Haw.). *Food Chem.* **2008**, *106*, 247–252. [CrossRef]
50. Hasan, M.T. *Official Methods of Analysis*, 15th ed.; Association of Official Analytical Chemists: Artington, VR, USA, 2015.
51. Sommer, I.; Schwartz, H.; Solar, S.; Sontag, G. Effect of gamma-irradiation on flavour 5′-nucleotides, tyrosine, and phenylalanine in mushrooms (*Agaricus bisporus*). *Food Chem.* **2010**, *123*, 171–174. [CrossRef]
52. Xiang, C.; Xu, Z.; Liu, J.; Li, T.; Yang, Z.; Ding, C. Quality, composition, and antioxidant activity of virgin olive oil from introduced varieties at Liangshan. *LWT Food Sci. Technol.* **2017**, *78*, 226–234. [CrossRef]
53. Li, X.; Lin, J.; Gao, Y.; Han, W.; Chen, D. Antioxidant activity and mechanism of Rhizoma Cimicifugae. *Chem. Cent. J.* **2012**, *6*, 140. [CrossRef]
54. Xia, Q.; Wang, L.; Xu, C.; Mei, J.; Li, Y. Effects of germination and high hydrostatic pressure processing on mineral elements, amino acids and antioxidants in vitro bioaccessibility, as well as starch digestibility in brown rice (*Oryza sativa* L.). *Food Chem.* **2017**, *214*, 533–542. [CrossRef] [PubMed]
55. Liu, K.; Xiao, X.; Wang, J.; Chen, C.Y.O.; Hu, H. Polyphenolic composition and antioxidant, antiproliferative, and antimicrobial activities of mushroom Inonotus sanghuang. *LWT Food Sci. Technol.* **2017**, *82*, 154–161. [CrossRef]
56. He, M.; Zeng, J.; Zhai, L.; Liu, Y.; Wu, H.; Zhang, R.; Li, Z.; Xia, E. Effect of in vitro simulated gastrointestinal digestion on polyphenol and polysaccharide content and their biological activities among 22 fruit juices. *Food Res. Int.* **2017**, *102*, 156–162. [CrossRef] [PubMed]
57. Hemalatha, P.; Bomzan, D.P.; Sathyendra Rao, B.V.; Sreerama, Y.N. Distribution of phenolic antioxidants in whole and milled fractions of quinoa and their inhibitory effects on α-amylase and α-glucosidase activities. *Food Chem.* **2016**, *199*, 330–338. [CrossRef]

8

Extraction of Carotenoids and Fat-Soluble Vitamins from *Tetradesmus Obliquus* Microalgae: An Optimized Approach by using Supercritical CO_2

Laura Chronopoulou [1], Chiara Dal Bosco [1], Fabrizio Di Caprio [1], Letizia Prosini [1], Alessandra Gentili [1,2], Francesca Pagnanelli [1] and Cleofe Palocci [1,2,*]

1 Chemistry Department, University of Rome La Sapienza, 00185 Rome, Italy
2 CIABC, University of Rome La Sapienza, 00185 Rome, Italy
* Correspondence: cleofe.palocci@uniroma1.it

Academic Editors: Alessandra Gentili and Chiara Fanali

Abstract: In recent years, great attention has been focused on rapid, selective, and environmentally friendly extraction methods to recover pigments and antioxidants from microalgae. Among these, supercritical fluid extraction (SFE) represents one of the most important alternatives to traditional extraction methods carried out with the use of organic solvents. In this study, the influence of parameters such as pressure, temperature, and the addition of a polar co-solvent in the SFE yields of carotenoids and fat-soluble vitamins from *T. obliquus* biomass were evaluated. The highest extraction of alpha-tocopherol, gamma-tocopherol, and retinol was achieved at a pressure of 30 MPa and a temperature of 40 °C. It was observed that overall, the extraction yield increased considerably when a preliminary step of sample pre-treatment, based on a matrix solid phase dispersion, was applied using diatomaceous earth as a dispersing agent. The use of ethanol as a co-solvent, under certain conditions of pressure and temperature, resulted in selectively increasing the yields of only some compounds. In particular, a remarkable selectivity was observed if the extraction was carried out in the presence of ethanol at 10 MPa and 40 °C: under these conditions, it was possible to isolate menaquinone-7, a homologous of vitamin K2, which, otherwise, cannot not recovered by using traditional extraction procedures.

Keywords: microalgae; *Scenedesmus*; supercritical fluid extraction; carotenoids; fat-soluble vitamins; antioxidants

1. Introduction

Aquatic species are promising sources of products for the fine chemicals industry, and this has aroused a growing interest toward such organisms for several applications such as the production of biofuels, the extraction of food additives or active ingredients for cosmetic formulations [1–3]. In particular, algae represent an attractive source for the extraction of vitamin K, carotenoids, and other fat-soluble vitamins.

Vitamin K is a family of structurally similar chemical compounds including phylloquinone (vitamin K1), which occurs in green plants, and menaquinones (vitamin K2 vitamers), which are predominantly of microbial origin [4,5]. Besides acting as a cofactor for the enzyme γ-glutamylcarboxylase, recent research has shown that vitamin K can protect against intracellular oxidative stress and cognitive decline [6–9]. Regarding the vitamin K content in common macroalgae, extremely variable concentrations of phylloquinone have been observed [10,11], however, it was not detected in *P. tricornutum* [11], while its concentration reached 750 µg/100 g in *Sargassum muticum* (commonly known as Japanese wireweed), which is a significantly higher value than that observed in terrestrial plants [10]. To the best of our knowledge, no information on the distribution of menaquinones has so far been reported.

Carotenoids are tetraterpenoids with numerous biological functions synthesized by plants, algae, fungi, and bacteria. They are essential for photosynthesis and, in general, for life in the presence of oxygen. Due to their chemical structure, based on a long chain of conjugated double bonds, these micronutrients are highly lipophilic, variously colored, and exhibit antioxidant properties [12]. Due to changes in lifestyle and the rising health consciousness of the average population, the demand for nutrient-rich supplements with health benefits has risen substantially. Carotenoids also have various medicinal properties and are widely used as preventatives against diseases such as cancer, diabetes, and cataracts [13]. Carotenoids are also currently used in food supplements, cosmetics, and pharmaceuticals [14].

Hydrocarbon carotenoids are referred to as carotenes, while their oxygenated derivatives are known as xanthophylls. Within the latter group, the oxygen atom may be present in the form of hydroxyl groups (as in the case of lutein) or as keto groups (as in the case of canthaxanthin) or as a combination of both (as in astaxanthin) [15]; other oxygenated functional groups occurring in xanthophylls are the epoxy group and carboxylic group. Moreover, all carotenoids having a polyene chain with 11 carbon atoms, and at least one unsubstituted β-ionone ring contributes to the synthesis of vitamin A by means of their conversion into retinol. Increasingly restrictive legislation regarding the origin of food preservatives (e.g., antioxidants and antimicrobials), coupled with the growing demand for natural compounds, have renewed the interest in biomass as a potential source for such compounds rather than resorting to chemical synthesis [16,17]. The best candidates for carotenoid biosynthesis are microalgae because they have many useful features from an industrial point of view: a high surface-to-volume ratio; high growth rates; different metabolic pathways; environmental adaptability; and the simplicity of screening and genetic manipulation [10]. Typical maximum growth rates of microalgae are between 0.5 and 2 d^{-1} that correspond to the duplication times of a few hours, which is much higher than the growth rates of terrestrial plants. Although microalgae are photosynthetic microorganisms, they can also grow in mixotrophic and heterotrophic conditions [18]. Microalgae generally have an optimal pH between 7–9 and an optimal temperature between 25–30 °C. Some carotenoids are synthesized by microalgae as primary metabolites to protect the photosystems from photodamage caused by excessive light exposition and to enlarge the harvesting spectrum of light (e.g., lutein and fucoxanthin). Others carotenoids such as astaxanthin and β-carotene are accumulated as secondary metabolites under stress conditions (e.g., nutrients starvation, osmotic shock, high temperature) to protect the cells from oxidative stresses [14,19]. Among the various species of microalgae, *Tetradesmus obliquus* (generally known as *Scenedesmus obliquus*), an ubiquitous microorganism in lakes and freshwater rivers, is one of the most studied strains for large scale industrial applications, thanks to its ability to grow well in the non-optimal conditions typical of industrial outdoor plants [20]. *T obliquus* can be cultivated with biomass productivities until 2 g/L d in photoautotrophic conditions [21] and until 6 g/L d in heterotrophic conditions [18]. This microalga appears to be a promising source of carotenoid-rich extracts [22]. However, to date, *T. obliquus* is still not exploited for the industrial production of carotenoids or vitamins [19]. One relevant limitation comes from the absence of extraction processes that are sufficiently selective and efficient [23]. Among the most widely used organic solvents for carotenoid extraction, there is hexane (particularly effective for carotenes), ethanol (advantageous for xanthophylls), and acetone. However, the use of organic solvents for extraction is not the preferred way to produce healthy foods due to the toxicity of solvent residues in food as well as the issues of environmental pollution. The development of an efficient extraction technique for the isolation of pharmaceutical compounds from natural resources is necessary.

Extraction using supercritical fluids (SFE) is currently considered an important green alternative to traditional methods. The properties of supercritical fluids (SFs) can be considered as intermediate between the ones of liquids and gases. Similarly to gases, SFs are highly compressible, but have high densities comparable to those of liquids. The combination of some of the properties of liquids with those of gases provides supercritical fluids with some very interesting features. For example, supercritical fluids can effuse through solid materials like a gas, but can also act like a liquid and dissolve substances. In SFE, the organic phase used in typical solid–liquid extractions (SLE) is substituted by a supercritical fluid. The manipulation of both the temperature and pressure of the fluid can solubilize the substance of interest in a complex matrix and selectively extract it. Compared to SLE, SFE is indeed simpler, faster, and more efficient but without consuming large quantities of organic solvents, which are both expensive and potentially harmful. Other immediate advantages of SFE compared to traditional extraction techniques are process flexibility due to the continuous modulation of the solvent power/selectivity of the supercritical fluid and the elimination of polluting organic solvents, which also prevents expensive post-processing of the extracts for solvent removal. CO_2 is the most commonly used supercritical fluid thanks to its non-toxicity, chemical inertia, low cost, and most importantly, low critical values. Its low critical temperature (below 32 °C) makes CO_2 ideal for the extraction of thermolabile compounds. For these reasons, the use of CO_2 as an extraction solvent has been successfully reported in the literature for the isolation of many compounds from various sources [24–28]. For example, supercritical CO_2 has been tested for the extraction of carotenoids and triglycerides from microalgae such as *Hematococcus pluvialis*, *Scenedesmus* sp., and *Chlorella* sp. in different previous works, by mainly using ethanol as the co-solvent, as reported in a recent review [29]. However, there is scarce information about the co-extraction of vitamins and carotenoids from microalgae. In this study, the possibility of extracting carotenoids and fat-soluble vitamins from *T. obliquus* by means of CO_2 in the supercritical phase was evaluated. The effect of several parameters such as the CO_2 physical variables, the addition of co-solvents (methanol and limonene), and an inert dispersing phase on the recovery of different carotenoids and fat-soluble vitamins was investigated. The supercritical fluid extraction was also compared in terms of yield and selectivity with conventional extraction methods.

2. Results and Discussion

2.1. Extraction via Matrix Solid-Phase Dispersion

According to the methodology reported in the literature [30,31], the HPLC-MS analysis of the extracts obtained by the matrix solid-phase dispersion (MSPD) showed that the most abundant carotenoids and fat-soluble vitamins in the algal biomass were lutein and α-tocopherol, respectively. The MSPD extraction, applied individually in accordance with that described in Section 3.4, allowed 10 different compounds to be isolated and identified via HPLC-MS.

2.2. Supercritical CO_2 ($SCCO_2$) Extraction

2.2.1. Evaluation of the Optimal Extraction Time

A series of preliminary SFE was performed on samples of *T. obliquus* at constant pressure (P) and temperature (T) values and by varying the extraction time between 1 and 4 h in order to determine its optimal value. In particular, by using a P_{CO2} of 30 MPa and a T_{CO2} of 50 °C, the percentage of extracted carotenoids was evaluated as a function of the incubation time. The 90-min extraction time was selected as the most appropriate one since, at higher incubation times, no increase in the amount of extracted material was recorded. Therefore, all subsequent extractions in the static approach were carried out for 90 min. For operations in the dynamic mode, an initial static extraction time of 90 min was used, followed by a dynamic extraction step for 10 min. In both cases, at the end of the process, CO_2 was withdrawn from the cell and the extract was recovered in 2 mL of ethanol.

2.2.2. Addition of Diatomaceous Earth as a Dispersing Phase

Before extraction, the algal biomass was mixed with diatomaceous earth in a 1:10 *w*/*w* proportion, as reported in the experimental section. This sample preparation was selected with the aim of obtaining a better yield and extraction reproducibility (data not shown). Indeed, mixing with diatomaceous earth can help to break the cell walls and membranes of the microalgae, exposing the cellular content to the action of the extracting fluid as well as increasing the contact area between the sample and the solvent.

2.3. Evaluation of the Influence of Pressure and Temperature on the Recovery of Carotenoids from T. Obliquus

2.3.1. Extraction Conditions and Extraction Variables

For all of the SFE and MSPD extraction procedures, a fixed quantity of diatomaceous earth/microalga mixture (equal to 0.200 g of sample, of which 0.01818 g was *T. obliquus* biomass) was used. Each experiment was repeated at least twice. Three different CO_2 pressure values were selected during the extractions (25, 30, and 35 MPa, respectively) and for each pressure, the extractions were performed at three different temperatures (40, 50, and 60 °C). The pressure and temperature values were selected on the basis of the literature data and taking into account the thermolability of the carotenoids. In general, as expected, it was observed that as the pressure of the supercritical CO_2 increases at constant temperature, the extraction yields increase as a consequence of an increase in the solvent power. In contrast, as the temperature increases at constant pressure, the solvent power of CO_2 decreases, and therefore carotenoid extraction yields are reduced.

Moreover, we investigated the effect of modifiers such as methanol (5% *v*/*v*) and limonene (5% *v*/*v*), on the composition of the extracts. As can be seen in Figures 1 and 2, the extractions carried out with the addition of limonene did not show a qualitative or quantitative improvement in the composition of the extract, except for the best extraction of phytofluene (a non-polar compound structurally very similar to limonene), while as expected, the addition of methanol allowed a better recovery of all the more polar carotenoids.

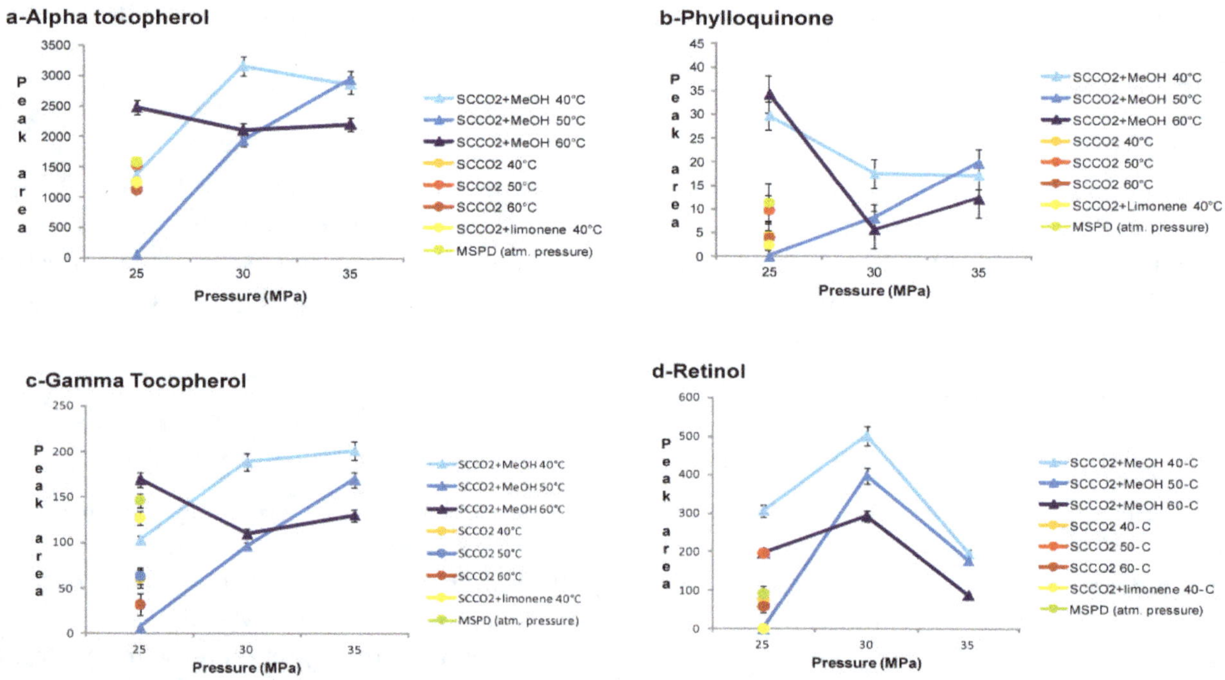

Figure 1. Extraction yields (average peak area ± standard error) of (**a**) alpha tocopherol, (**b**) phylloquinone, (**c**) gamma tocopherol, and (**d**) retinol obtained with SFE and MSPD.

2.3.2. $SCCO_2$ Extraction of Fat-Soluble Vitamins and Carotenoids from T. Obliquus in Comparison with MSPD

The supercritical fluid extraction of micronutrients from algal biomass in comparison with the solid/liquid extraction methodology is reported in Figures 1 and 2. We optimized the extraction conditions in $SCCO_2$ for the following compounds: α-tocopherol, canthaxantin, γ-tocopherol, lutein, phylloquinone, phytofluene, retinol, and menaquinone-7, whose structures are shown in Figures 3 and 4. In Figures 5 and 6, the mass spectra of each extracted compound are reported.

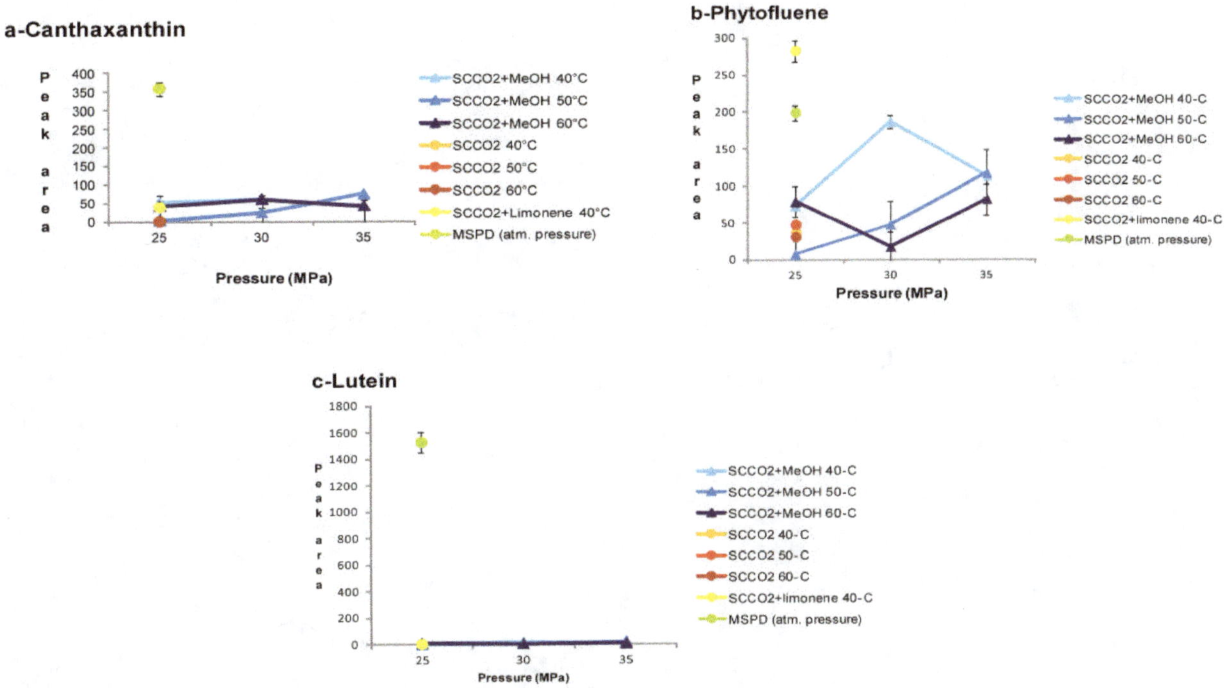

Figure 2. Extraction yields (average peak area ± standard error) of (**a**) canthaxanthin, (**b**) phytofluene, and (**c**) lutein obtained with SFE and MSPD.

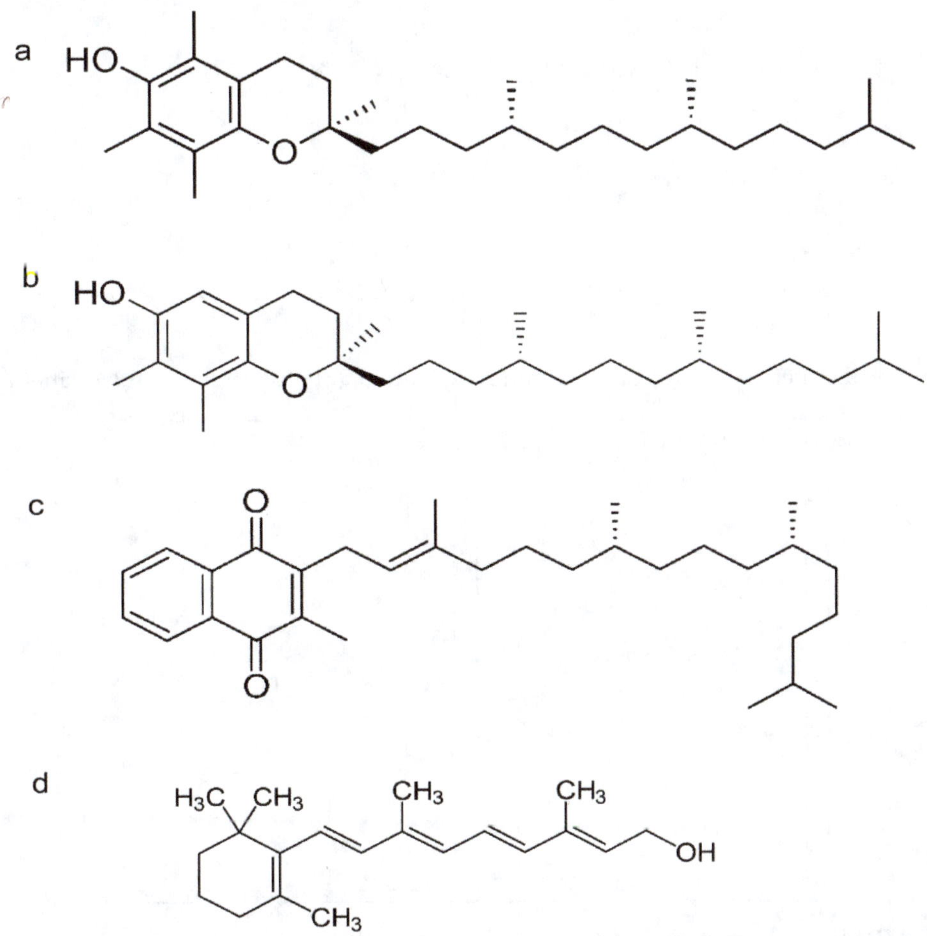

Figure 3. Chemical structures of: (**a**) alpha tocopherol, (**b**) gamma-tocopherol, (**c**) phylloquinone, and (**d**) retinol.

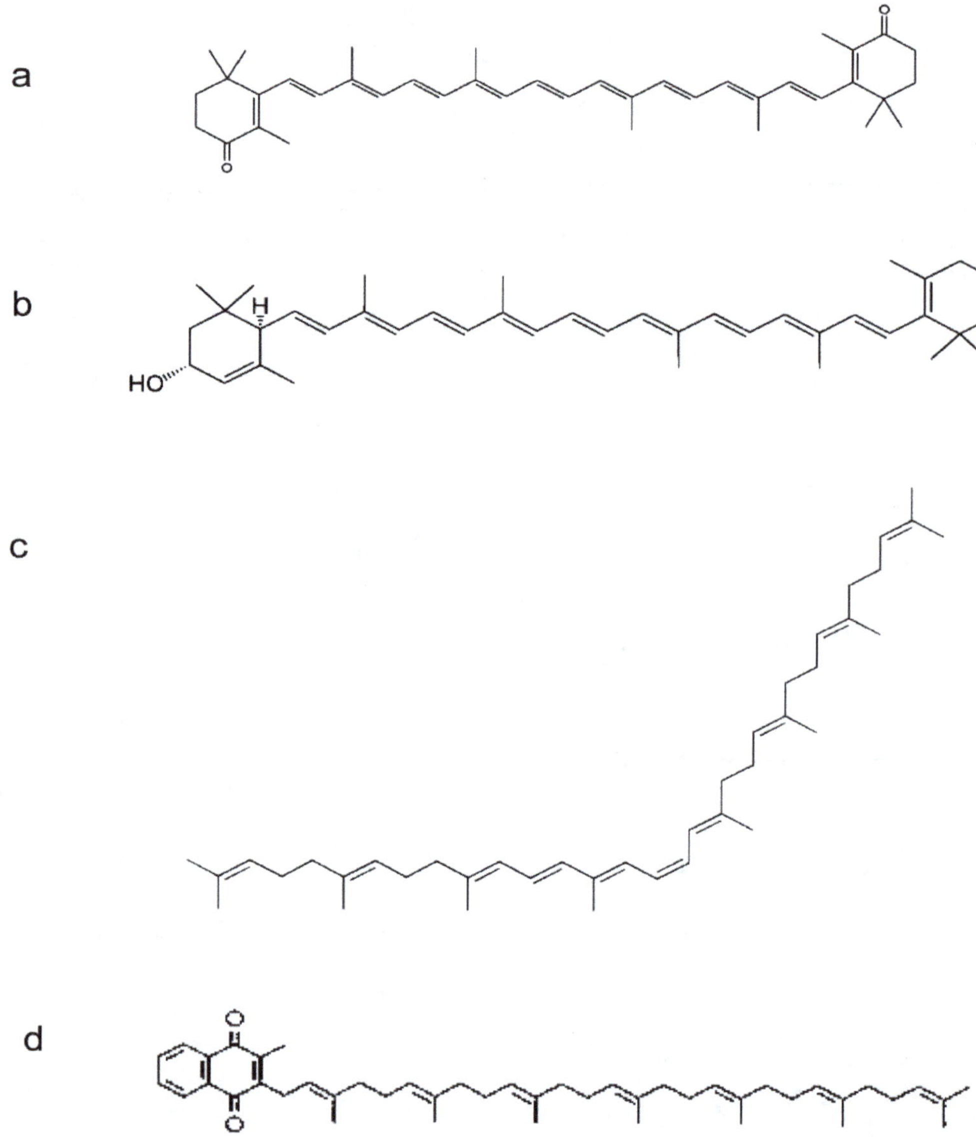

Figure 4. Chemical structures of: (**a**) canthaxanthin, (**b**) lutein, (**c**) phytofluene, and (**d**) menaquinone-7.

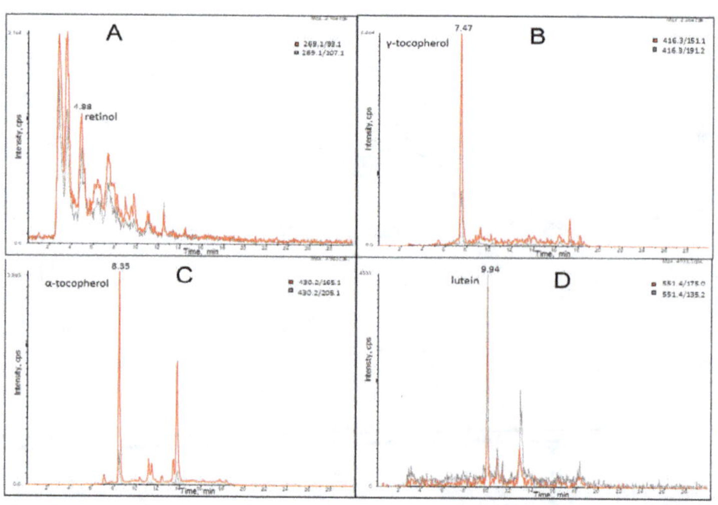

Figure 5. LC-MRM profile of retinol (**A**), gamma-tocopherol (**B**), alpha tocopherol (**C**) and lutein (**D**) extracted from a microalga sample by SFE. Extraction conditions: P_{CO2} = 35 MPa, T_{CO2} = 40 °C, 5% MeOH.

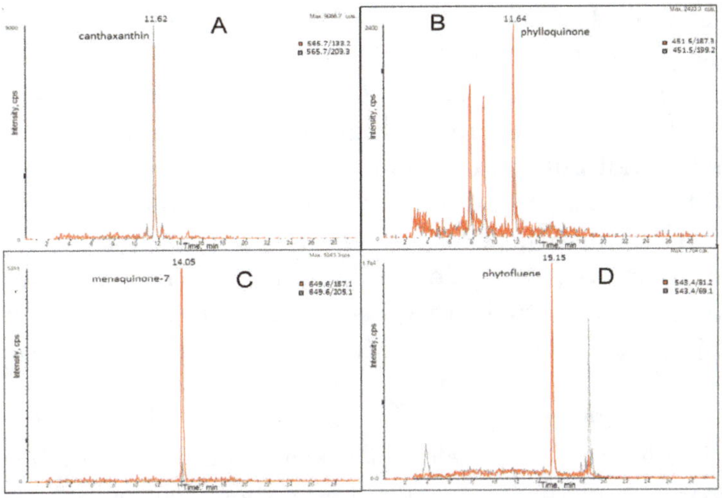

Figure 6. LC-MRM profile of canthaxanthin (**A**), phylloquinone (**B**), menaquinone-7 (**C**), and phytofluene (**D**) extracted from a microalga sample by SFE. Extraction conditions: A, B and D: $P_{CO2} = 35$ MPa, $T_{CO2} = 40$ °C, 5% MeOH. C: $P_{CO2} = 10$ MPa, $T_{CO2} = 40$ °C.

As reported in Figure 1a at P_{CO2} 30 MPa and T_{CO2} 40 °C with the addition of MeOH, the concentration of α-tocopherol detected in the extract was 61.8 μg/mL. The extraction with organic solvents allowed a lower recovery of α-tocopherol, with a concentration of 27.7 μg/mL. The molecular structure of α-tocopherol contains an oxygen atom that makes the molecule more polar than the carotenoids: this explains its higher SF extraction when adding MeOH as a cosolvent for CO_2.

Regarding alpha tocopherol extraction from different types of algae, a strong variability of its content has been reported in the literature and often very close taxa have very different alpha tocopherol contents. Such results point out the importance of growth conditions for obtaining higher quantities of this compound [32]. On this basis, it is very difficult to find useful extraction data to compare the efficiency of the extraction technique used.

Canthaxanthin is co-extracted with $SCCO_2$ in the same experimental conditions (Figure 2a). The extraction yields, however, were not quantitatively comparable with those obtained by MSPD. It is most likely that the chemical structure of canthaxanthin, which contains two carbonyl groups, makes it very poorly soluble in supercritical CO_2 (even at high density and in the presence of methanol). Indeed, the recovery of canthaxanthin in CO_2-only extraction was below the limit of detection, while better results were obtained by increasing the polarity of the solvent phase with the addition of MeOH.

Phylloquinone contains two carbonyl groups together with the presence of a long hydrocarbon chain in its molecular structure. This feature can be responsible for a greater solubility in supercritical CO_2 than in organic solvent mixtures. In the case of phylloquinone, the best extraction conditions were obtained at CO_2 pressure values of 25 MPa and T = 60 °C in the presence of MeOH (Figure 1b).

γ-tocopherol, one of the eight vitamers of vitamin E, showed a better extraction profile with $SCCO_2$, and the addition of MeOH at a temperature of 40 °C and pressure of both 35 and 30 MPa, produced quantitatively higher extraction yields than those obtained with the MSPD technique (Figure 1c). α-tocopherol and γ-tocopherol are structurally similar molecules, and despite being extracted under the same experimental conditions, their recovery was different due to the different relative quantities contained in the microalgae.

Retinol, or vitamin A, as a metabolite of provitamin A carotenoids, could be formed during the extraction procedure by increasing the extraction temperature. In fact, in almost all the $SCCO_2$ extraction conditions tested, a larger recovery of this vitamin was observed when compared to the MSPD technique (Figure 1d).

Phytofluene, a colorless carotenoid precursor, has a structure with 40 carbon atoms and five conjugated double bonds. It showed a better recovery profile with the MSPD technique compared to SFE, although the addition of limonene allowed a better extraction of this carotenoid with the SFE technique (Figure 2b). This may be due to its remarkably apolar structure, while further degradation occurs at temperatures above 40 °C and at higher CO_2 pressures.

Lutein, known as E161b in the European codification of food additives, was extracted in more significant amounts with the MSPD technique than with $SCCO_2$ (Figure 2c). Lutein contains two hydroxyl groups within the molecule and is a very polar compound. It is partially recovered in SFE extractions with thc addition of MeOH, but in lower relative yields.

Finally, extractions on the microalgae carried out in $SCCO_2$ at a pressure of 10 MPa and T = 40 °C allowed the selective extraction of menaquinone-7 (Figure 6), which was not detected in the MSPD extractions. $SCCO_2$ extractions at higher pressures and temperatures did not show the presence of menaquinone-7, confirming the hypothesis that it could be chemically degraded at high CO_2 pressure or temperature values.

Menaquinone-7, like the other menaquinones, has a bacterial origin. It is not usually synthesized from algae, although in the literature, its presence has been hypothesized in the microalgae of the genus *Scenedesmus* [33]. Moreover, the microalgae used in this study were grown in a non-sterile environment, therefore they may have contained the products of a unique system formed by the microbiota in symbiosis with microalgae [34]. It has been proven that several microalgae species cannot survive without such associated bacteria because these latter furnish essential vitamins (such as vitamin B_{12}) to the microalgae [35].

Our results demonstrate that SFE with $SCCO_2$ is a green method for the extraction of high purity thermolabile compounds such as carotenoids. However, the yield of polar carotenoids, as reported in the literature, is often low [36]. By optimizing some key parameters (use of entrainers), it is possible to improve the solubility of more polar analytes in $SCCO_2$. Table 1 reports a comparison of the obtained SFE extraction yields based on the peak areas obtained from the mass spectra with those obtained with MSPD. By varying the SFE conditions, we were able to obtain comparable and sometimes higher extraction yields than MSPD for alpha and gamma tocopherol, canthaxanthin, phylloquinone, phytofluene, retinol, and menaquinone-7.

Table 1. The SFE extraction yields of alpha tocopherol, phylloquinone, gamma-tocopherol, retinol, canthaxanthin, phytofluene, and lutein using different experimental conditions. The values reported are expressed as percentages of MSPD extraction yields ± standard error.

Compound	$SCCO_2$ 25 MPa			$SCCO_2$ + limonene	$SCCO_2$ + MeOH 25 MPa			$SCCO_2$ + MeOH 30 MPa			$SCCO_2$ + MeOH 35 MPa		
	40 °C	50 °C	60 °C		40 °C	50 °C	60 °C	40 °C	50 °C	60 °C	40 °C	50 °C	60 °C
Alpha-tocopherol	(77.71 ± 4.95)%	(97.3 ± 3.59)%	(71.12 ± 3.91)%	(78.73 ± 3.28)%	(86.25 ± 5.16)%	(2.88 ± 2.76)%	(156.95 ± 5.31)%	(200.15 ± 4.64)%	(122.39 ± 4.61)%	(133.44 ± 5.45)%	(180.73 ± 5.33)%	(185.79 ± 3.78)%	(139.39 ± 3.74)%
Phylloquinone	(38.83 ± 4.61)%	(86.50 ± 4.31)%	(34.77 ± 4.31)%	(21.45 ± 4.15)%	(262.14 ± 5.16)%	0	(302.74 ± 5.89)%	(154.46 ± 5.72)%	(71.67 ± 6.11)%	(50.31 ± 6.73)%	(151.81 ± 5.63)%	(174.76 ± 6.28)%	(108.56 ± 6.45)%
Gamma-tocopherol	(41.42 ± 3.76)%	(43.26 ± 3.73)%	(21.77 ± 5.71)%	(86.73 ± 3.82)%	(69.74 ± 3.16)%	(4.09 ± 3.16)%	(115.46 ± 3.81)%	(128.90 ± 3.45)%	(65.30 ± 3.18)%	(74.72 ± 3.62)%	(137.43 ± 3.92)%	(115.52 ± 3.75)%	(88.98 ± 3.49)%
Retinol	(81.44 ± 3.18)%	(212.56 ± 3.47)%	(62.03 ± 3.94)%	0	(332.72 ± [illegible].58)%	0	(213.75 ± 5.76)%	(543.00 ± 5.73)%	(430.97 ± 8.56)%	(317.32 ± 4.61)%	(213.00 ± 3.94)%	(192.93 ± 6.31)%	(94.78 ± 4.72)%
Canthaxanthin	0	0	0	(8.09 ± 2.98)%	(11.02 ± 3.11)%	(0.56 ± 0.44)%	(9.16 ± 3.26)%	(12.99 ± 3.45)%	(4.99 ± 2.31)%	(12.99 ± 4.23)%	(9.25 ± 3.62)%	(16.14 ± 3.15)%	(9.25 ± 3.52)%
Phytofluene	([illegible]8.24 ± 5.64)%	(23.50 ± 4.25)%	(15.21 ± 2.58)%	(142.65 ± 7.51)%	(36.43 ± [illegible].39)%	(94.04 ± 3.98)%	(56.80 ± 4.58)%	(3.34 ± 2.13)%	(23.50 ± 8.57)%	(58.92 ± 8.63)%	(39.46 ± 8.91)%	(8.44 ± 7.32)%	(41.13 ± 8.69)%
Lutein	0	0	0	0	(0.74 ± 0.53)%	0	(0.70 ± 0.45)%	(1.03 ± 0.74)%	(0.29 ± 0.25)%	(0.22 ± 0.20)%	(0.67 ± 0.64)%	(1.25 ± 0.50)%	(0.40 ± 0.40)%

3. Materials and Methods

3.1. Biomass Production

A strain of the microalgae *Tetradesmus obliquus* was maintained in the laboratory under phototrophic conditions as previously described [36]. *T. obliquus* is generally known as *Scenedesmus obliquus*, but it has been recently reclassified by Wynne and Hallan [37].

The biomass used for the extraction tests was produced by diluting 1 to 10 (*v/v*) microalgae from the maintenance flasks in two column photobioreactors (ø = 9 cm, h = 65 cm) with the cultivation medium. The initial biomass concentration was 0.05 g/L. The cultivation medium used was a "tap water based medium", which is a cultivation medium obtained by adding $NaNO_3$ and K_2HPO_4 to the local tap water; its exact chemical composition has been described in a previous work [38,39]. The photobioreactors were maintained under 24 h/24 constant illumination at 100 μmol $m^{-2}s^{-1}$, by means of cool-white florescent lamps and constant air feeding (0.5 L/min) at a room temperature of 27 ± 3 °C. After 15 days of cultivation, the produced microalgae biomass was harvested by centrifugation at 1370× *g* for 5 min and then the obtained pellet was freeze dried.

3.2. Chemicals and Solvents

Methanol, ethanol, 2-propanol, hexane, and acetone (HPLC grade) were purchased from Sigma (St. Louis, MO, USA). Dichloromethane and acetonitrile (analytical grade) were obtained from Chromasolv (Barcelona, Spain). Diatomaceous earth SPE-ED MATRIX 38 was purchased from Applied Separations (Allentown, PA, USA). Syringe-like polypropylene tubes (i.d. 26 mm, 75 mL capacity) and polyethylene frits were obtained from Alltech (Deerfield, IL, USA).

The following standards were purchased from Aldrich-Fluka-Sigma Chemical (St. Louis, MO, USA): retinol, ergocalciferol, δ-tocopherol, β-tocopherol, γ-tocopherol, cholecalciferol, α-tocopherol, menaquinone-4, menaquinone-7, phylloquinone, all-trans-lutein, all trans-zeaxanthin, all-trans-β-cryptoxanthin, and all-trans-β-carotene. Standards of α-tocotrienol, β-tocotrienol, δ-tocotrienol, and γ-tocotrienol were bought from LGC Standards (Middlesex, U.K.). Standards of 15-cis-phytoene, all-trans-phytoene, all-trans-phytofluene, 13-cis-β-carotene, 9-cis-β-carotene, all-trans-ζ-carotene, all-trans-γ-carotene, all-trans-lycopene, and 5-cis-lycopene were purchased from CaroteNature GmbH (Ostermundigen, Switzerland). All chemicals had a purity grade of >97%.

3.3. Biomass Pretreatment

The biomass was freeze dried and stored at −20 °C. Before extraction, the biomass was manually ground into a fine powder. When diatomaceous earth was used, it was mixed with the ground biomass at a 1:10 *w/w* ratio (algae:diatomaceous earth).

3.4. MSPD Extraction

The results obtained by SFE were compared with those from the MSPD extraction. The last one was performed as follows: 200 mg of sample (biomass and diatomaceous earth 1:10 *w/w*) was ground with a pestle into a ceramic mortar until an evenly colored powder was obtained. Subsequently, this powder was used for filling a syringe-like polypropylene tube previously prepared with a first layer of C18 sorbent (0.4 g). The resultant chromatographic bed was held among two polyethylene frits. Vacuum-assisted elution was conducted with 15 mL of methanol, 5 mL of 2-propanol, and 20 mL of hexane by collecting the analytes into a 50 mL falcon. Samples were then centrifuged at 6000 rpm for 10 min. The supernatant was poured out into a glass tube with a conical bottom (i.d. 2 cm) and evaporated up to dryness under a gentle flow of nitrogen in a water bath kept at 25 °C. Finally, the dry extract was dissolved in 2 mL of ethanol, sonicated for 2 min, and passed through a PTFE 0.45 μm filter. Forty microliters were injected into the chromatographic column for the LC-MS analysis.

3.5. *Supercritical Fluid Extraction*

Supercritical fluid extractions (SFE) were performed on a SFE 300 analytical extractor manufactured by Carlo Erba Instruments. A scheme of the extraction apparatus is shown in Figure 1. The extractions took place in a metal tubular reactor of 1 cm^3 where the samples were introduced. Cooled CO_2 was fed into the high-pressure reactor, pressurized at the desired target pressure by a syringe pump, and heated to the desired temperature with a system of recirculating air in the thermostated chamber where the reactor was located. In all of the experiments, a static extraction was followed by a 10-min dynamic extraction performed through the depressurization of $SCCO_2$ in 2 mL of ethanol.

For each extraction experiment, 200 mg of ground powder (containing biomass and diatomaceous earth 1:10 *w*/*w*) were placed in the extraction cell. The operating parameters were varied in the pressure range of 10–35 MPa and in the temperature range of 40–60 °C. A series of extractions was also performed in the presence of a modifier by adding 5% (*v*/*v*) methanol (T = 40, 50, and 60 °C; P = 25, 30, and 35 MPa) or limonene (T = 40 °C; P = 30 MPa) to the sample inside the extraction cell.

3.6. *Mass Spectrometry Experiments*

Analytes were detected by a 4000 Qtrap (AB SCIEX, Foster City, CA, USA.) mass spectrometer equipped with an atmospheric pressure chemical ionization (APCI) probe on a Turbo V source. A positive ionization mode was used, setting a needle current (NC) of 3 μA and a probe temperature of 450 °C. High-purity nitrogen was used as the curtain (40 psi) and collision (4 mTorr) gas, whereas air was the nebulizer (55 psi) and makeup (30 psi) gas. The preliminary calibration of Q1 and Q3 mass analyzers was conducted by infusing a polypropylene glycol solution at 10 μL/min. The unit mass resolution was established by maintaining a full width at half-maximum (fwhm) of approximately 0.7 ± 0.1 unit in each mass-resolving quadrupole. APCI−Q1−full scan spectra and product ion scan spectra of the analytes were acquired by working in flow injection analysis (1−10 ng injected, 1 mL/min flow rate).

3.7. *Liquid Chromatography*

Liquid chromatography (LC) was performed by a micro HPLC series 200 (PerkinElmer, Norwalk, CT, USA.) equipped with an autosampler, vacuum degasser, and column chiller. Analytes were separated on a ProntoSIL C30 column (4.6 mm × 250 mm, 3 μm) from Bischoff Chromatography (Leonberg, Germany), protected by a guard C30 column (4.0 mm × 10 mm, 5 μm), under non aqueous-reversed phase (NARP) conditions at 19 °C. The elution profile by using methanol (phase A) and 2-propanol/hexane (50:50, *v*/*v*; phase B) was as follows: 0−1 min, 0% B; 1−15 min, 0−75% B; 15−15.1 min, 75−99.5% B; and 15.1−30.1 min, 99.5% B. The mobile phase was entirely introduced into the MS detector at a flow rate of 1 mL/min. Phase B was also used to wash the autosampler injection device.

The separation and detection of MK-7 were confirmed by using a specific chromatographic method with increased efficiency in separating vitamin K homologues from interfering compounds. This method differs from the previous one for the use of two reversed-phase columns connected in series (SUPELCOSILTM C18, 4.6 mm × 50 mm, 5 mm, Supelco–Sigma–Aldrich, Bellefonte, PA, USA; and Alltima C18, 4.6 mm × 250 mm; 5 mm, Alltech, Deerfield, IL, USA).

4. Conclusions

In this study, the influence of parameters such as pressure, temperature, and the addition of a polar co-solvent on the SFE yields of carotenoids and fat-soluble vitamins from *T. obliquus* biomass was studied. The optimized extraction conditions revealed the possibility to substantially increase the yields of some compounds with respect to conventional solid–liquid extraction. In particular, by varying the SFE polarity, we were able to obtain comparable and sometimes higher extraction yields than MSPD for many low or medium-polar carotenoids and vitamins. We also obtained a remarkable selectivity (at 10 MPa and 40 °C) for the extraction of the compound menaquinone-7, whose extraction has been rarely achieved by using traditional procedures.

Author Contributions: Investigation, L.C., L.P., C.D.B., A.G., and F.D.C.; Writing—original draft preparation, L.C., L.P., A.G., and C.D.B.; Writing—review and editing, F.D.C., F.P., C.D.B., A.G., L.C., and C.P.; Supervision, C.P.

References

1. Sabu, S.; Bright-Singh, I.; Joseph, V. Molecular identification and comparative evaluation of tropical marine microalgae for biodiesel production. *Mar. Biotechnol.* **2017**, *19*, 328–344. [CrossRef] [PubMed]
2. Matsumoto, M.; Nojima, D.; Nonoyama, T.; Ikeda, K.; Maeda, Y.; Yoshino, T.; Tanaka, T. Outdoor cultivation of marine diatoms for year-round production of biofuels. *Mar. Drugs* **2017**, *15*, 94. [CrossRef] [PubMed]
3. Ahmed, A.B.; Adel, M.; Karimi, P.; Peidayesh, M. Pharmaceutical, cosmeceutical, and traditional applications of marine carbohydrates. *Adv. Food Nutr. Res.* **2014**, *73*, 197–220. [PubMed]
4. Nelsestuen, G.L.; Shah, A.M.; Harvey, S.B. Vitamin K-dependent proteins. *Vitam. Horm.* **2000**, *58*, 355–389. [PubMed]
5. Gentili, A.; Miccheli, A.; Tomai, P.; Baldassarre, M.E.; Curini, R.; Pérez-Fernández, V. Liquid chromatography–tandem mass spectrometry method for the determination of vitamin K homologues in human milk after overnight cold saponification. *J. Food Compos. Anal.* **2016**, *47*, 21–30. [CrossRef]
6. Diener, H.C.; Hart, R.G.; Koudstaal, P.J.; Lane, D.A.; Lip, G.Y.H. Atrial fibrillation and cognitive function: Jacc review topic of the week. *J. Am. Coll. Cardiol.* **2019**, *73*, 612–619. [CrossRef] [PubMed]
7. Brangier, A.; Ferland, G.; Rolland, Y.; Gautier, J.; Féart, C.; Annweiler, C. Vitamin K antagonists and cognitive decline in older adults: A 24-month follow-up. *Nutrients* **2018**, *10*, 666. [CrossRef]
8. Vos, M.; Esposito, G.; Edirisinghe, J.N.; Vilain, S.; Haddad, D.M.; Slabbaert, J.R.; Van Meensel, S.; Schaap, O.; De Strooper, B.; Meganathan, R.; et al. Vitamin K2 is a mitochondrial electron carrier that rescues pink1 deficiency. *Science* **2012**, *336*, 1306–1310. [CrossRef] [PubMed]
9. Gentili, A.; Cafolla, A.; Gasperi, T.; Bellante, S.; Caretti, F.; Curini, R.; Pérez-Fernández, V. Rapid, high performance method for the determination of vitamin K1, menaquinone-4 and vitamin K1 2, 3-epoxide in human serum and plasma using liquid chromatography-hybrid quadrupole linear ion trap mass spectrometry. *J. Chromatogr. A* **2014**, *1338*, 102–110. [CrossRef]
10. Roeck-Holtzhauer, Y.D.; Quere, I.; Claire, C. Vitamin analysis of five planktonic microalgae and one macroalga. *J. Appl. Phycol.* **1991**, *3*, 259–264. [CrossRef]
11. Tarento, T.D.C.; McClure, D.D.; Vasiljevski, E.; Schindeler, A.; Dehghani, F.; Kavanagh, J.M. Microalgae as a source of vitamin K1. *Algal Res.* **2018**, *36*, 77–87. [CrossRef]
12. Flores-Hidalgo, M.; Torres-Rivas, F.; Monzon-Bensojo, J.; Escobedo-Bretado, M.; Glossman-Mitnik, D.; Barraza-Jimenez, D. Electronic Structure of Carotenoids in Natural and Artificial Photosynthesis. In *Carotenoids*; Cvetkovic, D.J., Nikolic, G.S., Eds.; IntechOpen: London, UK, 2017; pp. 17–33. [CrossRef]
13. Mein, J.R.; Lian, F.; Wang, X.D. Biological activity of lycopene metabolites: Implications for cancer prevention. *Nutr. Rev.* **2008**, *66*, 667–683. [CrossRef] [PubMed]
14. Sathasivam, R.; Ki, J.S. A Review of the Biological Activities of Microalgal Carotenoids and Their Potential Use in Healthcare and Cosmetic Industries. *Mar. Drugs* **2018**, *16*, 26. [CrossRef] [PubMed]
15. Higuera-Ciapara, I.; Félix-Valenzuela, L.; Goycoolea, F.M. Astaxanthin: A review of its chemistry and applications. *Crit. Rev. Food Sci. Nutr.* **2006**, *46*, 185–196. [CrossRef] [PubMed]
16. Gordon, H.T.; Bauernfeind, J.C. Carotenoids as food colorants. *Crit. Rev. Food Sci. Nutr.* **1982**, *18*, 59–97. [CrossRef] [PubMed]
17. Macías-Sánchez, M.D.; Mantell Serrano, C.; Rodríguez Rodríguez, M.; Martínez de la Ossa, E.; Lubián, L.M.; Montero, O. Extraction of carotenoids and chlorophyll from microalgae with supercritical carbon dioxide and ethanol as cosolvent. *J. Sep. Sci.* **2008**, *31*, 1352–1362. [CrossRef] [PubMed]
18. Di Caprio, F.; Altimari, P.; Iaquaniello, G.; Toro, L.; Pagnanelli, F. Heterotrophic cultivation of T. obliquus under non-axenic conditions by uncoupled supply of nitrogen and glucose. *Biochem. Eng. J.* **2019**, *145*, 127–136. [CrossRef]
19. Ambati, R.R.; Gogisetty, D.; Aswathanarayana, R.G.; Ravi, S.; Bikkina, P.N.; Bo, L.; Yuepeng, S. Industrial potential of carotenoid pigments from microalgae: Current trends and future prospects. *Crit. Rev. Food Sci.* **2019** *59*, 1880–1902. [CrossRef]

20. Wu, Y.H.; Hu, H.Y.; Yu, Y.; Zhang, T.Y.; Zhu, S.F.; Zhuang, L.L.; Zhang, X.; Lu, Y. Microalgal species for sustainable biomass/lipid production using wastewater as resource: A review. *Renew. Sust. Energ. Rev.* **2014**, *33*, 675–688. [CrossRef]
21. Breuer, G.; Lamers, P.P.; Martens, D.E.; Draaisma, R.B.; Wijffels, R.H. Effect of light intensity, pH, and temperature on triacylglycerol (TAG) accumulation induced by nitrogen starvation in *Scenedesmus obliquus*. *Bioresour. Technol.* **2013**, *143*, 1–9. [CrossRef]
22. Amaro, H.M.; Fernandes, F.; Valentão, P.; Andrade, P.B.; Sousa-Pinto, I.; Malcata, F.X.; Guedes, A.C. Effect of Solvent System on Extractability of Lipidic Components of Scenedesmusobliquus (M2-1) and Gloeothece sp. on Antioxidant Scavenging Capacity Thereof. *Mar. Drugs.* **2015**, *13*, 6453–6471. [CrossRef] [PubMed]
23. Postma, P.R.; 't Lam, G.P.; Barbosa, M.J.; Wijffels, R.H.; Eppink, M.H.M.; Olivieri, G. Microalgal Biorefinery for Bulk and High-Value Products: Product Extraction Within Cell Disintegration. In *Handbook of Electroporation*; Miklavcic, D., Ed.; Springer International Publishing: Berlin, Germany, 2016; pp. 1–20. [CrossRef]
24. De Andrade Lima, M.; Kestekoglou, I.; Charalampopoulos, D.; Chatzifragkou, A. Supercritical Fluid Extraction of Carotenoids from Vegetable Waste Matrices. *Molecules* **2019**, *24*, 466. [CrossRef] [PubMed]
25. Mo, Z.Z.; Lin, Z.X.; Su, Z.R.; Zheng, L.; Li, H.L.; Xie, J.H.; Xian, Y.F.; Yi, T.G.; Huang, S.Q.; Chen, J.P. *Angelica sinensis* Supercritical Fluid CO_2 Extract Attenuates D-Galactose-Induced Liver and Kidney Impairment in Mice by Suppressing Oxidative Stress and Inflammation. *J. Med. Food* **2018**, *21*, 887–898. [CrossRef] [PubMed]
26. Chronopoulou, L.; Agatone, A.; Palocci, C. Supercritical CO2 extraction of oleanolic acid from grape pomace. *Int. J. Food Sci. Tech.* **2013**, *48*, 1854–1860. [CrossRef]
27. Molino, A.; Larocca, V.; Di Sanzo, G.; Martino, M.; Casella, P.; Marino, T.; Karatza, D.; Musmarra, D. Extraction of Bioactive Compounds Using Supercritical Carbon Dioxide. *Molecules* **2019**, *24*, 782. [CrossRef] [PubMed]
28. Tyśkiewicz, K.; Dębczak, A.; Gieysztor, R.; Szymczak, T.; Rój, E. Determination of fat- and water-soluble vitamins by supercritical fluid chromatography: A review. *J. Sep. Sci.* **2018**, *41*, 336–350. [CrossRef] [PubMed]
29. Gallego, R.; Bueno, M.; Herrero, M. Sub- and supercritical fluid extraction of bioactive compounds from plants, food-by-products, seaweeds and microalgae: An update. *TrAC, Trends Anal. Chem.* **2019**, *116*, 198–213. [CrossRef]
30. Gentili, A.; Dal Bosco, C.; Fanali, S.; Fanali, C. Large-scale profiling of carotenoids by using non aqueous reversed phase liquid chromatography—Photodiode array detection—Triple quadrupole linear ion trap mass spectrometry: Application to some varieties of sweet pepper (*Capsicum annuum* L.). *J. Pharm. Biomed. Anal.* **2019**, *164*, 759–767. [CrossRef]
31. Gentili, A.; Caretti, F.; Ventura, S.; Pérez-Fernández, V.; Venditti, A.; Curini, R. Screening of carotenoids in tomato fruits by using liquid chromatography with diode array–linear ion trap mass spectrometry detection. *J. Agric. Food Chem.* **2015**, *63*, 7428–7439. [CrossRef]
32. Mudimu, O.; Koopmann, I.K.; Rybalka, N.; Friedl, T.; Schulz, R.; Bilger, W. Screening of microalgae and cyanobacteria strains for α-tocopherol content at different growth phases and the influence of nitrate reduction on α-tocopherol production. *J. Appl. Phycol.* **2017**, *29*, 2867–2875. [CrossRef]
33. Schwender, J.; Seemann, M.; Lichtenthaler, H.K.; Rohmer, M. Biosynthesis of isoprenoids (carotenoids, sterols, prenyl side-chains of chlorophylls and plastoquinone) via a novel pyruvate/glyceraldehyde 3-phosphate non-mevalonate pathway in the green alga Scenedesmus obliquus. *Biochem. J.* **1996**, *316*, 73–80. [CrossRef] [PubMed]
34. Lian, J.; Wijffels, R.H.; Smidt, H.; Sipkema, D. The effect of the algal microbiome on industrial production of microalgae. *Microb. Biotechnol.* **2018**, *11*, 806–818. [CrossRef] [PubMed]
35. Croft, M.T.; Lawrence, A.D.; Raux-Deery, E.; Warren, M.J.; Smith, A.G. Algae acquire vitamin B12 through a symbiotic relationship with bacteria. *Nature* **2005**, *438*, 90–93. [CrossRef] [PubMed]
36. Pour, H.S.R.; Tavakoli, O.; Sarrafzadeh, M.H. Experimental optimization of SC-CO2 extraction of carotenoids from *Dunaliella salina*. *J. Supercrit. Fluids* **2017**, *121*, 89–95. [CrossRef]
37. Di Caprio, F.; Scarponi, P.; Altimari, P.; Iaquaniello, G.; Pagnanelli, F. The influence of phenols extracted from olive mill wastewater on the heterotrophic and mixotrophic growth of *Scenedesmus* sp. *J. Chem. Technol. Biotechnol.* **2018**, *93*, 3619–3626. [CrossRef]

38. Wynne, M.J.; Hallan, J.K. Reinstatement of Tetradesmus, G.M. Smith (Sphaeropleales, Chlorophyta). *Feddes Repertorium* **2016**, *126*, 83–86. [CrossRef]
39. Di Caprio, F.; Altimari, P.; Pagnanelli, F. Integrated microalgae biomass production and olive mill wastewater biodegradation: Optimization of the wastewater supply strategy. *Chem. Eng. J.* **2018**, *349*, 539–546. [CrossRef]

9

Alternative Ultrasound-Assisted Method for the Extraction of the Bioactive Compounds Present in Myrtle (*Myrtus communis* L.)

Ana V. González de Peredo [1], Mercedes Vázquez-Espinosa [1], Estrella Espada-Bellido [1], Marta Ferreiro-González [1], Antonio Amores-Arrocha [2], Miguel Palma [1], Gerardo F. Barbero [1,*] and Ana Jiménez-Cantizano [2]

1 Department of Analytical Chemistry, Faculty of Sciences, University of Cadiz, Agrifood Campus of International Excellence (ceiA3), IVAGRO, 11510 Puerto Real, Cadiz, Spain; ana.velascogope@uca.es (A.V.G.d.P.); mercedes.vazquez@uca.es (M.V.-E.); estrella.espada@uca.es (E.E.-B.); marta.ferreiro@uca.es (M.F.-G.); miguel.palma@uca.es (M.P.)

2 Department of Chemical Engineering and Food Technology, Faculty of Sciences, University of Cadiz, Agrifood Campus of International Excellence (ceiA3), IVAGRO, 11510 Puerto Real, Cadiz, Spain; antonio.amores@uca.es (A.A.-A.); ana.jimenezcantizano@uca.es (A.J.-C.)

* Correspondence: gerardo.fernandez@uca.es

Academic Editors: Alessandra Gentili and Chiara Fanali

Abstract: The bioactive compounds in myrtle berries, such as phenolic compounds and anthocyanins, have shown a potentially positive effect on human health. Efficient extraction methods are to be used to obtain maximum amounts of such beneficial compounds from myrtle. For that reason, this study evaluates the effectiveness of a rapid ultrasound-assisted method (UAE) to extract anthocyanins and phenolic compounds from myrtle berries. The influence of solvent composition, as well as pH, temperature, ultrasound amplitude, cycle and solvent-sample ratio on the total phenolic compounds and anthocyanins content in the extracts obtained were evaluated. The response variables were optimized by means of a Box-Behnken design. It was found that the double interaction of the methanol composition and the cycle, the interaction between methanol composition and temperature, and the interaction between the cycle and solvent-sample ratio were the most influential variables on the extraction of total phenolic compounds (92.8% methanol in water, 0.2 s of cycle, 60 °C and 10:0.5 mL:g). The methanol composition and the interaction between methanol composition and pH were the most influential variables on the extraction of anthocyanins (74.1% methanol in water at pH 7). The methods that have been developed presented high repeatability and intermediate precision (RSD < 5%) and the bioactive compounds show a high recovery with short extraction times. Both methods were used to analyze the composition of the bioactive compounds in myrtle berries collected from different locations in the province of Cadiz (Spain). The results obtained by UAE were compared to those achieved in a previous study where microwave-assisted extraction (MAE) methods were employed. Similar extraction yields were obtained for phenolic compounds and anthocyanins by MAE and UAE under optimal conditions. However, UAE presents the advantage of using milder conditions for the extraction of anthocyanins from myrtle, which makes of this a more suitable method for the extraction of these degradable compounds.

Keywords: anthocyanins; bioactive compounds; Box–Behnken design; ultrasound-assisted extraction; myrtle; *Myrtus communis* L.; phenolic compounds

1. Introduction

People's diet is currently improving with a growing demand for healthy food, such as vegetables and fruit. Some berries have been particularly demanded mainly due to their phenolic composition [1,2]. Phenolic compounds are quite prone to oxidation due to their high content in double bonds and hydroxyl groups [3]. This characteristic provides them with a substantial capacity to prevent the oxidation of free radicals, i.e., chemically unstable species that may damage lipid cells, proteins and DNA [4]. Therefore, numerous studies support a positive association between the consumption of berries, which are rich in phenolic compounds, and the prevention against some diseases, such as cardiovascular or neurodegenerative diseases [5]. For that reason, the phenolic compounds extracted from berries are being used by the food, cosmetic and pharmaceutical industries to replace synthetic antioxidants [6,7].

Myrtle (*Myrtus communis* L.), is a widely spread plant throughout the Mediterranean area and the Middle East [8]. This evergreen shrub produces dark blue edible berries of different shapes in most of the characterized ecotypes [9]. These berries are rich in antioxidant compounds, in a considerably greater degree than most other fruit types [10]. Specifically, myrtle berries have a high content of phenolic compounds and anthocyanins [11]. The major phenolic compounds which can be identified in myrtle berries are quercetin 3-*O*-galactoside, quercetin 3-*O*-rhamnoside, myricetin 3-*O*-rhamnoside, quercetin 3-*O*-glucoside, ellagic acid and myricetin [12,13]. The major anthocyanins which can be identified in myrtle berries are delphinidin 3,5-*O*-diglucoside, delphinidin 3-*O*-glucoside, cyanidin 3-*O*-galactoside, cyanidin 3-*O*-glucoside, cyanidin 3-*O*-arabinoside, petunidin 3-*O*-glucoside, delphinidin 3-*O*-arabinoside, peonidin 3-*O*-glucoside, malvidin 3-*O*-glucoside, petunidin 3-*O*-arabinoside and malvidin 3-*O*-arabinoside [14,15]. These compounds, as above mentioned, have exhibited some potentially positive properties for human health, such as antidiabetic, anti-inflammatory, anti-cancer and antioxidant properties [16]. These medicinal properties have meant new fields where myrtle can be used. Consequently, it is currently used in the perfume, cosmetics, healthcare and food industries [17]. In spite of myrtle's broad potential and the fact that it grows in many extensive areas, its intensive exploitation takes place mainly in Sardinia. In this area, "Mirto" liquor is produced by macerating myrtle leaves and berries [18].

In order to determine the quality control of these beneficial compounds, the development of fast and efficient methods of extraction and analysis is required. Several studies have been carried out on the antioxidant activity and phenolic compounds in the extracts obtained from myrtle berries [19]. Most of these studies use maceration as the extraction technique, while few studies have been found that employ new extraction methods to obtain the extract from myrtle berries [20]. Compared to the commonly used extraction methods, some novel extraction techniques, such as ultrasound-assisted extraction (UAE), microwave-assisted extraction (MAE) or pressurized-liquid extraction (PLE) greatly reduce time, costs and volume of solvent, and also improve the quality of the extracts [21]. In a previous study recently published by the authors, MAE was used for the determination of total phenolic compounds and total anthocyanins in myrtle berries [22]. Based on the results obtained, it was concluded that MAE is an eco-friendlier and easier to use a technique for the extraction of both, phenolic compounds and anthocyanins, from myrtle berries. In this study, UAE is presented as an alternative extraction method, since UAE offers extraction yields comparable to those obtained by means of MAE, but exceeds MAE in terms of the number of solvent used, easiness, and economic cost. The use of ultrasounds is supported by the phenomenon of cavitation. This phenomenon makes of UAE a widely used method, since it breaks cell walls and releases the target compounds out of their natural matrices [23]. For the reason, UAE has been used for the extraction of the antioxidant compounds found in fruits matrices, such as papayas [24], mulberries [25], oranges [26] or sugarcane [27]. With regard to myrtle, some articles have been found in literature where UAE is used to extract biological compounds from myrtle berries [28,29], but it has not yet been carried out a thorough optimization and development of this technique for the specific extraction of phenolic compounds and anthocyanins. This would be very convenience, since the correct applicability would improve the analysis of the raw material, which would enhance the quality, for example, of liquor, as the main product obtained from myrtle. Moreover, the correct applicability of the method would allow to track down the evolution of the fruit's chemical composition during its maturity process. Reliable maturity indices would be useful to establish optimum harvesting periods, etc.

The aim of the present study is, therefore, to determine optimum conditions for the efficient extraction methods to obtain the greatest possibly yields of substances with antioxidant activities, anthocyanins and phenolic compounds from myrtle berries. Furthermore, the UAE results were compared to those obtained by MAE.

2. Results and Discussion

2.1. Development of the UAE Method

The Box-Behnken design was applied to the optimization of the variables that mainly influence UAE with regards to total phenolic and total anthocyanins yields. The main variables that affect UAE efficiency are solvent composition, temperature, amplitude of the ultrasound, cycle, pH, and solvent-sample ratio [30]. Table 1 shows such influential variables and their corresponding values studied in this work.

Table 1. Influential variables with their corresponding values studied in this work.

Variables	Studied Ranges	
	Phenolic Compounds	**Anthocyanins**
Temperature (°C)	10, 35, 60	10, 35, 60
Amplitude (%)	30, 50, 70	30, 50, 70
Cycle (s)	0.2, 0.45, 0.7	0.2, 0.45, 0.7
pH	2, 4.5, 7	2, 4.5, 7
Solvent-sample ratio (mL/0.5 g)	10, 15, 20	10, 15, 20
Solvent composition (% methanol in water)	50, 75, 100	25, 50, 75

Analysis of variance (ANOVA) was applied to the set of results in order to evaluate the effect of the different factors on their response and the possible interactions between them. Tables 2 and 3 show the results obtained from this analysis.

Table 2. ANOVA for the quadratic model adjusted to the extraction of total phenolic compounds.

	Source	Coefficient	Sum of Squares	Degrees of Freedom	Mean Square	*F*-Value	*p*-Value
	Model		2940.04	27	108.89	2.31	0.0179
Methanol	X_1	−1.15	31.97	1	31.97	0.6787	0.4175
Temperature	X_2	−1.01	24.50	1	24.50	0.5202	0.4772
Amplitude	X_3	0.0166	0.0066	1	0.0066	0.0001	0.9906
Cycle	X_4	−4.17	418.01	1	418.01	8.88	0.0062
pH	X_5	−0.8371	16.82	1	16.82	0.3571	0.5553
Ratio	X_6	0.5498	7.26	1	7.26	0.1541	0.6979
Methanol × Temperature	X_1X_2	6.75	364.53	1	364.53	7.74	0.0099
Methanol × Amplitude	X_1X_3	1.01	8.14	1	8.14	0.1729	0.6809
Methanol × Cycle	X_1X_4	−2.00	63.93	1	63.93	1.36	0.2545
Methanol × pH	X_1X_5	4.38	153.56	1	153.56	3.26	0.0826
Methanol × Ratio	X_1X_6	−0.7961	5.07	1	5.07	0.1077	0.7454
Temperature × Amplitude	X_2X_3	−0.7882	4.97	1	4.97	0.1055	0.7479
Temperature × Cycle	X_2X_4	0.6373	3.25	1	3.25	0.0690	0.7949
Temperature × pH	X_2X_5	2.87	132.23	1	132.23	2.81	0.1058

Table 2. *Cont.*

	Source	Coefficient	Sum of Squares	Degrees of Freedom	Mean Square	*F*-Value	*p*-Value
Temperature × Ratio	X_2X_6	−2.36	44.72	1	44.72	0.9495	0.3388
Amplitude × Cycle	X_3X_4	−4.26	144.96	1	144.96	3.08	0.0911
Amplitude × pH	X_3X_5	−1.14	10.47	1	10.47	0.2222	0.6413
Amplitude × Ratio	X_3X_6	1.87	56.01	1	56.01	1.19	0.2855
Cycle × pH	X_4X_5	−2.64	55.80	1	55.80	1.18	0.2863
Cycle × Ratio	X_4X_6	6.71	359.66	1	359.66	7.64	0.0104
pH × Ratio	X_5X_6	1.68	22.66	1	22.66	0.4812	0.4940
Methanol × Methanol	X_1^2	−8.06	668.48	1	668.48	14.19	0.0009
Temperature × Temperature	X_2^2	−0.2020	0.4197	1	0.4197	0.0089	0.9255
Amplitude × Amplitude	X_3^2	1.79	32.81	1	32.81	0.6967	0.4115
Cycle × Cycle	X_4^2	1.87	36.12	1	36.12	0.7669	0.3892
pH × pH	X_5^2	2.96	90.01	1	90.01	1.91	0.1786
Ratio × Ratio	X_6^2	−2.20	49.57	1	49.57	1.05	0.3144
Residual		45.85	1224.48	26	47.10		
Lack of Fit			1075.70	21	51.22	1.72	0.2858
Pure Error			148.78	5	29.76		
Total			4164.52	53			

Table 3. ANOVA for the quadratic model adjusted to the extraction of total anthocyanins.

	Source	Coefficient	Sum of Squares	Degrees of Squares	Mean Square	*F*-Value	*p*-Value
	Model		2514.45	27	93.13	1.47	0.1631
Methanol	X_1	4.80	553.02	1	553.02	8.75	0.0065
Temperature	X_2	−2.05	100.49	1	100.49	1.59	0.2185
Amplitude	X_3	−1.88	85.15	1	85.15	1.35	0.2563
Cycle	X_4	−2.47	146.43	1	146.43	2.32	0.1400
pH	X_5	1.13	30.92	1	30.92	0.4892	0.4905
Ratio	X_6	−0.1120	0.3008	1	0.3008	0.0048	0.9455
Methanol × Temperature	X_1X_2	−5.59	250.28	1	250.28	3.96	0.0572
Methanol × Amplitude	X_1X_3	−4.68	174.94	1	174.94	2.77	0.1082
Methanol × Cycle	X_1X_4	−2.18	76.24	1	76.24	1.21	0.2821
Methanol × pH	X_1X_5	6.51	338.73	1	338.73	5.36	0.0288
Methanol × Ratio	X_1X_6	0.0864	0.0598	1	0.0598	0.0009	0.9757
Temperature × Amplitude	X_2X_3	−0.1454	0.1691	1	0.1691	0.0027	0.9591
Temperature × Cycle	X_2X_4	−1.16	10.77	1	10.77	0.1704	0.6831
Temperature × pH	X_2X_5	−3.14	157.76	1	157.76	2.50	0.1262
Temperature × Ratio	X_2X_6	0.8116	5.27	1	5.27	0.0834	0.7751
Amplitude × Cycle	X_3X_4	3.73	111.16	1	111.16	1.76	0.1963
Amplitude × pH	X_3X_5	−1.63	21.28	1	21.28	0.3368	0.5667
Amplitude × Ratio	X_3X_6	−1.04	17.31	1	17.31	0.2739	0.6052
Cycle × pH	X_4X_5	−0.7232	4.18	1	4.18	0.0662	0.7990
Cycle × Ratio	X_4X_6	0.2305	0.4249	1	0.4249	0.0067	0.9353
pH × Ratio	X_5X_6	−0.9561	7.31	1	7.31	0.1157	0.7365
Methanol × Methanol	X_1^2	−1.95	39.12	1	39.12	0.6191	0.4385
Temperature × Temperature	X_2^2	2.23	50.98	1	50.98	0.8068	0.3773
Amplitude × Amplitude	X_3^2	2.89	86.14	1	86.14	1.36	0.2536

Table 3. *Cont.*

	Source	Coefficient	Sum of Squares	Degrees of Squares	Mean Square	*F*-Value	*p*-Value
Cycle × Cycle	X_4^2	−1.17	14.04	1	14.04	0.2221	0.6414
pH × pH	X_5^2	2.03	42.24	1	42.24	0.6683	0.4211
Ratio × Ratio	X_6^2	−1.64	27.72	1	27.72	0.4386	0.5136
Residual		23.34	1643.02	26	63.19		
Lack of Fit			1469.28	21	69.97	2.01	0.2248
Pure Error			173.75	5	34.75		
Total			4157.47	53			

This information was supplemented with Pareto Charts (Figure 1). Pareto charts show each effect and combination of effects by a bar in decreasing order of significance. From a graphical point of view, this allows to visualize the influencing variables and their degree of influence.

(a)
(b)

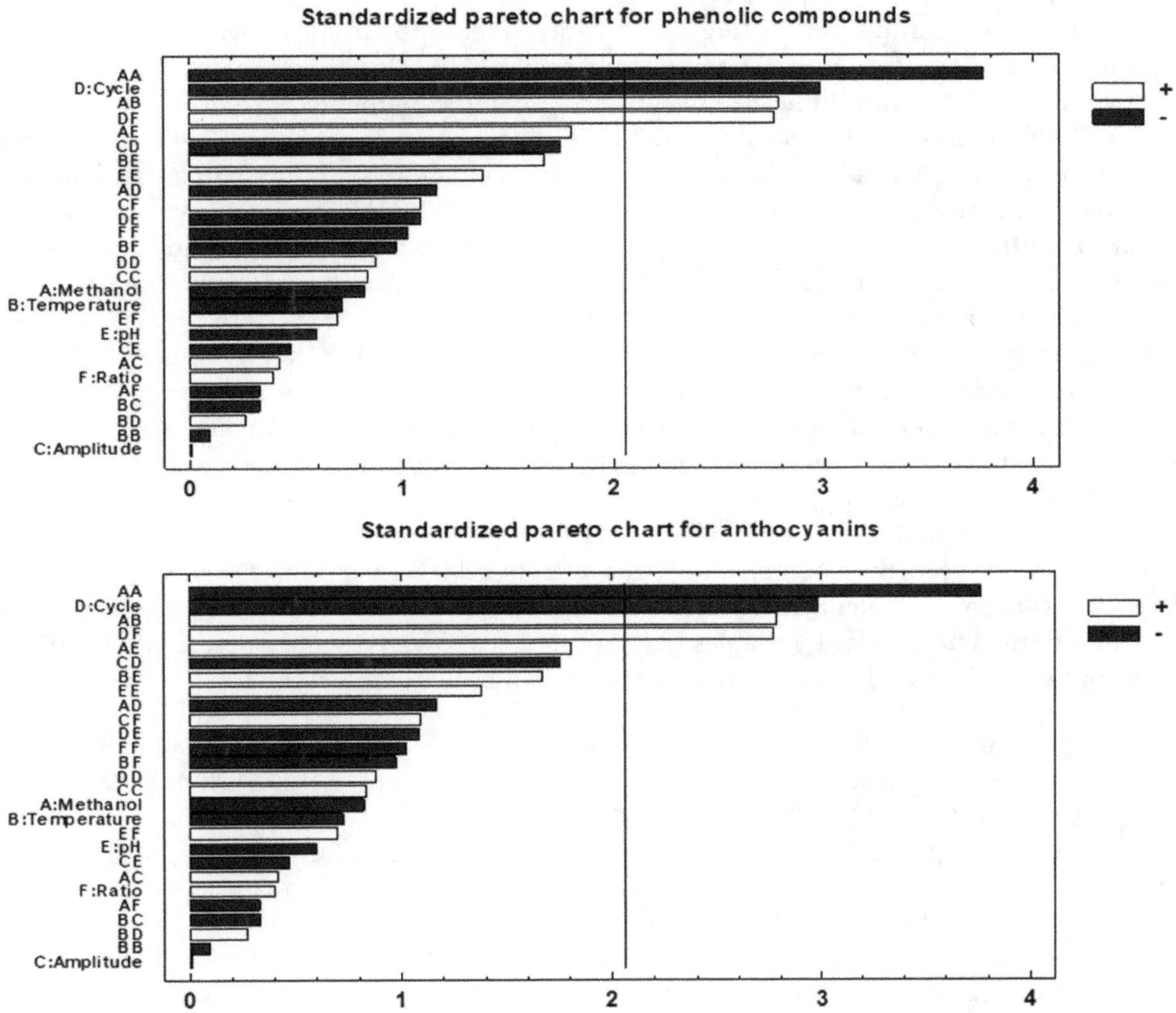

Figure 1. Pareto charts of the standardized effects: (**a**) Total phenolic compounds and (**b**) total anthocyanins.

In the case of total phenolic compounds (Figure 1a), cycle (X_4) was the only linear term which had a significant effect, with a p-value lower than 0.01. Its effect on the response variable was negative ($b_4 = -4.17$). Numerous studies show that the cycle is an influential variable since the use of ultrasound cycles (pulse) improves the extraction of certain compounds of interest, such as the phenolic compounds in natural matrices [31–33]. The negative effect means that a decrease in the cycle increases the extraction of phenolic compounds. This may be due to the negative chemical and physical effects of cavitation [34]. The negative effect is often due to the reactions of free radicals formed during the sonication with molecules in the medium [35], which accelerates the degradation process of phenolic compounds. In addition, the solvent composition had a significant quadratic influence (X_1^2) on the response variable (p-value < 0.01). The solvent composition is an important variable since it is necessary to extract the phenolic compounds with solvents of similar polarity [36]. Specifically, X_1^2 showed a negative effect ($b_{11} = -8.06$). With regards to interactions between factors, minor interactions between methanol and temperature (X_1X_2) (p-value < 0.01) and between cycle and solvent-sample ratio (X_4X_6) (p-value < 0.05) were observed. Both interactions showed positive coefficients ($b_{12} = 6.75$ and $b_{46} = 6.70$).

In the case of anthocyanins (Figure 1b), solvent composition (X_1) was the only linear term that had a significant effect, with a p-value lower than 0.01. Its effect on the response variable was positive ($b_1 = 4.80$), which indicates that an increase in the methanol percentage in the solvent favored the anthocyanins content in the extract. Many pieces of research have been found in the literature which shows that hydroalcoholic mixtures are more efficient than pure solvents for the extraction of moderately polar molecules, such as phenolic compounds [36]. Phenolic compounds have a moderate polarity, so they are not extracted adequately when pure water mixtures are used (high polarity). The use of methanol increases the solubility of phenolic compounds and the use of water in a lower percentage helps the desorption of the solute from the sample [37]. With regard to the interactions between factors, a minor interaction between methanol and pH (X_1X_5) (p-value < 0.05) was observed with a positive coefficient ($b_{15} = 6.51$). These results agree with the bibliographic data [38] where the concentration of organic solvent used and the pH are influential variables on the extraction and stability of anthocyanins from vegetable matrices. Which regard to quadratic effects, non-significant interactions were obtained (p-value > 0.05).

The polynomial Equations (1) and (2) for anthocyanins and total phenolic compounds were obtained from the coefficients of the effects and interactions (Tables 2 and 3). Therefore, two second-order mathematical models were obtained to predict the Y_{TA} and Y_{TP} response values as a function of the independent variables. Lack of fit test showed p-values greater than 0.05 for phenolic compounds and for anthocyanins which means that both models fit well.

$$\begin{gathered} Y_{TA}\ (\mathrm{mg{\cdot}g^{-1}}) = 23.3422 + 4.80026{\cdot}X_1 - 2.04623{\cdot}X_2 - 1.88357{\cdot}X_3 - 2.47009{\cdot}X_4 + 1.13497{\cdot}X_5 - \\ 0.111961{\cdot}X_6 - 1.95032{\cdot}X_1^2 - 5.59331{\cdot}X_1X_2 - 4.67624{\cdot}X_1X_3 - 2.18294{\cdot}X_1X_4 + 6.50698{\cdot}X_1X_5 + \\ 0.0864269{\cdot}X_1X_6 + 2.22638{\cdot}X_2^2 - 0.145373{\cdot}X_2X_3 - 1.16027{\cdot}X_2X_4 - 3.14005{\cdot}X_2X_5 + 0.811562{\cdot}X_2X_6 \\ + 2.89386{\cdot}X_3^2 + 3.72761{\cdot}X_3X_4 - 1.63114{\cdot}X_3X_5 - 1.04011{\cdot}X_3X_6 - 1.16818{\cdot}X_4^2 - 0.72317{\cdot}X_4X_5 + \\ 0.230451{\cdot}X_4X_6 + 2.02637{\cdot}X_5^2 - 0.956103{\cdot}X_5X_6 - 1.64156{\cdot}X_6^2 \end{gathered} \tag{1}$$

$$\begin{gathered} Y_{TP}\ (\mathrm{mg{\cdot}g^{-1}}) = 45.8533 - 1.15407{\cdot}X_1 - 1.0103{\cdot}X_2 + 0.0166375{\cdot}X_3 - 4.17338{\cdot}X_4 - 0.837125{\cdot}X_5 + \\ 0.549829{\cdot}X_6 - 8.06168{\cdot}X_1^2 + 6.75029{\cdot}X_1X_2 + 1.00895{\cdot}X_1X_3 - 1.99896{\cdot}X_1X_4 + 4.38119{\cdot}X_1X_5 \\ -0.796125{\cdot}X_1X_6 - 0.201989{\cdot}X_2^2 - 0.78815{\cdot}X_2X_3 + 0.637275{\cdot}X_2X_4 + 2.87481{\cdot}X_2X_5 - 2.3643{\cdot}X_2X_6 \\ + 1.78604{\cdot}X_3^2 - 4.25675{\cdot}X_3X_4 - 1.14384{\cdot}X_3X_5 + 1.87097{\cdot}X_3X_6 + 1.87383{\cdot}X_4^2 - 2.6411{\cdot}X_4X_5 + \\ 6.70505{\cdot}X_4X_6 + 2.95825{\cdot}X_5^2 + 1.68306{\cdot}X_5X_6 - 2.19536{\cdot}X_6^2 \end{gathered} \tag{2}$$

Both mathematical models can be reduced by omitting the insignificant terms (p-value > 0.05). The Equations (3) and (4) of the two reduced models were expressed as follows:

$$Y_{TA}\ (\mathrm{mg{\cdot}g^{-1}}) = 23.3422 + 4.80026{\cdot}X_1 - 6.60698{\cdot}X_1X_5, \tag{3}$$

$$Y_{TP}\ (\text{mg}\cdot\text{g}^{-1}) = 45.8533 - 4.17338\cdot X_4 - 8.06168\cdot X_{12} + 6.75029\cdot X_1X_2 + 6.70505\cdot X_4X_6 \quad (4)$$

The trends outlined above were recorded in three-dimensional (3D) surface plots using the fitted model in order to improve our understanding of both, the main and the interaction effects, of the most influential parameters. The combined effects of cycle-methanol, methanol–temperature and cycle-ratio on the total phenolic compounds recovery are represented in Figure 2a–c. The combined effect of solvent composition and pH on the total anthocyanins recovery is represented in Figure 2b.

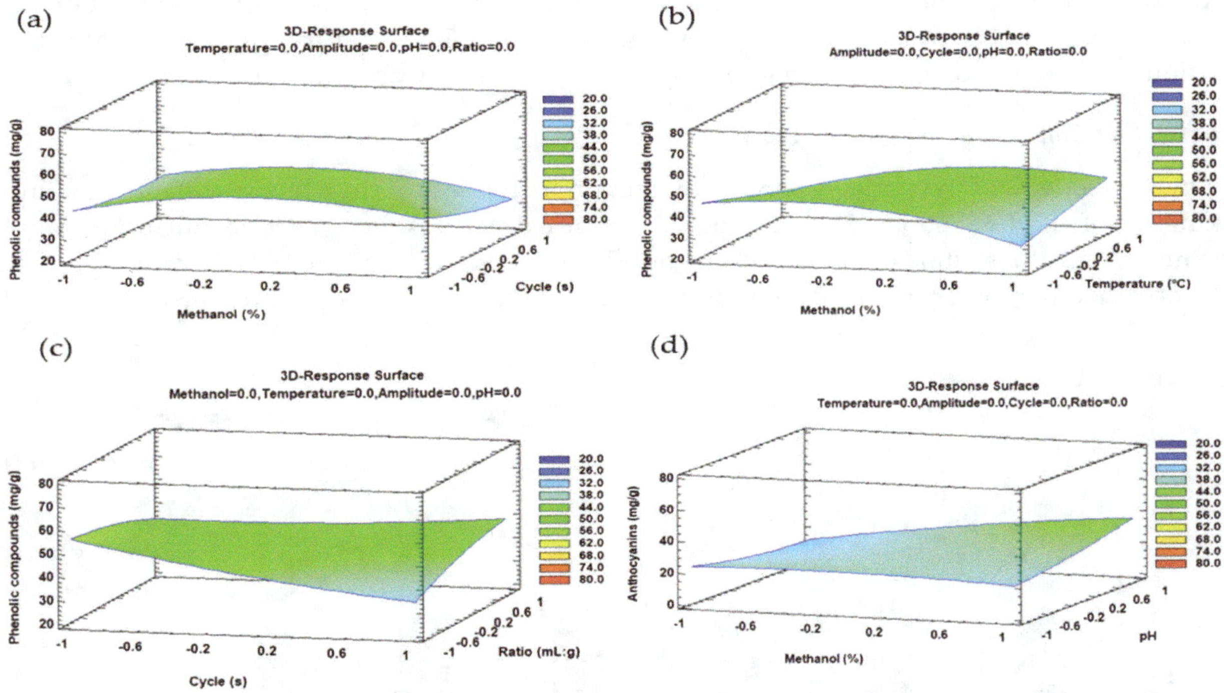

Figure 2. 3D-surface plots of the Box–Behnken design to represent the influence of: (**a**) Solvent composition and cycle on the total phenolic compounds; (**b**) solvent composition-temperature on the total phenolic compounds; (**c**) cycle-ratio on the total phenolic compounds; (**d**) solvent composition-pH on the total anthocyanins.

2.2. Optimal Conditions

According to the experimental design, the ideal UAE conditions to extract the phenolic compounds were as follows: 92.8% methanol in water as a solvent, 60 °C extraction temperature, 65.48% ultrasound amplitude, 0.2 s cycles, pH 6.8, and 10:0.5 mL:g solvent-sample ratio. With regard to the temperature, no higher temperatures were verified, since they might imply a greater degradation of the compounds of interest and a high loss of methanol that would affect the solvent-sample ratio [39]. With respect to pH, an almost neutral value was determined as optimal, since different research shows that acidified solvents may enhance the formation of free radicals in aqueous solutions because of their higher concentration of H^+ or thermal treatment [32], which would hinder the recovery of the phenolic compounds [35].

With regards to anthocyanins, optimum UAE conditions were as follows: 74.1% methanol in water solvent, 10 °C extraction temperature, 30% ultrasound amplitude, 0.3 s cycles, pH 7, and 18:0.5 mL:g as the solvent-sample ratio. With respect to temperature, the lowest end of the range studied (10 °C) was determined as the optimum value. Although anthocyanins are also phenolic compounds, they are more thermally sensitive than other phenolic compounds. High temperatures can diminish the recovery of anthocyanins due mainly to oxidation, cleavage of covalent bonds or an increase in oxidation reactions as a result of the thermal treatment [40]. With respect to the solvent pH, neutral pH was found to be optimum for the extraction of anthocyanins. Although pH between 1 and 3 usually generates stable conformation for anthocyanins, there are many articles in the literature where the highest extraction yields take place with a higher pH (3–7) [34]. This behavior might be the effect of different factors on anthocyanins stability (light, temperature, extraction time, etc.), which may turn them into other compounds [41]. Specifically, some authors affirm that ultrasound can promote the degradation of the anthocyanins because of the radical hydroxyl ($OH^{\bullet}$) and *hydrogen peroxide* (H_2O_2) produced inside the cavitation bubbles when subjected to conditions, such as high ultrasonic power, high amplitude, low temperature and long treatment time [42]. No higher pH was checked for anthocyanins since this may cause unstable structures as a result of basic hydrolysis [43].

In conclusion, for both, phenolic compounds and anthocyanins, maximum extractions were obtained when the solvent had a high percentage of methanol and neutral pH. Specifically, for the extraction of phenolic compounds a higher range of percentages was required.

2.3. Extraction Time

Once the effects of the variables on the extraction methods and the optimal values were known, the kinetics of the extractions was studied. Several extractions were carried out under optimal ultrasound conditions while extraction time varied between 2, 5, 10, 15, 20, and 25 min. The average results obtained (n = 3) for phenolic compounds and for anthocyanins are represented in Figure 3.

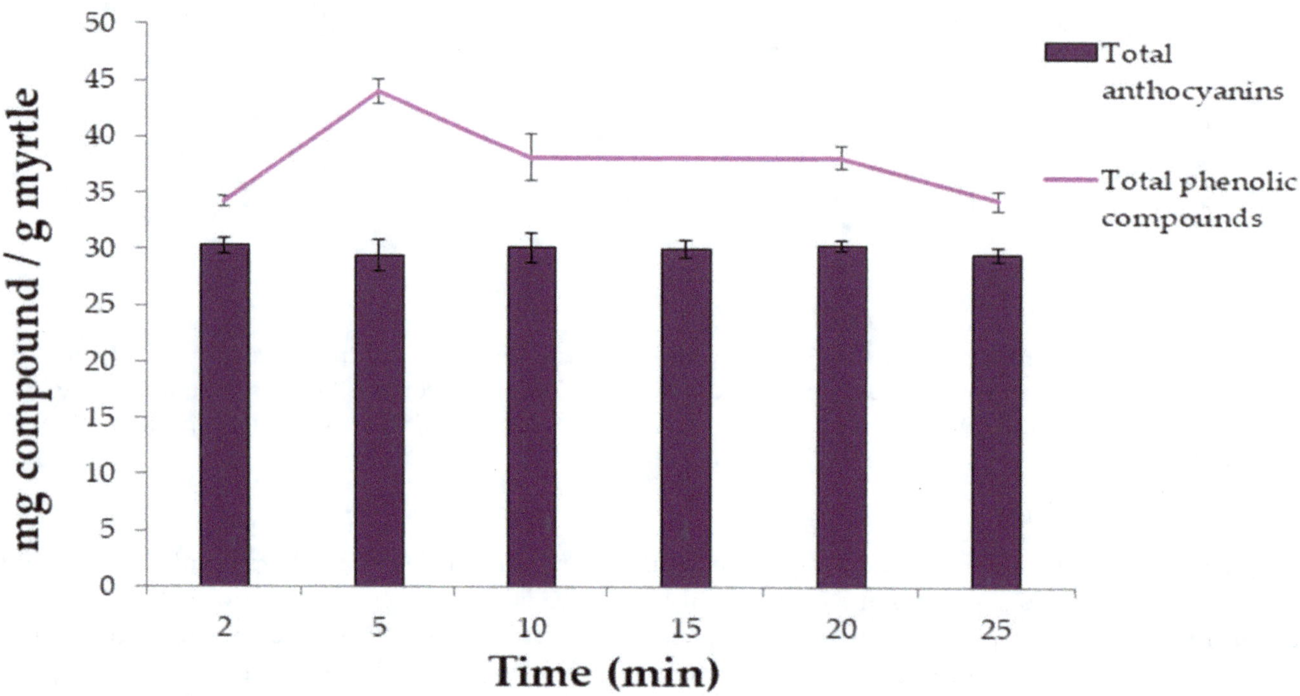

Figure 3. Recovery of anthocyanins ($mg{\cdot}g^{-1}$) and total phenolic compounds ($mg{\cdot}g^{-1}$) using different extraction times (n = 3).

It can be seen that large recoveries are achieved for both types of bioactive compounds and that long extraction times are not required. The Phenolic compounds present their maximum extraction at 5 min. However, longer extraction times led to lower recoveries, probably due to degradation of the phenolic compounds [25]. With respect to the anthocyanins, 2 min was determined as their optimum extraction time, since it exhibited the same yields as with longer times, while saving both, time and costs.

2.4. Repeatability and Intermediate Precision of UAE Methods

The precision of the extraction methods was evaluated in terms of repeatability (intra-day) and intermediate precision (inter-day). Repeatability was evaluated by performing 10 extractions under the same conditions on the same day. Intermediate precision was evaluated by performing 10 additional extractions on each one of the following two days. Altogether, 30 extractions were carried out under optimal extraction conditions to evaluate the precision of the extraction method for phenolic compounds and for anthocyanins. This method is employed in numerous studies [25,44]. The results were expressed by the coefficient of variation (CV) of the means. The repeatability results obtained were: 2.95% for phenolic compounds and 2.23% for anthocyanins. The intermediate precision results were: 4.66% for phenolic compounds and 4.15% for anthocyanins. As it can be seen, all the results are within acceptable limits (±10%) according to AOAC [45] and supported the accuracy—with diversions lower than 5.0%—of the extraction methods for total anthocyanins and total phenolic compounds.

2.5. Application of the Developed Methods to Ecotypes from Two Locations

In order to determine the applicability of the developed methods, once they had been optimized, they were applied to a new set of samples. Specifically, 14 ecotypes of myrtle were evaluated. 8 ecotypes from Puerto Real region (My-1, My-2, My-3, My-4, My-5, My-6, My -7, and My-8) and 6 from San José del Valle region (My-9, My-10, My-11, My-12, My-13, and My-14). The phenolic compounds were extracted from the 14 ecotypes in duplicate by applying the UAE method according to the optimum conditions previously determined. This should ensure the greatest possible yields. The quantification of the total phenolic compounds content in the extracts was carried out by Folin-Ciocalteau reagent.

The anthocyanins compounds were also extracted from the 14 samples according to the optimal conditions determined for the developed UAE method. The anthocyanins content in the extracts was quantified by UHPLC-UV-vis. The total anthocyanins content is the result of adding up each separate anthocyanin content. The average extraction and quantification results are shown in Table 4.

2.6. Analysis by Conglomerate

As a consequence of the chemical results obtained, it can be observed that there are differences between the average values obtained from the different myrtle ecotypes. To objectively study if these visual differences are related to the origin of the ecotypes, a comparative chemometric study was carried out using all the average values. Specifically, the data matrix $D_{13 \times 14}$ ($D_{variables \times ecotypes}$) (Table 4) was evaluated using an exploratory tool, i.e., Hierarchical Cluster Analysis (HCA). Ward's method and square Euclidean distance were employed and the variables for the differentiation were: The total phenolic compounds ($mg \cdot g^{-1}$) from each experiment; each individual anthocyanin content, 11 anthocyanins ($mg \cdot g^{-1}$), and total anthocyanins ($mg \cdot g^{-1}$). Myrtle is a shrub that grows better in warm and humid areas, and requires rich, humid soils. These characteristics match those of Puerto Real, which is near the sea with humid and sandy soils. San José del Valle has drier climate conditions and its clay soil is not so fertile [46,47].

These differences may lead to differences between the maturation processes of the autochthonous ecotypes in Puerto Real and San José del Valle, and consequently, to variations in their bioactive composition. The results of the analysis are graphically represented as a dendrogram in Figure 4. An obvious differentiation of the samples into two groups can be observed: Cluster A, includes only the ecotypes from Puerto Real, and Cluster B, only includes the ecotypes from San José del Valle. Therefore, based on their tendency to fall into a particular group in accordance with their origin, it can be said that phenolic compounds and anthocyanins contents in each ecotype is related to the berries' geographical area of origin. Specifically, the ecotypes in Cluster A, Puerto Real, present a total phenolic compounds and anthocyanins content greater than the ecotypes in Cluster B, from San José del Valle. The differences can be attributed to the different climatic and soil conditions above mentioned.

Table 4. Extract concentrations (mg·g^{-1}) of total anthocyanins and total phenolic compounds (n = 3) obtained from different myrtle ecotypes by means of ultrasound-assisted method (UAE).

Compounds [1]	Myrtle ecotypes from Puerto Real								Myrtle ecotypes from San José del Valle					
	MY-1	MY-2	MY-3	MY-4	MY-5	MY-6	MY-7	MY-8	MY-9	MY-10	MY-11	MY-12	MY-13	MY-14
D-3,5-diGl	0.166 ± 0.006	0.415 ± 0.016	0.404 ± 0.015	0.504 ± 0.020	0.377 ± 0.014	0.375 ± 0.016	0.183 ± 0.008	0.103 ± 0.0034	0.127 ± 0.004	0.065 ± 0.002	0.079 ± 0.0023	0.176 ± 0.0067	0.113 ± 0.003	0.134 ± 0.004
Del-3-Glu	10.552 ± 0.161	16.309 ± 0.534	10.305 ± 0.358	11.114 ± 0.384	12.557 ± 0.481	12.258 ± 0.432	8.315 ± 0.326	5.929 ± 0.165	3.667 ± 0.152	1.566 ± 0.008	2.145 ± 0.084	3.823 ± 0.124	2.987 ± 0.121	3.346 ± 0.135
Cy-3-Ga	0.150 ± 0.006	0.382 ± 0.008	0.216 ± 0.009	0.359 ± 0.007	0.319 ± 0.012	0.371 ± 0.0135	0.176 ± 0.007	0.392 ± 0.012	0.050 ± 0.001	0.052 ± 0.002	0.068 ± 0.0018	0.1757 ± 0.005	0.077 ± 0.002	0.036 ± 0.001
Cy-3-Gl	1.794 ± 0.084	3.004 ± 0.15	1.126 ± 0.023	1.241 ± 0.036	1.613 ± 0.03	1.313 ± 0.0578	1.946 ± 0.034	0.791 ± 0.043	0.852 ± 0.024	0.293 ± 0.01	0.301 ± 0.016	0.512 ± 0.025	0.452 ± 0.013	0.786 ± 0.032
Cy-3-Ar	0.092 ± 0.003	0.126 ± 0.002	0.076 ± 0.002	0.130 ± 0.002	0.112 ± 0.004	0.170 ± 0.006	0.061 ± 0.14	0.081 ± 0.0025	0.170 ± 0.003	0.040 ± 0.001	0.410 ± 0.021	0.657 ± 0.027	0.549 ± 0.23	0.188 ± 0.006
Pet-3-Gl	6.217 ± 0.110	11.007 ± 0.38	7.173 ± 0.314	8.691 ± 0.301	8.329 ± 0.297	10.242 ± 0.392	5.538 ± 0.0423	5.082 ± 0.176	6.810 ± 0.235	1.580 ± 0.018	1.676 ± 0.065	2.123 ± 0.085	1.823 ± 0.067	6.643 ± 0.114
Del-3-Ara	0.840 ± 0.005	2.378 ± 0.102	1.635 ± 0.062	1.775 ± 0.067	1.917 ± 0.064	2.009 ± 0.081	1.335 ± 0.021	1.562 ± 0.035	1.833 ± 0.076	0.367 ± 0.006	0.324 ± 0.009	0.679 ± 0.021	0.454 ± 0.014	1.234 ± 0.033
Peo-3-Gl	1.030 ± 0.017	1.179 ± 0.043	0.659 ± 0.019	0.689 ± 0.021	0.610 ± 0.023	0.816 ± 0.029	0.755 ± 0.34	0.481 ± 0.019	0.473 ± 0.015	0.305 ± 0.017	0.357 ± 0.012	0.921 ± 0.032	0.564 ± 0.013	0.446 ± 0.012
Mal-3-Gl	9.834 ± 0.314	16.691 ± 0.641	16.050 ± 0.65	18.301 ± 0.614	12.748 ± 0.768	23.817 ± 1.032	8.244 ± 0.273	9.072 ± 0.348	6.987 ± 0.246	5.195 ± 0.068	5.453 ± 0.185	7.679 ± 0.23	6.456 ± 0.241	5.979 ± 0.263
Pet-3-Ar	0.269 ± 0.009	0.551 ± 0.034	0.448 ± 0.014	0.587 ± 0.013	0.482 ± 0.028	0.751 ± 0.0013	0.332 ± 0.012	0.583 ± 0.015	0.438 ± 0.017	0.197 ± 0.002	0.206 ± 0.008	0.657 ± 0.023	0.454 ± 0.012	0.446 ± 0.012
Mal-3-Ar	0.226 ± 0.004	0.302 ± 0.009	0.402 ± 0.012	0.436 ± 0.121	0.296 ± 0.015	0.598 ± 0.019	0.240 ± 0.009	0.395 ± 0.014	0.185 ± 0.005	0.221 ± 0.001	0.244 ± 0.012	0.846 ± 0.032	0.679 ± 0.031	0.165 ± 0.006
Total anthocyanins	31.170 ± 1.023	52.346 ± 1.124	38.493 ± 3.516	43.827 ± 2.557	43.274 ± 1.904	60.252 ± 0.002	35.207 ± 1.175	32.903 ± 1.162	21.591 ± 0.92	10.381 ± 0.037	11.263 ± 0.431	18.249 ± 2.30	14.608 ± 1.964	19.403 ± 1.97
Total Phenolic compounds	70.747 ± 1.433	86.439 ± 3.125	64.7586 ± 0.914	60.034 ± 0.44	72.155 ± 3.174	76.124 ± 1.170	76.070 ± 3.417	69.815 ± 2.947	49.824 ± 0.362	52.66 ± 0.038	45.789 ± 1.824	51.278 ± 2.051	49.103 ± 0.192	50.897 ± 1.564

[1] Del-3,5-diGl, delphinidin 3,5-*O*-diglucoside; Del-3-Glu, delphinidin 3-*O*-glucoside; Cy-3-Ga, cyanidin 3-*O*-galactoside; Cy-3-Gl, cyanidin 3-*O*-glucoside; Cy-3-Ar, cyanidin 3-*O*-arabinoside; Pet-3-Gl, petunidin 3-*O*-glucoside; Del-3-Ara, delphinidin 3-*O*-arabinoside; Peo-3-Gl, peonidin 3-*O*-glucoside; Mal-3-Gl, malvidin 3-*O*-glucoside; Pet-3-Ar, petunidin 3-*O*-arabinoside; Mal-3-Ar, malvidin 3-*O*-arabinoside.

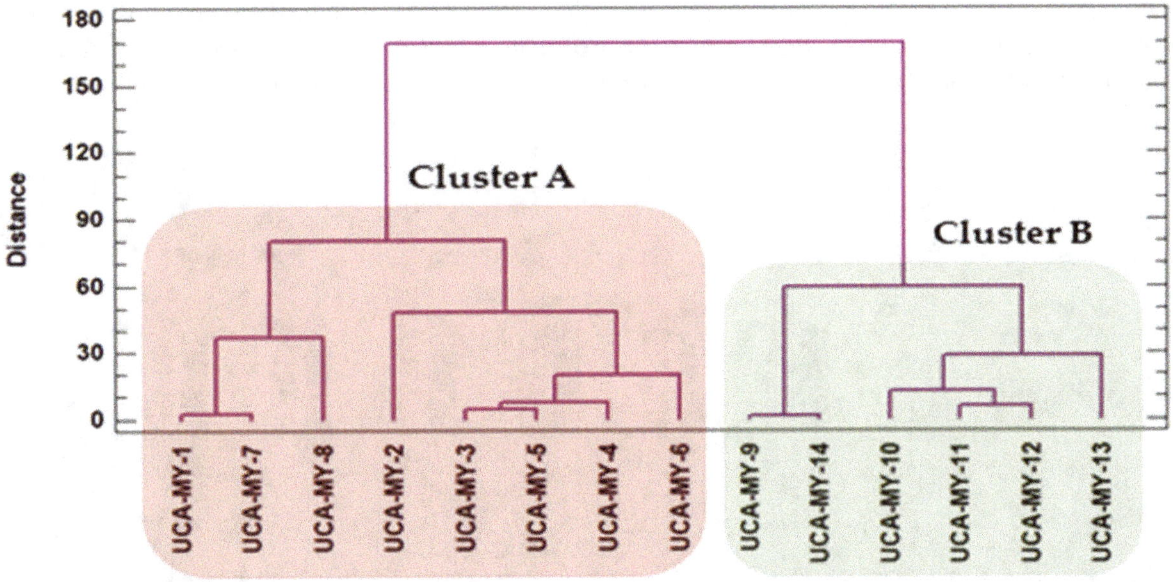

Figure 4. Dendrogram representing the bioactive compounds in 14 myrtle ecotypes according to the matrix of the results from Hierarchical Cluster Analysis (HCA) (Table 3).

2.7. Comparison Study: UAE vs. MAE

As above mentioned, the greatest phenolic compounds and anthocyanins yields using UAE were obtained under the following optimal conditions: methanol:water solvent ratio 92.8% v/v at 60 °C for phenolic compounds and methanol:water solvent ratio 74.1% v/v at 10 °C for anthocyanins.

In comparison with traditionally used methods for the extraction of the phenolic compounds in myrtle berries, UAE achieves a greater recovery of the compounds of interest, while using less solvent and in a shorter time, with the consequent cost reduction [11,17]. This increased effectiveness with greater extraction yields of both, total phenolic compounds and total anthocyanins, could be based on the phenomenon of cavitation, which breaks cell walls and releases the compounds of interest from the myrtle berries' matrices.

With the purpose of rounding up this study, the results obtained using UAE were compared to those achieved by MAE in previous work. For that purpose, the same number of samples were run under optimum conditions and later on analyzed [22]. The total phenolic compounds and total anthocyanins content extracted from myrtle berries at different times using UAE and MAE are shown in Figure 5.

With respect to the phenolic compounds (Figure 5a), a similar trend is observed in both extraction methods. The phenolic compounds yield increases until the maximum extraction value is reached at 5 min for UAE and at 15 min for MAE. From then on, the quantity of the extract begins to decrease. The optimum time for UAE, 5 min, indicates that UAE degrades phenolic compounds faster than MAE and the recovery is also lower. When compared to UAE, as recently reported by Ghafoor et al. [48,49], MAE obtains extracts with a substantially greater content of phenolic compounds than the one obtained by UAE.

With respect to the anthocyanins (Figure 5b), their content levels are very similar in MAE and UAE extracts. In addition, the extraction time required to achieve good yields (2 min) is low for both methods, which would considerably reduce costs when operating at an industrial scale. At the optimal time of two minutes, the anthocyanins extraction is slightly higher when MAE is used. However, at longer times, MAE extracts have lower anthocyanins content [50]. MAE optimal operating temperature (50 °C) makes anthocyanins, thermally labile, begin to degrade before.

Additionally, both techniques were applied to different myrtle ecotypes (Figure 6). As above noted, MAE stands out as a more efficient method for the extraction of the phenolic compounds in myrtle berries. With respect to anthocyanins, although some particular ecotypes produced greater yields by MAE, most of them produced greater yields when UAE was employed. Anthocyanins are extremely susceptible to degradation and the combination of high pressure and temperature that is employed for MAE would enhance such degradation and affect negatively their recovery.

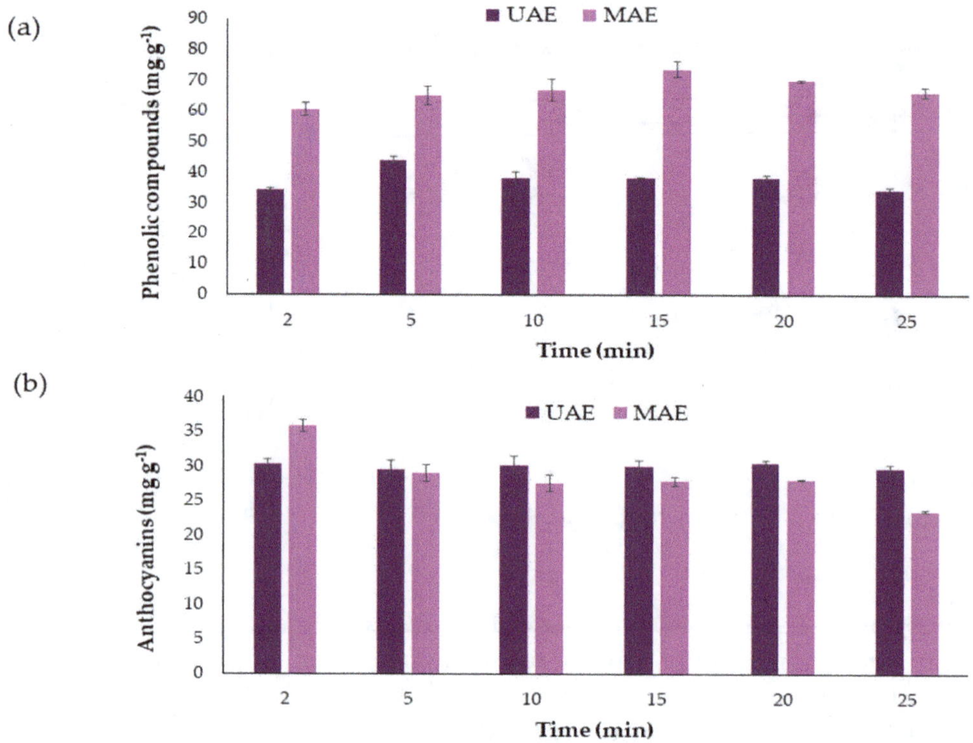

Figure 5. Comparison between the implementation of UAE and microwave-assisted extraction (MAE) to myrtle berries: (**a**) To extract total phenolic compounds; (**b**) to extract anthocyanins.

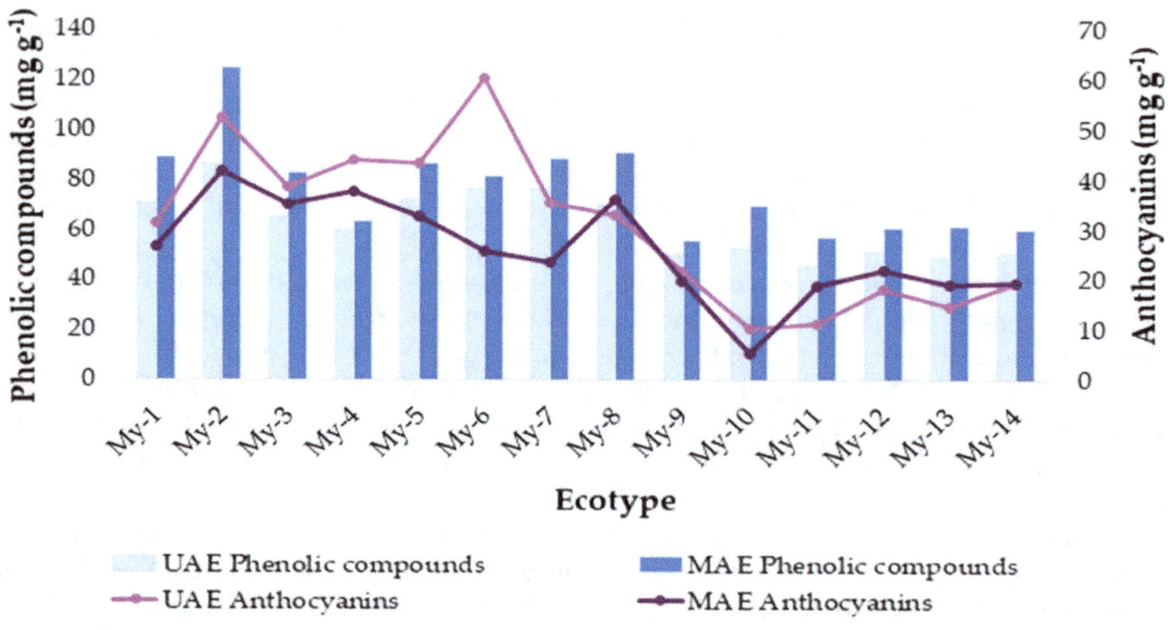

Figure 6. Concentration of total phenolic compounds and total anthocyanins (n = 3) in UAE and MAE extracts from several myrtle berries ecotypes.

Therefore, the UAE should be seriously considered as the preferred method for the extraction of the anthocyanins in myrtle berries.

3. Materials and Methods

3.1. Plant Materials

The biological materials used for this study were different myrtle berries (14 ecotypes) collected by the authors from two different areas in Cadiz province during their optimum ripeness stages in 2016. The first collection area was Puerto Real (eight ecotypes). This area is characterized by its humid climate due to its proximity to the sea. The second collection area was San José del Valle (6 ecotypes), also within Cadiz province, but located inland at 50 km from the coast. This location has a drier climate and its soils have a lower water content. The guidelines described by M., Mulas and M.R., Cani [47] were applied to characterize the morphology of both, leaves and berries, to confirm that the samples had been collected from different ecotypes. The samples were subjected to a pretreatment to improve the contact surface with the solvent [51]. First, the seeds were separated from the pulp. Secondly, the pulp was lyophilized in a Virtis Benchtop K freeze dryer (SP Cientific, New York, United States) and crushed by means of a regular spice grinder. Finally, the samples were stored in a freezer at −20 °C prior to analysis.

3.2. Chemicals and Solvents

The solvents used for the extraction were a mix of methanol and water at different concentration levels and with different pH. The methanol (Fischer Scientifics, Loughborough, United Kingdom) was HPLC grade. Ultra-pure water was obtained from a Milli-Q water purification system (EMD Millipore Corporation, Bedford, MA, United States). The pH adjustment of the solvents was done by means of hydrochloric acid and sodium hydroxide, both analytical grade and purchased from Panreac Química S.A.U. (Castellar del Valles, Barcelona, Spain). For the Folin-Ciocalteau spectrophotometric method, anhydrous sodium carbonate (Panreac Química S.A.U., Castellar del Valles) and Folin–Ciocalteu (Merck KGaA, EMD Millipore Corporation, Darmstadt, Germany) were employed. For the HPLC analyses, methanol (Fischer Scientific, Loughborough, United Kingdom) and formic acid (Scharlau, Barcelona, Spain) were used. These solvents were degassed and filtered through a 0.22 μm membrane (Nylon Membrane Filter, FILTER-LAB, Barcelona, Spain) before being used. The standard for the phenolic compounds was gallic acid and the standard for anthocyanins was cyanidin chloride. Both standards were purchased from Sigma-Aldrich Chemical Co. (St. Louis, MO, USA).

3.3. Ultrasound-Assisted Extraction Procedure

To extract the total phenolic compounds and the total anthocyanins from the myrtle berries, UAE was used. A UP200S probe (Hielscher Ultrasound Technology, Berlin, Germany) was employed, coupled to a processor that allows adjusting the amplitude and the cycle. For the adjustment of the temperature, a thermostatic bath (Frigiterm-10, Selecta, Barcelona, Spain) was employed. The temperature, the cycle, and the amplitude were selected for each extraction according to the experiment. About 0.5 g of the lyophilized and homogenized sample was weighed in a Falcon tube and the corresponding volume of solvent was added depending on the experiment. The Falcon tube was placed in a double vessel through which the water from the thermostatic bath circulated. The initial extraction time set was 10 min, followed by a sample cooling time. After that time, the extracts were centrifuged (7500 rpm, 5 min) and the supernatants were placed in a 25 mL volumetric flask. The precipitates from the extraction were redissolved in 5 mL of the same extraction solvent and centrifuged again under the same conditions. The new supernatants were placed in the volumetric flask and it was completed with the same solvent. The final extracts were stored at −20 °C for their correct conservation until further analysis. The UAE conditions set for the extractions were: Solvent composition (50–100% methanol in water for phenolic compounds and 25–75% for anthocyanins), temperature (10–60 °C), amplitude (30–70%), cycle (0.2–0.7 s), pH (2–7) and solvent-sample ratio (10:0.5–20:0.5 mL:g).

3.4. Determining the Content of Total Phenolic Compounds by Folin-Ciocalteau Essay

The total phenolic compounds content in myrtle berry was determined by adapted/modified Folin-Ciocalteau (FC) method [52]. This method has been previously used by many researchers to determine the total phenolic compounds content [53,54]. It is based on a redox reaction in a basic medium that gives rise to a complex of blue coloration with a wide absorption up to 765 nm. The extracts were filtered using a 0.45 μm nylon syringe filter (Membrane Solutions, Dallas, United States). The protocol of the method is the following: 250 μL of the previously filtered extract was transferred to a 25 mL volumetric flask. After this, 12.5 mL of water, 1.25 mL of Folin-Ciocalteau reagent and 5 mL of a 20% aqueous sodium carbonate solution were added. Finally, the flask was made up with water, and after 30 min the absorbance was measured at the maximum. All the extracts were analyzed in duplicate. The range of absorbance obtained for the studied samples was 0.4-1.4. The equipment used to measure the absorbance was a Helios Gamma (γ) Unicam UV-vis Spectrophotometer (Thermo Fisher Scientific, Waltham, MA, United States). The calibrated curve for the quantification was constructed based on a reference standard gallic acid pattern under the same conditions as the extracts [55]. The computer application used to process the data was Microsoft Office Excel 2013. The following regression equation $y = 0.0010x + 0.0059$ and the following correlation coefficient $R^2 = 0.9999$ were obtained. The linear range of work was 100–2600 mg L^{-1}. The results are expressed in milligrams of gallic acid equivalent per gram of lyophilized weight.

3.5. Identification of Anthocyanins by UHPLC-QToF-MS

Ultra-high performance liquid chromatography (UHPLC) coupled to quadrupole-time-of-flight mass spectrometry (QToF-MS) (Xevo G2 QToF, Waters Corp., Milford, MA, United States) was used to identify the anthocyanins in the UAE extracts. The column employed was a reverse-phase C18 analytical column with 1.7 μm particle size, 2.1 mm × 100 mm (ACQUITY UPLC CSH C18, Waters). The mobile phase was 2% formic acid–water solution (phase A) and methanol solution (phase B). The studied bioactive compounds were determined by employing the UHPLC-QToF-MS method described in a previously research [22]. The individual anthocyanins were identified based on their retention time and molecular weight. The following eleven anthocyanins were identified in the samples: delphinidin 3,5-*O*-diglucoside ($m/z = 627.1561$), delphinidin 3-*O*-glucoside ($m/z = 465.1033$), cyanidin 3-*O*-galactoside ($m/z = 449.1084$), cyanidin 3-*O*-glucoside ($m/z = 449.1084$), cyanidin 3-*O*-arabinoside ($m/z = 419.0978$), petunidin 3-*O*-glucoside ($m/z = 479.1189$), delphinidin 3-*O*-arabinoside ($m/z = 435.0927$), peonidin 3-*O*-glucoside ($m/z = 463.1240$), malvidin 3-*O*-glucoside ($m/z = 493.1346$), petunidin 3-*O*-arabinoside ($m/z = 449.1084$) and malvidin 3-*O*-arabinoside ($m/z = 463.1240$). Before their identification, all the UAE extracts were filtered through a 0.20 μm nylon syringe filter (Membrane Solutions, Dallas, TX, United States). The anthocyanins structures are shown in Figure 7.

3.6. Determination of Anthocyanins by UHPLC-UV-Vis System

For the separation and quantification of the anthocyanins present in UAE extracts from myrtle berries, an Elite UHPLC LaChrom Ultra System (Hitachi, Tokyo, Japan) was used. The UHPLC system consists of an L-2420U UV-Vis detector, an L-2200U autosampler, an L-2300 column oven set at 50 °C and two L-2160 U pumps. The column used was a "Fused Core" C18 with 2.6 μm particle size, 2.1 mm × 100 mm (Phenomenex Kinetex, Torrance, CA, United States). The mobile phase consisted of a 5% formic acid–water solution (phase A) and a methanol solution (phase B). The studied bioactive compounds were determined by employing the UHPLC-UV-Vis method described in previous research [22]. Before their analysis, all the UAE extracts were filtered through a 0.20 μm nylon syringe filter (Membrane Solutions, Dallas, TX, United States) and diluted in Milli-Q water. The individual anthocyanins present in myrtle extracts were quantified in cyanidin equivalents by means of a regression curve of anthocyanidin standard, cyanidin chloride

($y = 252640.4136x - 28462.4337$; $R^2 = 0.9999$). The standards with a known concentration were prepared between 0.06 and 35 mg·L^{-1}. The limit of detection (LOD) (0.196 mg·L^{-1}) and the limit of quantification (LOQ) (0.653 mg·L^{-1}) were calculated as three and ten times respective to the standard deviation of the blank divided by the slope of the calibration curve. Assuming that the 11 anthocyanins have similar absorbance, and taking into account the molecular weight of each anthocyanin, a calibration curve was plotted for each anthocyanin present in myrtle, which allowed to quantify the compounds of interest. All the analyses were carried out in duplicate. Figure 8 shows the HPLC chromatogram that represents the eleven anthocyanins detected in the analyses.

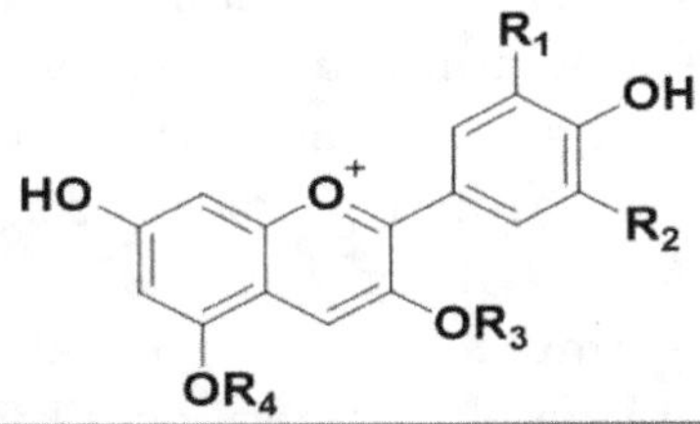

Anthocyanins present in myrtle berries	R1	R2	R3	R4
Delphinidin 3,5-*O*-diglucoside	OH	OH	$C_6H_{11}0_5$	$C_6H_{11}0_5$
Delphinidin 3-*O*-glucoside	OH	OH	$C_6H_{11}0_5$	H
Cyanidin 3-*O*-galactoside	OH	H	$C_6H_{11}0_5$	H
Cyanidin 3-*O*-glucoside	OH	H	$C_6H_{11}0_5$	H
Cyanidin 3-*O*-arabinoside	OH	H	$C_5H_90_4$	H
Petunidin 3-*O*-glucoside	OH	H	$C_{12}H_{21}0_9$	H
Delphinidin 3-*O*-arabinoside	OH	OH	$C_5H_90_4$	H
Peonidin 3-*O*-glucoside	OCH_3	H	$C_6H_{11}0_5$	H
Malvidin 3-*O*-glucoside	OCH_3	OCH_3	$C_6H_{11}0_5$	H
Petunidin 3-*O*-arabinoside	OCH_3	OH	$C_5H_90_4$	H
Malvidin 3-*O*-arabinoside	OCH_3	OCH_3	$C_5H_90_4$	H

Figure 7. Radicals of the different anthocyanins present in myrtle berries.

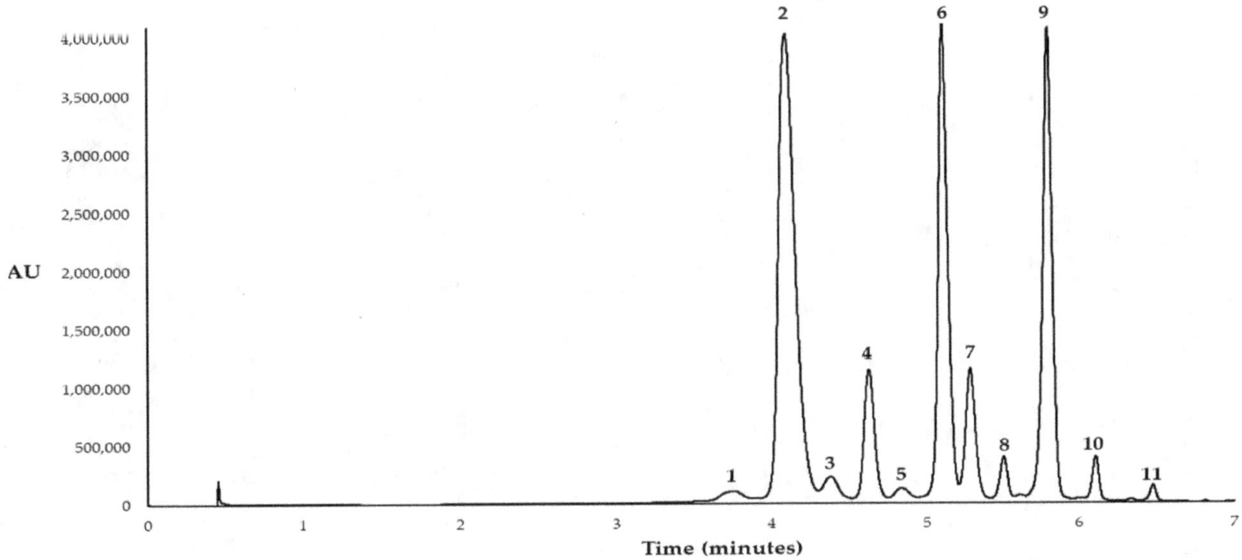

Figure 8. Chromatogram of the 11 anthocyanins identified in the UAE extracts from myrtle berries. Peak assignment: (1) Delphinidin 3,5-*O*-diglucoside; (2) delphinidin 3-*O*-glucoside; (3) cyanidin 3-*O*-galactoside; (4) cyanidin 3-*O*-glucoside; (5) cyanidin 3-*O*-arabinoside; (6) petunidin 3-*O*-glucoside; (7) delphinidin 3-*O*-arabinoside; (8) peonidin 3-*O*-glucoside; (9) malvidin 3-*O*-glucoside; (10) petunidin 3-*O*-arabinoside; (11) malvidin 3-*O*-arabinoside.

3.7. Application of Box-Behnken Design (BBD) to the Optimization of the Extraction Methods

In order to optimize the extraction variables, a response surface experiment (RSM) known as Box-Behnken (BBD) was carried out [56]. Box-Behnken design (BBD) is an independent rotatable quadratic design with no embedded factorial or fractional factorial points. The variable combinations are at the midpoint of the edges and at the center of the space [57]. It is useful because it allows one to avoid carrying out experiments under extreme conditions and, therefore, the possibility of deceiving results [58]. When this statistical experiment design is employed in conjunction with a response surface methodology (RSM) the effects of six independent factors on each response can be studied. The independent factors studied were: Solvent composition (% methanol in water) (X_1), solvent pH (X_2), extraction temperature (X_3), ultrasound amplitude (X_4), cycle (X_5), and a solvent-sample ratio (X_6). For each variable, there are three levels, coded as –1 (low), 0 (central point or middle), and +1 (high). Specifically, the studied ranges were as follows: Solvent composition: 50, 75, 100% for phenolic compounds and 25, 50, 75% for anthocyanins; temperature: 10, 35, 60 °C; amplitude: 30, 50, 70%; cycle: 0.2, 0.45, 0.7 s; pH: 2, 4.5, 7 and solvent-sample ratio: 10:0.5, 15:0.5, 20:0.5 mL:g. The ranges for the study were selected taking into account previous experiences by the research team. The response variables studied were: The experimental results for total phenolic compounds (Y_{TP}, mg·g^{-1}) and the experimental results for total anthocyanins (Y_{TA}, mg·g^{-1}). The design consisted of 54 treatments with six repetitions at the center point. All the trials were performed in random order. The whole experimental design matrix used can be seen in Table 5. My-9 from San Jose del Valle was the myrtle sample used for the optimization procedure.

Table 5. Experimental and predicted values for total phenolic compounds and total anthocyanins contents based on Box–Behnken design.

Run	Factors						Responses			
	X_1	X_2	X_3	X_4	X_5	X_6	Y_{TP} (mg·g^{-1})		Y_{TA} (mg·g^{-1})	
							Experimental	Predicted	Experimental	Predicted
1	0	0	−1	0	−1	−1	55.4367	51.0831	20.7595	23.8541
2	0	0	1	0	−1	−1	48.6499	49.6621	29.8821	25.4294
3	0	0	−1	0	1	−1	51.4343	48.3304	29.0606	31.2985
4	0	0	1	0	1	−1	41.6667	42.334	27.9193	26.3493
5	0	0	−1	0	−1	1	46.1649	45.0747	27.6777	27.6226
6	0	0	1	0	−1	1	47.6107	51.1375	25.6501	25.0375
7	0	0	−1	0	1	1	50.4893	49.0542	28.4150	31.2426
8	0	0	1	0	1	1	45.7652	50.5417	23.6023	22.1329
9	0	−1	0	−1	−1	0	57.1168	57.3752	28.5256	24.7846
10	0	1	0	−1	−1	0	44.7735	48.3304	26.8602	29.2928
11	0	−1	0	1	−1	0	41.9739	53.0361	27.755	23.6113
12	0	1	0	1	−1	0	41.3703	46.5404	26.6112	23.4784
13	0	−1	0	−1	1	0	57.8351	55.2335	27.0548	34.781
14	0	1	0	−1	1	0	71.3186	57.688	27.1787	26.729
15	0	−1	0	1	1	0	41.3184	40.33	28.5542	30.715
16	0	1	0	1	1	0	48.1604	45.3335	18.8744	18.0219
17	−1	0	−1	−1	0	0	37.8811	41.5155	14.3354	19.5394
18	1	0	−1	−1	0	0	39.2436	41.1874	19.2532	42.8583
19	−1	0	1	−1	0	0	48.7908	48.0444	18.0553	17.6695
20	1	0	1	−1	0	0	43.5972	51.7521	21.1111	22.2834
21	−1	0	−1	1	0	0	54.6703	45.6802	17.6946	11.5099
22	1	0	−1	1	0	0	35.7746	37.3562	20.6987	26.097
23	−1	0	1	1	0	0	37.9611	35.1821	10.1680	24.5504
24	1	0	1	1	0	0	33.6931	30.8939	20.6241	20.4326
25	0	−1	−1	0	0	−1	39.6751	44.4043	25.5771	30.4887
26	0	1	−1	0	0	−1	42.5789	48.6886	26.0578	25.0639
27	0	−1	1	0	0	−1	43.7849	42.272	27.3046	29.0925
28	0	1	1	0	0	−1	42.0191	43.4037	24.5787	23.0862
29	0	−1	−1	0	0	1	47.4523	46.4907	27.6042	30.7219
30	0	1	−1	0	0	1	39.3819	41.3177	28.7060	28.5433
31	0	−1	1	0	0	1	58.3748	51.8422	25.7965	25.1653
32	0	1	1	0	0	1	48.6688	43.5167	28.9420	22.4052
33	−1	−1	0	0	−1	0	58.0268	57.5556	22.9897	19.5292
34	1	−1	0	0	−1	0	41.2425	32.9846	19.5373	27.3024

Table 5. *Cont.*

Run	Factors						Responses			
	X_1	X_2	X_3	X_4	X_5	X_6	Y_{TP} (mg·g^{-1})		Y_{TA} (mg·g^{-1})	
							Experimental	Predicted	Experimental	Predicted
35	−1	1	0	0	−1	0	38.5507	36.2848	24.6490	32.9035
36	1	1	0	0	−1	0	46.8626	38.7149	20.2517	18.3034
37	−1	−1	0	0	1	0	35.7902	41.3694	17.7105	15.0653
38	1	−1	0	0	1	0	34.6257	34.3231	17.7143	48.8664
39	−1	1	0	0	1	0	20.7714	31.5978	19.0511	15.8794
40	1	1	0	0	1	0	48.5130	51.5526	19.2534	27.3073
41	−1	0	0	−1	0	−1	49.3362	46.1577	18.9945	14.4979
42	1	0	0	−1	0	−1	49.0123	49.4397	21.8380	28.2914
43	−1	0	0	1	0	−1	30.0802	28.3987	16.72389	13.4627
44	1	0	0	1	0	−1	24.1848	23.6849	20.7428	18.5244
45	−1	0	0	−1	0	1	35.7748	35.4395	16.4343	13.6402
46	1	0	0	−1	0	1	33.0203	35.537	19.5057	27.7794
47	−1	0	0	1	0	1	44.0929	44.5007	14.9678	13.5268
48	1	0	0	1	0	1	34.2591	36.6024	19.4501	18.9343
49	0	0	0	0	0	0	37.2240	45.8533	27.1667	23.3422
50	0	0	0	0	0	0	52.1910	45.8533	27.2028	23.3422
51	0	0	0	0	0	0	47.5538	45.8533	17.7051	23.3422
52	0	0	0	0	0	0	50.5994	45.8533	14.0658	23.3422
53	0	0	0	0	0	0	43.5576	45.8533	26.5943	23.3422
54	0	0	0	0	0	0	43.9939	45.8533	27.3184	23.3422

The results for total phenolic compounds and total anthocyanins contents were entered into a polynomial equation. The response of the total phenolic compounds and the anthocyanins obtained in each of the experiments was entered into a second-order polynomial equation in order to correlate the relationship between the independent variables and the response (Equation (5)):

$$Y = \beta_0 + \sum_{i=1}^{k} \beta_i X_i + \beta_{ii} X_i^2 + \sum_{i} \sum_{i=1}^{k} \beta_{ij} X_i X_j + r \quad (5)$$

where Y is the predicted response (Y_{TP} and Y_{TA}); β_0 is the model constant; X_i and X_j are the independent variables; β_i are the linear coefficients; β_{ij} are the coefficients corresponding to the interactions; βii are the quadratic coefficients and r is the pure error sum of squares.

Design Expert software 11 (Trial Version, Stat Ease Inc., Minneapolis, MN, USA) was the software employed for experimental design, the data analysis, and the model building. The statistical significance of the model, lack of fit, and regression terms were evaluated based on the analysis of variance (ANOVA).

The results of applying the extraction method to different myrtle ecotypes were studied using a multivariate analysis, hierarchical clustering analysis (HCA). Ward's method and the Euclidean square distance, were employed. Statgraphic Centurion XVII (Statgraphics Technologies, Inc., The Plains, VA, United States) was the software used.

4. Conclusions

This work has successfully developed quick and effective methods to extract bioactive compounds, such as anthocyanins and total phenolics from myrtle (Myrtus communis L.) pulp. A thorough search in the relevant literature showed that this is the first study in which UAE has been optimized for the extraction of phenolic compounds and anthocyanins from myrtle berries. The following optimal UAE conditions have been determined to extract the phenolic compounds: 92.8% methanol in water, 6.8 pH, 60 °C temperature, 65.48% ultrasound amplitude, 0.2 s cycle, and 10:0.5 as the optimum solvent-sample ratio. With regards to anthocyanins, optimal UAE conditions were: 74.1% methanol in water, 7 pH, 10 °C temperature, 30% ultrasound amplitude, 0.3 s cycle, and 18:0.5 as the optimum solvent-sample ratio. The optimum extraction times were only 5 and 2 min for phenolic compounds and anthocyanins, respectively. Both extraction methods presented satisfactory intra-day repeatability and inter-day repeatability (CV < 5%). The methods were applied to 14 different myrtle ecotypes.

The hierarchical cluster analysis (HCA), showed a correlation between the bioactive composition (total phenolic compounds and total and individual anthocyanins contents in the extracts) and the ecotypes' geographical area of origin. In conclusion, the results have indicated that UAE is a feasible alternative to conventional methods for the extraction of valuable components from myrtle berries. These results would mean a substantial improvement at the industrial level, since they would allow the manufacturers to quickly determine the quality of the raw materials and save costs. Furthermore, UAE results were compared to those achieved by MAE. The proposed UAE method proved to be an effective procedure to extract the bioactive compounds in myrtle berries, and a particularly efficient alternative for the extraction of anthocyanins.

Author Contributions: Conceptualization, E.E.-B. and G.F.B.; methodology, A.V.G.d.P. and M.V.-E.; software, M.F.-G.; formal analysis, A.V.G.d.P., M.V.-E and A.A.-A.; investigation, A.V.G.d.P. and M.V.-E.; resources, A.J.-C. and M.P.; data curation, E.E.-B., M.F.-G. and G.F.B.; writing—original draft preparation, A.V.G.d.P. and M.V.-E.; writing—review and editing, G.F.B. and E.E.-B.; supervision, E.E.-B. and G.F.B.; project administration, G.F.B. and E.E.-B.

References

1. Tian, Y.; Puganen, A.; Alakomi, H.L.; Uusitupa, A.; Saarela, M.; Yang, B. Antioxidative and antibacterial activities of aqueous ethanol extracts of berries, leaves, and branches of berry plants. *Food Res. Int.* **2018**, *106*, 291–303. [CrossRef] [PubMed]
2. Li, D.; Ye, Q.; Jiang, L.; Luo, Z. Effects of nano-TiO_2 packaging on postharvest quality and antioxidant activity of strawberry (Fragaria × ananassa Duch.) stored at low temperature. *J. Sci. Food Agric.* **2017**, *97*, 1116–1123. [CrossRef] [PubMed]
3. Łata, B.; Trampczynska, A.; Paczesna, J. Cultivar variation in apple peel and whole fruit phenolic composition. *Sci. Hortic.* **2009**, *121*, 176–181. [CrossRef]
4. Ahmad, N.; Zuo, Y.; Lu, X.; Anwar, F.; Hammed, S. Anwar. Characterization of free and conjugated phenolic compounds in fruits of selected wild plants. *Food Chem.* **2016**, *190*, 80–89. [CrossRef] [PubMed]
5. Bamba, B.S.B.; Shi, J.; Tranchant, C.C.; Xue, S.J.; Forney, C.F.; Lim, L.-T. Influence of Extraction Conditions on Ultrasound-Assisted Recovery of Bioactive Phenolics from Blueberry Pomace and Their Antioxidand Activity. *Molecules* **2018**, *23*, 1685. [CrossRef] [PubMed]
6. Wannes, A.; Mhamdi, B.; Sriti, J.; Ben Jemia, M.; Ouchikh, O.; Hamdaoui, G.; Kchouk, M.E.; Marzouk, B. Antioxidant activities of the essential oils and methanol extracts from myrtle (*Myrtus communis* var. italica L.) leaf, stem and flower. *Food Chem. Toxicol.* **2010**, *48*, 1362–1370. [CrossRef] [PubMed]
7. Limwachiranon, J.; Jiang, L.; Huang, H.; Sun, J.; Luo, Z. Improvement of phenolic compounds extraction from high-starch lotus (*Nelumbo nucifera* G.) seed kernels using glycerol: New insights to amylose/amylopectin—Phenolic relationships. *Food Chem.* **2019**, *274*, 933–941. [CrossRef] [PubMed]
8. Pereira, P.C.; Cebola, M.-J.; Bernardo-Gil, M.G. Evolution of the Yields and Composition of Essential Oil from Portuguese Myrtle (*Myrtus comunis* L.) through the Vegetative Cycle. *Molecules* **2009**, *14*, 3094–3105. [CrossRef] [PubMed]
9. Yildirim, F.; San, B.; Yildirim, A.N.; Polat, M.; Erciŝli, S. Mineral Composition of Leaves and Fruit in Some Myrtle (*Myrtus communis* L.). *Genotypes. Erwerbs Obstbau* **2015**, *57*, 149–152. [CrossRef]
10. Kuskoski, E.M.; Asuero, A.G.; Troncoso, A.M.; Mancini-Filho, J.; Fett, R. Aplicación de diversos métodos químicos para determinar actividad antioxidante en pulpa de frutos. *Ciênc. Tecnol. Aliment.* **2005**, *25*, 726–732. [CrossRef]
11. Maldini, M.; Chessa, M.; Petretto, G.L.; Montoro, P.; Rourke, J.P.; Foddai, M.; Nicoletti, M.; Pintore, G. Profiling and Simultaneous Quantitative Determination of Anthocyanins in Wild *Myrtus communis* L. Berries from Different Geographical Areas in Sardinia and their Comparative Evaluation. *Phytochem. Anal.* **2016**, *27*, 249–256. [CrossRef] [PubMed]
12. Babou, L.; Hadidi, L.; Grosso, C.; Zaidi, F.; ValentÃ£o, P.; Andrade, P.B. Study of phenolic composition and antioxidant activity of myrtle leaves and fruits as a function of maturation. *Eur. Food Res. Technol.* **2016**, *242*, 1447–1457. [CrossRef]
13. Tuberos, C.I.G.; Rosa, A.; Bifulco, E.; Melis, M.P.; Atseri, A.; Pirisi, F.M.; Dessi, M.A. Chemical composition and antioxidant activities of *Myrtus communis* L. berries extracts. *Food Chem.* **2010**, *123*, 1242–1251. [CrossRef]

14. Scorrano, S.; Lazzoi, M.R.; Mergola, L.; Di Bello, M.P.D.; Del Sole, R.; Vasapollo, G. Anthocyanins Profile by Q-TOF LC/MS in *Myrtus communis* Berries from Salento Area. *Food Anal. Methods* **2017**, *10*, 2404–2411. [CrossRef]
15. Tuberoso, C.I.G.; Melis, M.; Angioni, A.; Pala, M.; Cabras, P. Myrtle hydroalcoholic extracts obtained from different selections of *Myrtus communis* L. *Food Chem.* **2007**, *101*, 806–811. [CrossRef]
16. Dahmoune, F.; Nayak, B.; Moussi, K.; Remini, H.; Madani, K. Optimization of microwave-assisted extraction of polyphenols from *Myrtus communis* L. leaves. *Food Chem.* **2015**, *166*, 585–595. [CrossRef] [PubMed]
17. Messaoud, C.; Boussaid, M. *Myrtus communis* berry color morphs: A comparative analysis of essential oils, fatty acids, phenolic compounds, and antioxidant activities. *Chem. Biodiv.* **2011**, *8*, 300–310. [CrossRef] [PubMed]
18. Serreli, G.; Jerković, I.; Gil, K.A.; Marijanović, Z.; Pacini, V.; Tuberoso, C.I.G. Phenolic Compounds, Volatiles and Antioxidant Capacity of White Myrtle Berry Liqueurs. *Plant Foods Hum. Nutr.* **2017**, *72*, 205–210. [CrossRef] [PubMed]
19. Aidi Wannes, W.; Marzouk, B. Differences between myrtle fruit parts (*Myrtus communis* var. *italica*) in phenolics and antioxidant contents. *J. Food Biochem.* **2013**, *37*, 585–594. [CrossRef]
20. Pereira, P.; Cebola, M.J.; Oliveira, M.C.; Bernardo-Gil, M.G. Supercritical fluid extraction vs conventional extraction of myrtle leaves and berries: Comparison of antioxidant activity and identification of bioactive compounds. *J. Supercrit. Fluids.* **2016**, *113*, 1–9. [CrossRef]
21. Picó, Y. Ultrasound-assisted extraction for food and environmental samples. *TrAC Trends Anal. Chem.* **2013**, *43*, 84–99. [CrossRef]
22. González de Peredo, A.V.; Vázquez-Espinosa, M.; Espada-Bellido, E.; Jiménez-Cantizano, A.; Ferreiro-González, M.; Amores-Arrocha, A.; Barbero, G.F. Development of new analytical microwave-assisted extraction methods for bioactive compounds from myrtle (*myrtus communis* L.). *Molecules* **2018**, *23*, 2992. [CrossRef] [PubMed]
23. Mason, T.J.; Paniwnyk, L.; Lorimer, K.P. The uses of ultrasound in food technology. *Ultrason. Sonochem.* **1996**, *3*, S253–S260. [CrossRef]
24. Briones-Labarca, V.; Plaza-Morales, M.; Giovagnoli-Vicuña, C.; Jamett, F. High hydrostatic pressure and ultrasound extractions of antioxidant compounds, sulforaphane and fatty acids from Chilean papaya (*Vasconcellea pubescens*) seeds: Effects of extraction conditions and methods. *LWT Food Sci. Technol.* **2015**, *60*, 525–534. [CrossRef]
25. Espada-Bellido, E.; Ferreiro-González, M.; Carrera, C.; Palma, M.; Barroso, C.G.; Barbero, G.F. Optimization of the ultrasound-assisted extraction of anthocyanins and total phenolic compounds in mulberry (*Morus nigra*) pulp. *Food Chem.* **2017**, *219*, 23–32. [CrossRef] [PubMed]
26. Khan, M.K.; Abert-Vian, M.; Fabiano-Tixier, A.S.; Dangles, O.; Chemat, F. Ultrasound-assisted extraction of polyphenols (*flavanone glycosides*) from orange (*Citrus sinensis* L.) peel. *Food Chem.* **2010**, *119*, 851–858. [CrossRef]
27. Feng, S.; Luo, Z.; Tao, B.; Chen, C. Ultrasonic-assisted extraction and purification of phenolic compounds from sugarcane (*Saccharum officinarum* L.) rind. *LWT Food Sci. Technol.* **2015**, *60*, 970–976. [CrossRef]
28. Sarais, G.; D'Urso, G.; Lai, C.; Pirisi, F.M.; Pizza, C.; Montoro, P. Targeted and untargeted mass spectrometric approaches in discrimination between *Myrtus communis* cultivars from Sardinia region. *J. Mass Spectrom.* **2016**, *51*, 704–715. [CrossRef] [PubMed]
29. Pereira, P.; Cebola, M.J.; Oliveira, M.C.; Bernardo Gil, M.G. Antioxidant capacity and identification of bioactive compounds of *Myrtus communis* L. extract obtained by ultrasound-assisted extraction. *J. Food Sci. Technol.* **2017**, *54*, 4362–4369. [CrossRef] [PubMed]
30. Bridgers, E.N.; Chinn, M.S.; Truong, V.D. Extraction of anthocyanins from industrial purple-fleshed sweetpotatoes and enzymatic hydrolysis of residues for fermentable sugars. *Ind. Crops Prod.* **2010**, *32*, 613–620. [CrossRef]
31. Hashemi, S.M.B.; Michiels, J.; Joris, A.Y.; Hosseini, M. Kolkhoung (*Pistacia khinjuk*) kernel oil quality is affected by different parameters in pulsed ultrasound-assisted solvent extraction. *Ind. Crops Prod.* **2015**, *70*, 28–33. [CrossRef]
32. Kazemi, M.; Karim, R.; Mirhosseini, H.; Abdul Hamid, A. Optimization of pulsed ultrasound-assisted technique for extraction of phenolics from pomegranate peel of Malas variety: Punicalagin and hydroxybenzoic acids. *Food Chem.* **2016**, *206*, 156–166. [CrossRef] [PubMed]

33. You, Q.; Yin, X.; Ji, C. Pulsed counter-current ultrasound-assisted extraction and characterization of polysaccharides from *Boletus edulis*. *Carbohydr. Polym.* **2014**, *101*, 379–385. [CrossRef] [PubMed]
34. Machado, A.P.D.F.; Pereira, A.L.D.; Barbero, G.F.; Martínez, J. Recovery of anthocyanins from residues of *Rubus fruticosus*, *Vaccinium myrtillus* and *Eugenia brasiliensis* by ultrasound assisted extraction, pressurized liquid extraction and their combination. *Food Chem.* **2017**, *231*, 1–10. [CrossRef] [PubMed]
35. Kidak, R.; Ince, N.H. Ultrasonic destruction of phenol and substituted phenols: A review of current research. *Ultrason. Sonochem.* **2006**, *13*, 195–199. [CrossRef] [PubMed]
36. Machado, A.P.D.F.; Pasquel-Reátegui, J.L.; Barbero, G.F.; Martínez, J. Pressurized liquid extraction of bioactive compounds from blackberry (*Rubus fruticosus L.*) residues: A comparison with conventional methods. *Food Res. Int.* **2015**, *77*, 675–683. [CrossRef]
37. Mustafa, A.; Turner, C. Pressurized liquid extraction as a green approach in food and herbal plants extraction: A review. *Anal. Chim. Acta* **2011**, *703*, 8–18. [CrossRef] [PubMed]
38. Cavalcanti, R.N.; Santos, D.T.; Meireles, M.A.A. Non-thermal stabilization mechanisms of anthocyanins in model and food systems-An overview. *Food Res. Int.* **2011**, *44*, 499–509. [CrossRef]
39. Carrera, C.; Ruiz-Rodriguez, A.; Palma, M.; Barroso, C.G. Ultrasound assisted extraction of phenolic compounds from grapes. *Anal. Chim. Acta* **2012**, *732*, 100–104. [CrossRef] [PubMed]
40. Pereira, D.T.V.; Tarone, A.G.; Cazarin, C.B.B.; Barbero, G.F.; Martínez, J. Pressurized liquid extraction of bioactive compounds from grape marc. *J. Food Eng.* **2019**, *240*, 105–113. [CrossRef]
41. Tiwari, B.K.; O'Donnell, C.P.; Cullen, P.J. Effect of non thermal processing technologies on the anthocyanin content of fruit juices. *Trends Food Sci. Technol.* **2009**, *20*, 137–145. [CrossRef]
42. Tiwari, B.K.; Patras, A.; Brunton, N.; Cullen, P.J.; O'Donnell, C.P. Effect of ultrasound processing on anthocyanins and color of red grape juice. *Ultrason. Sonochem.* **2009**, *17*, 598–604. [CrossRef] [PubMed]
43. Fleschhut, J.; Kratzer, F.; Rechkemmer, G.; Kulling, S.E. Stability and biotransformation of various dietary anthocyanins in vitro. *Eur. J. Nutr.* **2006**, *45*, 7–18. [CrossRef] [PubMed]
44. Stipcovich, T.; Barbero, G.F.; Ferreiro-González, M.; Palma, M.; Barroso, C.G. Fast analysis of capsaicinoids in Naga Jolokia extracts (*Capsicum chinense*) by high-performance liquid chromatography using fused core columns. *Food Chem.* **2018**, *239*, 217–224. [CrossRef] [PubMed]
45. AOAC International (Ed.) AOAC Peer Verified Methods Program. In *Manual on Policies and Procedures*; AOAC International: Arlingt, MD, USA, 1998.
46. González-Varo, J.P.; Albaladejo, R.G.; Aparicio, A. Mating patterns and spatial distribution of conspecific neighbours in the Mediterranean shrub *Myrtus communis* (Myrtaceae). *Plant Ecol.* **2009**, *203*, 207–215. [CrossRef]
47. Mulas, M. Germplasm Evaluation of Spontaneous Myrtle (*Myrtus communis* L.) for Cultivar Selection and Crop Development. *J. Herbs Spices Med. Plants* **1999**, *6*, 31–49. [CrossRef]
48. Ghafoor, K.; Choi, Y.H.; Jeon, J.Y.; Jo, I.H. Optimization of ultrasound- assisted extraction of phenolic compounds, antioxidants, and anthocyanins from grape (*Vitis vinifera*) seeds. *J. Agric. Food Chem.* **2009**, *57*, 4988–4994. [CrossRef] [PubMed]
49. Casazza, A.; Aliakbarian, B.; Mantegna, S.; Cravotto, G.; Perego, P. Extraction of phenolics from Vitis vinifera wastes using non-conventional techniques. *J. Food Eng.* **2010**, *100*, 50–55. [CrossRef]
50. Wijngaard, H.; Hossain, M.B.; Rai, D.K.; Brunton, N. Techniques to extract bioactive compounds from food by-products of plant origin. *Food Res. Int.* **2012**, *46*, 505–513. [CrossRef]
51. Tušek, A.J.; Benković, M.; Cvitanović, A.B.; Valinger, D.; Jurina, T.; Kljusurić, J.G. Kinetics and thermodynamics of the solid-liquid extraction process of total polyphenols, antioxidants and extraction yield from Asteraceae plants. *Ind. Crops Prod.* **2016**, *91*, 205–214. [CrossRef]
52. Singleton, V.L.; Orthofer, R.; Lamuela-Raventós, R.M. Analysis of total phenols and other oxidation substrates and antioxidants by means of folin-ciocalteu reagent. *Methods Enzymol.* **1999**, *299*, 152–178. [CrossRef]
53. Vallverdú-Queralt, A.; Medina-Remón, A.; Andres-Lacueva, C.; Lamuela-Raventos, R.M. Changes in phenolic profile and antioxidant activity during production of diced tomatoes. *Food Chem.* **2011**, *126*, 1700–1707. [CrossRef] [PubMed]
54. Li, Y.; Skouroumounis, G.K.; Elsey, G.M.; Taylor, D.K. Microwave-assistance provides very rapid and efficient extraction of grape seed polyphenols. *Food Chem.* **2011**, *129*, 570–576. [CrossRef] [PubMed]
55. Batista, Â.G.; Ferrari, A.S.; Da Cunha, D.C.; da Silva, J.K.; Cazarin, C.B.B.; Correa, L.C.M.; Prado, M.A.; de Carvalho-Silva, L.B.; Esteves, E.A.; Junior, M.R.M. Polyphenols, antioxidants, and antimutagenic effects

of *Copaifera langsdorffii* fruit. *Food Chem.* **2016**, *197*, 1153–1159. [CrossRef] [PubMed]
56. Zhou, Y.; Zheng, J.; Gan, R.-Y.; Zhou, T.; Xu, D.-P.; Li, H.-B. Optimization of Ultrasound-Assisted Extraction of Antioxidants from the Mung Bean Coat. *Molecules* **2017**, *22*, 638. [CrossRef] [PubMed]
57. Maran, J.; Manikandan, S.; Thirugnanasambandham, K.; Vigna Nivetha, C.; Dinesh, R. Box-Behnken design based statistical modeling for ultrasound-assisted extraction of corn silk polysaccharide. *Carbohydr. Polym.* **2013**, *92*, 604–611. [CrossRef] [PubMed]
58. Ferreira, S.L.C.; Bruns, R.E.; Ferreira, H.S.; Matos, G.D.; David, J.M.; Brandao, G.C.; da Silva, E.G.P.; Portugal, L.A.; dos Reis, P.S.; Souza, A.S.; et al. Box-Behnken design: An alternative for the optimization of analytical methods. *Anal. Chim. Acta* **2007**, *597*, 179–186. [CrossRef] [PubMed]

10

^{1}H-NMR Profiling and Carbon Isotope Discrimination as Tools for the Comparative Assessment of Walnut (*Juglans regia* L.) Cultivars with Various Geographical and Genetic Origins

Raluca Popescu [1], Roxana Elena Ionete [1,*], Oana Romina Botoran [1], Diana Costinel [1], Felicia Bucura [1], Elisabeta Irina Geana [1], Yazan Falah Jadee 'Alabedallat [2] and Mihai Botu [2,3]

1 National Research and Development Institute for Cryogenics and Isotopic Technologies-ICSI Rm. Valcea, ICSI Analytics Group, 4 Uzinei Street, RO-240050 Râmnicu Vâlcea, Romania; raluca.popescu@icsi.ro (R.P.); oana.dinca@icsi.ro (O.R.B.); diana.costinel@icsi.ro (D.C.); felicia.bucura@icsi.ro (F.B.); irina.geana@icsi.ro (E.I.G.)

2 University of Craiova, Faculty of Horticulture, Department of Horticulture and Food Science, 13 Al. I. Cuza Street, RO-200585 Craiova, Romania; yazan.fja@gmail.com (Y.F.J.'A.); btmihai2@yahoo.com (M.B.)

3 University of Craiova, Fruit Growing Research Station (SCDP) Vâlcea, 464 Calea lui Traian Street, RO-240273 Râmnicu Vâlcea, Romania

* Correspondence: roxana.ionete@icsi.ro

Academic Editors: Alessandra Gentili and Chiara Fanali

Abstract: The aim of the study was to investigate the differences between walnut genotypes of various geographical and genetic origins grown under the same or different environmental conditions. The biological material analyzed consisted in walnut kernels of 34 cultivars, nine advanced selections, and six hybrids harvested in 2015 and 2016, summing up to a total of 64 samples. The walnuts, walnut oil, and residue were characterized in respect to their chemical (proximate composition—fat, protein, nutritional value, fatty acids profile by ^{1}H-NMR) and carbon-13 isotopic composition. The data was used to statistically discriminate the cultivars according to composition, geographical area of origin, and year of harvest, comparing the Romanian cultivars, selections, and hybrids with the internationally available ones.

Keywords: δ^{13}C-IRMS; fatty acids composition; ^{1}H-NMR; walnut varieties

1. Introduction

Rich in omega-3 fats, proteins, and with a higher number of antioxidants than most other foods, walnuts are recognized as an important component of a healthy diet. The walnut kernels' composition contains between 52 and 77% fats, 12 ÷ 25% protein, 5 ÷ 24% carbohydrates, quinones, tannins, minerals, and vitamins (A, B1, B2, P, and C), that may significantly differ with the genotype, growth location, and environmental conditions [1]. Compared to other types of nuts, they are rich in terms of polyunsaturated fatty acids (PUFA), such as omega-6 and omega-3 [2], the major fatty acids found in walnut being linoleic (C18:2), oleic (C18:1), α-linolenic (C18:3), palmitic (C16:0), and stearic acids (C18:0) [3,4].

Consumers are becoming increasingly aware of food quality and authenticity of the nutritional composition and health-promoting components. In this regard, walnuts have generated considerable interest since it was confirmed by various survey studies [5–7] that their consumption can lower the cholesterol, improve the arterial function (by decreasing total and LDL-cholesterol, and increasing

HDL-cholesterol), reduce inflammation, and decrease the likelihood of diabetes and neurological diseases. The beneficial components for health found in walnuts are the polyunsaturated fatty acids, phytosterols, proteins, biogenic amines (melatonin and serotonin), dietary fiber, folate, tannins, polyphenols, and minerals (magnesium, potassium, calcium, etc.) [1,2].

Covering a wide and diverse expanse of land in Central Asia [8], the walnut has spread in China, India, Europe, and America, and is nowadays cultivated in almost all countries with temperate climates. The walnut belongs to the Juglandaceae family that contains six genera for the temperate region, with the most important two being *Juglans* and *Pterocarya*. The *Juglans* genus contains 12 species, from which the high feasibility in obtaining new cultivars and parent stock are the *Juglans regia* L. and *Juglans nigra* [9]. Persian walnut (*Juglans regia* L.) is mostly cultivated in Asia (with China as the top producer), the U.S., and Europe. Of the European countries, Romania is in the top three for the yield/hectare and in the top 10 for the production and export [10], cultivating different varieties chosen according to their high yield, disease resistance, or product quality.

Evaluating and certifying the identity/origin and quality of walnuts, and their byproduct, the oil, is a challenge for the food industry, regulatory bodies, and consumers. There are various developed methods to differentiate vegetable oils, including nuts oil, and the most recent are the nuclear magnetic resonance-based metabolomics [11,12] and stable isotopes investigation [13–15], along with chemometrics [16]. Generally, the studies on walnuts investigated the major and minor compounds, such as fatty acids, sterols, polyphenols, volatiles, and minerals [17,18], or assessed the oxidative stability, antioxidant, and antimicrobial activity [1,19,20] to determine the variations given by the location, environmental conditions, cultivars, and technological processing [21–23]. Although the main purpose of isotopic studies is to identify the botanical and geographical origins of different food products, only a few were applied to walnuts and walnut oil [24–26], but lacked reference to the variation given by the cultivar/genotype, geographical origin, or year of production (environmental conditions—temperature, rainfall, and light).

As a consequence, the aim of the present work was to investigate the differences between walnut genotypes of various geographical and genetic origins grown under the same or different environmental conditions. The analyzed biological material consisted in walnut kernels of 34 cultivars, nine advanced selections, and six hybrids from the breeding program, harvested in 2015 and 2016. The walnuts, walnut oil, and residue were characterized in terms of chemical proximate composition (fats and protein content, nutritional value, fatty acids profile by ^{1}H-NMR) and carbon-13 isotopic composition. The data were used to statistically discriminate the cultivars according to the composition, geographical area of origin and year of harvest, comparing the Romanian cultivars, selections, and hybrids, with the internationally available ones.

2. Results and Discussion

2.1. Geographical and Year of Harvest Discrimination

Chemical composition of the kernel, oil, and residue for the walnut samples is highlighted in Table 1. The samples are grouped according to the year of harvest and geographical origin of the walnut cultivar, with an emphasis on the Romanian cultivars and the selections and hybrids developed at the SCDP Râmnicu Vâlcea (Romania).

The walnut kernels contained mainly fats (69.0%), with a mean concentration of crude protein of 20.3% and a 10.7% content of carbohydrates, resulting in a 700 kcal/100 g kernel energy. The extracted oil, with a mean of 874 kcal energy, was composed mainly of linolenic acid (as sum of ω-6 fatty acids)—55.6% molar, followed by 22.7% monounsaturated fatty acids (mainly oleic), 11.1% linoleic acid (as sum of ω-3 fatty acids), and 10.6% saturated fatty acids, giving a mean unsaturation degree (IV–iodine value) of 139.

Table 1. Composition of the walnut kernel, oil, and residue.

Geographical Origin/Composition	All Samples (n = 64)	Germplasm Collection of Fruit Growing Research Station (SCDP) Vâlcea, Grown to RO-VL1							RO-VL2 Local Selections (n = 3)	RO-DB Red Kernel (n = 1)	GR (n = 2)		PRC (n = 2)	
		USA Cultivars (n = 18)	FR Cultivars (n = 7)	RO Cultivars (n = 20)	RO Selections (n = 5)	RO Hybrids (n = 6)	Samples Collected in Both 2015 and 2016 (n = 30)							
							2015	2016			GR-1	GR-2	PRC-1	PRC-2
kernel														
Total fat-k (%)	69.0 ± 7.6	66.6 ± 7.1	73.6 ± 6.1	71.0 ± 9.3	66.3 ± 2.3	66.5 ± 3.9	66.4 ± 4.0	73.4 ± 10.7	66.7 ± 4.0	79.6	72.0	60.6	79.4	61.1
Protein-k (%)	20.3 ± 6.4	19.8 ± 7.9	19.7 ± 5.9	20.1 ± 6.3	22.8 ± 4.9	26.0 ± 2.2	26.2 ± 4.6	13.3 ± 2.2	18.1 ± 4.5	16.7	12.9	17.4	14.6	15.6
Carbohydrates-k (%)	10.7 ± 8.0	13.5 ± 9.8	6.7 ± 1.8	8.9 ± 8.3	10.8 ± 5.9	7.5 ± 5.5	7.3 ± 5.7	10.2 ± 0.9	15.2 ± 0.9	3.7	15.1	22.1	6.0	23.2
Energy-k (kcal/100 g)	700 ± 22	703 ± 24	699 ± 19	699 ± 22	693 ± 9	702 ± 33	696 ± 21	709 ± 21	693 ± 18	713	730	698	704	711
oil														
SFA-o (%molar)	10.6 ± 0.9	10.2 ± 0.8	10.9 ± 0.7	10.7 ± 1.0	10.5 ± 0.7	10.8 ± 0.8	10.7 ± 0.7	10.2 ± 0.9	11.1 ± 0.7	13.4	11.0	12.3	8.9	9.3
Oleic-o (%molar)	22.7 ± 5.2	23.6 ± 5.5	21.3 ± 3.4	23.1 ± 4.2	24.7 ± 5.6	18.8 ± 3.5	19.9 ± 3.2	26.1 ± 5.2	16.1 ± 2.4	20.9	19.9	20.2	38.5	31.2
Linoleic-o (%molar)	55.6 ± 4.6	55.2 ± 5.0	56.3 ± 2.1	55.1 ± 4.0	54.8 ± 3.9	59.2 ± 5.3	58.0 ± 3.3	52.7 ± 4.4	60.5 ± 1.5	54.1	55.2	55.7	43.4	50.6
Linolenic-o (%molar)	11.1 ± 1.6	11.1 ± 1.6	11.5 ± 1.7	11.0 ± 1.4	10.1 ± 2.0	11.2 ± 2.1	11.4 ± 2.0	11.0 ± 1.0	12.4 ± 0.4	11.6	14.0	11.8	9.3	8.9
IV-o	139 ± 4	138 ± 4	140 ± 4	138 ± 4	137 ± 6	142 ± 2	141 ± 4	136 ± 4	144 ± 1.3	134	141	138	126	131
PUFA (%molar)	66.7 ± 4.8	66.2 ± 5.0	67.8 ± 3.3	66.2 ± 4.0	64.9 ± 5.2	70.4 ± 3.4	69.4 ± 3.3	63.7 ± 4.7	72.9 ± 1.7	65.7	69.2	67.5	52.6	59.5
UFA/SFA	8.49 ± 0.83	8.9 ± 0.8	8.2 ± 0.6	8.38 ± 0.78	8.58 ± 0.62	8.30 ± 0.66	8.43 ± 0.63	8.87 ± 0.96	8.07 ± 0.6	6.46	8.13	7.12	10.2	9.8
PUFA/SFA	6.32 ± 0.55	6.5 ± 0.5	6.3 ± 0.5	6.21 ± 0.62	6.20 ± 0.39	6.55 ± 0.57	6.55 ± 0.57	6.26 ± 0.51	6.61 ± 0.3	4.90	6.31	5.48	5.92	6.43
Energy-o (kcal/100 g)	874 ± 69	900 ± 56	848 ± 60	854 ± 71	892 ± 11	886 ± 11	886 ± 18	857 ± 104	887 ± 9	816	903	903	744	902
residue														
Protein-r (%)	41.3 ± 6.7	39.8 ± 6.8	41.7 ± 2.6	40.2 ± 7.6	43.7 ± 5.3	40.7 ± 6.0	42.5 ± 7.9	38.7 ± 6.0	52.3 ± 6.2	34.7	39.2	44.1	45.3	46.2
Energy-r (kcal/100 g)	303 ± 17	307 ± 23	293 ± 6	303 ± 18	300 ± 8	310 ± 16	312 ± 23	295 ± 11	304 ± 7	312	286	289	297	300

k—kernel, o—oil, r—residue.

The fatty acids composition allowed the estimation of the different nutritional fractions: PUFA—polyunsaturated fatty acids, UFA/SFA—unsaturated fatty acids/ saturated fatty acids, PUFA/SFA—polyunsaturated fatty acids/saturated fatty acids. The results as a general composition for the walnuts and walnuts oil and for specific cultivars were in agreement with other published data [1,17,20,22,23,27].

Considering the walnuts grown at the Fruit Growing Research Station (SCDP) Vâlcea, the Romanian selections and hybrids tend to present the extreme values for this set of samples. The Romanian hybrids registered high protein content (26.0%) and low fats (66.5%) for the kernel and oil, with a low composition in oleic acid (18.8% molar), and high linoleic acid (59.2% molar), that led to increased values of PUFA (70.4 % molar) and IV (142) compared to the other groups of walnuts.

The Romanian selections were also found to have a low-fat content (66.3%), with a higher oleic concentration (24.7% molar) and lower linoleic (54.8% molar) and linolenic (10.1% molar) concentrations, resulting in low PUFA. The international and Romanian cultivars had similar compositions, the values being intermediate between the Romanian hybrids and the Romanian selections, with the exception of the higher fatty content for the French cultivars (mean of 73.6%) and higher PUFA/SFA and UFA/SFA ratios for the USA cultivars.

For harvest year, discriminations were considered only among the cultivars from the Fruit Growing Research Station (SCDP) Vâlcea sampled both in 2015 and 2016, namely 15 cultivars, from which 10 were international and five Romanian ones. The classification of the samples was 100% correct, with the $\delta^{13}C$ of the residue (which is the carbon-13 fingerprint of the proteins and carbohydrates), protein content of the kernel and polyunsaturated fatty acids concentration, and the iodine value as the most important factors for discrimination. The walnuts from 2015 had a lower content of fats with a higher degree of unsaturation than the ones harvested in 2016, together with a higher content of proteins. Also, since in 2015 there were higher mean temperatures and precipitation in the maturation period (July–September) than in 2016 [28], a possible explanation could be drawn that increased temperatures and precipitation are conductive to the formation of proteins in the kernels and lower quantities of fat with higher concentrations of polyunsaturated fatty acids.

Discriminant analysis of walnuts grown in different locations is presented in Figure 1. For all samples, the first two most important factors summed 83.59% of the variance (F1 = 46.25% and F2 = 37.35%), with 100% correct classification of the samples.

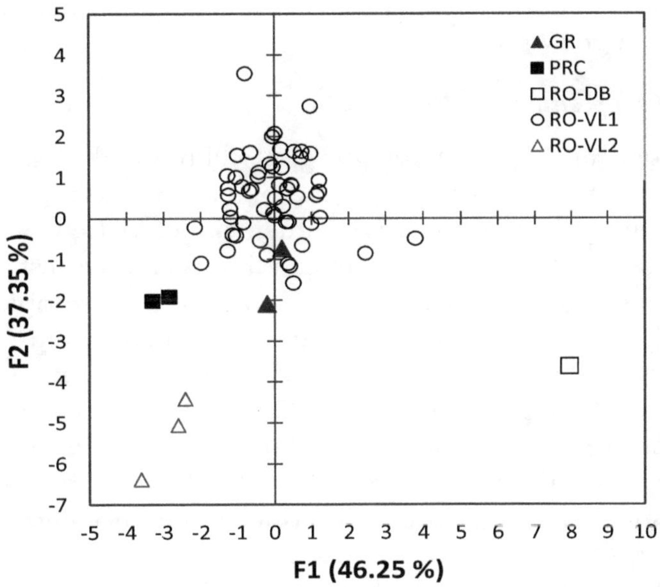

A

Figure 1. *Cont.*

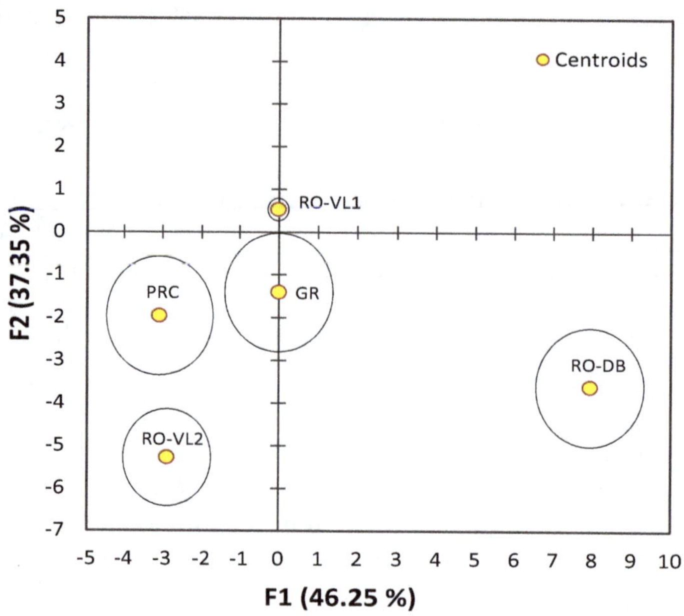

Figure 1. Discriminant analysis of walnuts according to the geographical origin of the walnut samples: (**A**) samples distribution by groups; (**B**) group centroids as mean discriminant scores for each group in the dependent variable for each of the discriminant functions.

The most important parameters for the geographical discrimination of walnuts were the $\delta^{13}C$ of the kernel, oil, and residue, the SFA, oleic concentrations, together with the indices IV, UFA/SFA, and PUFA/SFA. Since the number of samples grown in different locations than RO-VL1 was low, no general conclusions could be drawn. As a general observation, the two samples from China presented high oleic composition (38.5 and, respectively, 31.2% molar) compared with the other cultivars, resulting in a low IV of 129.

Other distinct compositions were observed for the walnut with red kernels (RO-DB), which had high fatty content (79.6%) with higher SFA (13.4% molar), and for the Romanian selections RO-VL2 ($n = 3$) which had low oleic and high linoleic/linolenic concentrations (mean of 60.5% and, respectively, 12.4% molar) resulting in a high IV (144).

2.2. Stable Carbon Isotopic Composition

The carbon stable isotopes are important for biological, cultivar and climatic differentiation [24,29,30]. Variation of $\delta^{13}C$ values between different plants and plant products may be attributed, besides the metabolic pathway, to geographical origin, environmental factors (temperature, humidity, rainfall, light exposure, water availability), cultivar, growth conditions and harvest time [26].

The variation in carbon-13 abundance in food is the result of isotopic fractionation that takes place in plants [25]. During the photosynthesis, the light (^{12}C) and heavier (^{13}C) isotopes of carbon will be discriminated, the processes being reflected in the isotopic compositions of the plant components. Carbon dioxide-fixing pathways in plants are the C3 (Calvin cycle), C4 (Hatch-Stack cycle), and CAM (Crassulacean acid metabolism), with the mention that all trees use the C3 metabolism. Since walnut is a C3 plant, the isotopic fingerprint of the kernel and oil should be in the range −34‰ to −22‰ [24]. Studies done on walnut and walnut oil samples produced in France and China, reported values of −28.67 ± 0.25‰ [25] and, respectively, −27.5‰ [24–26] for the $\delta^{13}C$.

In this study, the range of variation for the $\delta^{13}C$ of the kernel was of 7.6 delta units (between −29.9 and −22.3‰), while for the oil and residue was about six delta units, between −30.1 and −23.9‰ and, respectively, between −28.3 and −21.9‰. The oil was depleted in carbon-13 compared to the kernel with approx. 0.6‰, while the residue was enriched in carbon-13 compared to the kernel with approx. 1.9‰ (Table 2).

Table 2. Stable carbon composition (mean values) of the investigated walnut samples.

		δ^{13}C-Kernel (‰)	δ^{13}C-Oil (‰)	δ^{13}C-Residue (‰)
All samples (n = 64)		−26.8 ± 1.5	−27.4 ± 1.2	−24.9 ± 1.4
Samples collected in both years	2015 (n = 15)	−27.3 ± 1.4	−27.6 ± 0.8	−25.6 ± 1.3
	2016 (n = 15)	−26.9 ± 1.4	−27.1 ± 0.7	−24.5 ± 0.7
Samples growth to SCDP	USA cultivars (n = 18)	−27.4 ± 1.2	−27.5 ± 0.8	−25.3 ± 1.1
	RO selections (n = 5)	−27.1 ± 1.3	−28.3 ± 1.2	−25.2 ± 1.3
	RO cultivars (n = 20)	−27.0 ± 1.2	−27.4 ± 0.9	−25.2 ± 1.4
	RO hybrids (n = 6)	−26.9 ± 0.9	−28.0 ± 1.3	−24.5 ± 1.4
	FR cultivars (n = 7)	−26.8 ± 1.4	−27.8 ± 0.8	−25.0 ± 0.7
Samples from Greece	GR-2 2016	−26.3	−27.3	−24.1
	GR-1 2015	−26.3	−25.8	−22.7
Samples from China	PRC-2 (Zhong Lin no.1)	−25.6	−25.2	−22.1
	PRC-1 (Jin Bo Feng no.1)	−24.0	−26.0	−22.4
Samples from Romania (other than SCDP)	RO-DB red kernel selection (n = 1)	−25.1	−25.4	−23.6
	RO-VL (n = 3)	−23.3 ± 1.0	−24.8 ± 0.8	−22.7 ± 0.7
Samples classified by groups	C1 (n = 23)	−27.0 ± 1.7	−27.6 ± 1.3	−25.4 ± 1.6
	C2 (n = 25)	−26.6 ± 1.2	−27.4 ± 1.2	−24.7 ± 1.3
	C3 (n = 6)	−26.8 ± 1.4	−27.4 ± 1.1	−24.8 ± 1.1
	C4 (n = 8)	−26.3 ± 1.5	−26.8 ± 0.7	−24.1 ± 1.0
	C5 (n = 2)	−28.2	−26.5	−24.0

In photosynthesis, C3 plants transform the carbon dioxide ($\delta^{13}C = -8‰$) in carbohydrates which will be further used to form other classes of compounds, including proteins and fatty acids, processes that will take place as a general rule with a 1‰ depletion in carbon-13 for proteins and 6‰ for lipids, since the ^{12}C isotope is favored for the reactions that take place [31]. The study done by Guo et al. [26] went in even more details and showed the differences between the δ^{13}C of individual fatty acids, which decreased slightly (0.7‰) in the elongation of $C_{18:0}$ from $C_{16:0}$, increased significantly (1.3‰) in the first desaturation (from $C_{18:0}$ to $C_{18:1}$) and kept stable in the further desaturation from $C_{18:1}$ to $C_{18:2}$.

The statistical analysis of the data (both isotopic and compositional) showed that the δ^{13}C of the kernel, oil, and residue were important in the geographical discrimination (with p-values < 0.0001), δ^{13}C of the residue (containing mostly the proteins and carbohydrates of the walnut) was important in differentiating the year of harvest (p-value = 0.004), while no carbon-13 fingerprint was important in discriminating the walnuts according to composition. These correlations indicate an influence of the location and year of harvest on the carbon isotopic composition of the walnuts. It can be mentioned that if the discrimination is done between cultivars from a restricted area, the isotopic composition is influenced by the biological parameters (metabolism) rather than the climatic factors, though no strong correlations were found between the isotopic fingerprint and composition of the walnuts.

The mean carbon-13 values of the different walnut groups are also presented in Table 2. The samples from 2016 were slightly more enriched in carbon-13 than the ones produced in 2015, with differences of about 0.4, 0.5 and 1.1‰ for the kernels, oil and residue. The trend is opposite than in other findings in which the isotopic fingerprint is enriched in the warmer years, though the two samples from Greece (Tripoli area) showed a more enriched isotopic composition for the oil and residue in the warmer year (2015).

Concerning the geographical origin of the cultivars, a trend of isotope enrichment was observed, from the walnuts produced in the Fruit Growing Research Station (SCDP) Vâlcea (kernel of −27‰, oil of −28‰ and residue of −25‰), to the ones produced in Greece, followed by the red kernel walnut from Romania–Dambovişa county and walnuts from China, ending with the three walnut selections from Vâlcea county that registered the highest δ^{13}C values (−23.3‰ for kernel, −24.8‰ for oil, and −22.7‰ for residue). The data are also presented as a distribution in Figure 2.

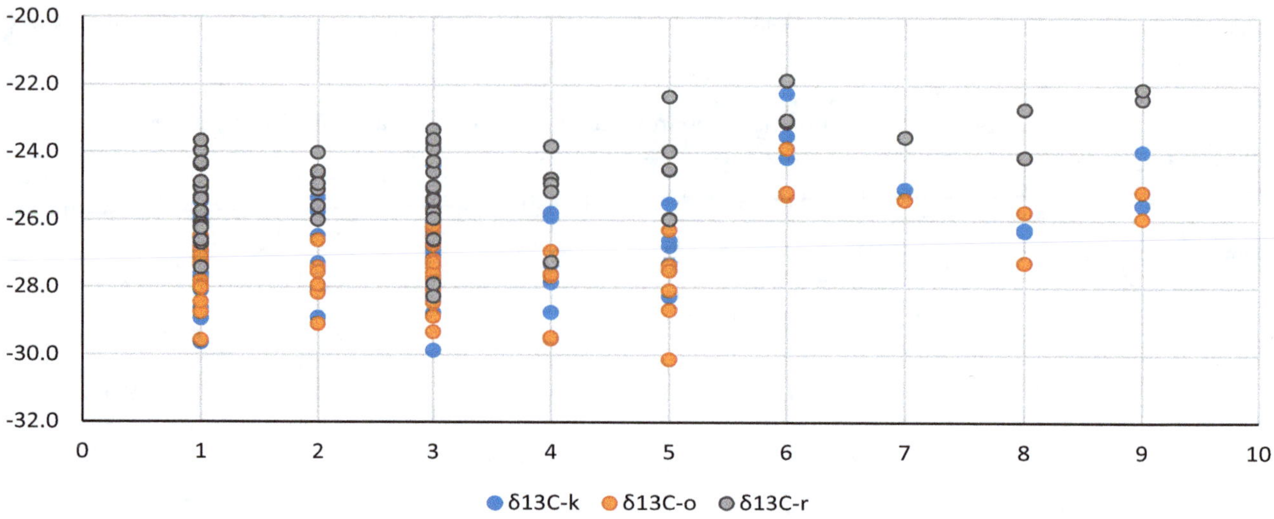

Figure 2. Carbon-13 composition of the kernel, oil and residue taking into account the geographical origin of the cultivars (1 = SCDP-USA cultivars, 2 = SCDP-FR cultivars, 3 = SCDP-RO cultivars, 4 = SCDP-RO selections, 5 = SCDP-RO hybrids, 6 = RO-VL, 7 = RO-DB, 8 = GR, 9 = PRC).

Using only the isotopic values, the walnuts can be differentiated according to the geographical origin, with the exception of the two samples from Greece which were classified in the RO–SCDP Vâlcea group, while the year of harvest discrimination gave only a 73.33% correct classification of the samples (11 out of 15). If the chemical composition is taken into account, groups C5 and C4 were the most enriched in carbon-13 (with the exception of C5 kernel), while groups C1, C2, and C3 were the most depleted.

2.3. Composition Discrimination

Nonspecific statistical analysis of the composition and isotopic data by agglomerative hierarchical clustering (AHC) classified the groups as presented in Table 3. The most important factors to differentiate these groups were the fat content, the energy of the oil, the oleic and linoleic concentrations, and the iodine value.

Groups C1 and C2 were the most similar, with the next level of similarity clustering groups C1 and C2 with C3, followed by group C5, while group C4 had the most distinct composition. Groups C1–C3 had the most unsaturated composition (high polyunsaturated acids concentrations resulting in high IV values), with group C1 being distinct due to the high protein content and group C3 due to the higher SFA content; group C5 (formed of Hartley and Vina cultivars harvested in 2016) had walnuts with low fat and protein compositions, but high in SFA and linolenic acid. Group C4, where Adams 10, Lara, one Chinese cultivar (Jin Bo Feng no.1), and other five Romanian cultivars all harvested in 2016 were found, had a walnut composition high in fats with increased oleic acid concentrations.

Group C1 is comprised mainly from samples of 2015 harvest; groups C3–C5 are formed of walnuts harvested in 2016, while group C2 is a mixture of 2015 and 2016 samples. For the cultivar Wilson Franquette, the walnuts sampled in both years of harvest were classified in the same group, indicating a metabolism that is less influenced by the environmental conditions. For the other cultivars, however, the walnuts obtained in different years were classified in different groups, more or less related to each other due to the difference in composition given by the production year.

Concerning the Romanian walnut cultivars, selections and hybrids, the majority of them were classified in the high unsaturation groups (C1–C3), with the exception of five of them produced in 2016 which had a composition high in fats and oleic acid (group C4).

Table 3. Classification of the walnuts according to composition.

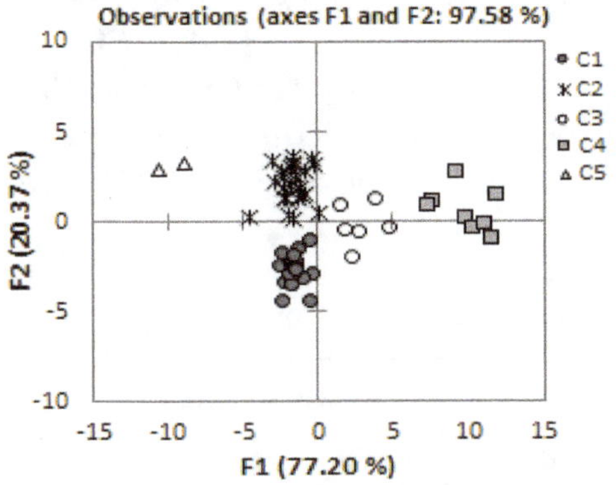

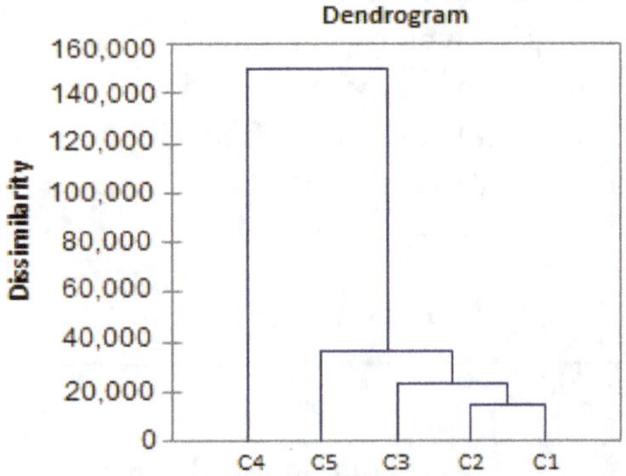

Group Labels	Group Components			
	SCDP–Cultivars	SCDP–Romanian Selections	SCDP–Romanian Hybrids	Other Geographical Origin
C1 (n = 23)	USA: Chase D9-2016, Hartley 2015, Pedro 2015, Tehama 2015, Vina 2015 FR: Fernette 2015, Fernor 2015, Lara 2015, RO: Argesan 2015, Germisara 2015, Muscelean 2015, Orastie 2015, Roxana 2015, Sibisel 44-2015, Sibisel Precoce 2015, Valcris-2015	Munteanu 2015, VL206S 2015, Vladesti 2015	HC 02-2015, UC 4/12-2015, T 2/3-2015	Fumureni 2-2016 (RO-VL2)
	C1 compositional characteristics: high protein, linoleic, linolenic, IV			
C2 (n = 25)	USA: Adams 10-2015, Chase D9-2015, Howe 2016, Idaho 2016, Payne 2016, Pedro 2016, Serr 2016, Wilson Franquette 2015/2016 RO: Germisara 2016, Mihaela 2015, Muscelean 2016, Timval 2015, Valcor 2016, Velnita 2015	Ignat Tudor 2015, VL 301B-2015	C2-2015, HC3390-2015, UC 3/1-2015	Brezoi-2015 (RO-VL2), Fumureni 1-2016 (RO-VL2), GR 1-2015, GR 2-2016, PRC 2-2016 (Zhong Lin no.1 cultivar)
	C2 compositional characteristics: high linoleic, linolenic, IV			
C3 (n = 6)	USA: Payne 2015 FR: Ferjean 2016, Fernette 2016, Fernor 2016 RO: Jupanesti 2015	-	-	Targoviste 1-red kernel 2016 (RO-DB)
	C3 compositional characteristics: high SFA, linoleic, linolenic, IV			
C4 (n = 8)	USA: Adams 10-2016, FR: Lara 2016 RO: Roxana 2016, Sibisel 44-2016, Timval 2016, Unival 2016, Valmit 2016	-	-	PRC 1-2016 (Jin Bo Feng no.1 cultivar)
	C4 compositional characteristics: high fat, oleic, UFA/ SFA; low linoleic, linolenic, IV			
C5(n = 2)	USA: Hartley 2016, Vina 2016	-	-	-
	C5 compositional characteristics: low fat, proteins; high SFA, linolenic			

3. Materials and Methods

3.1. Plant Material and Collection Site

The main biological material analyzed in this study consisted of walnut kernels of fruits harvested from the germplasm collection of the Fruit Growing Research Station (SCDP) belonging to the University of Craiova and located in Râmnicu Vâlcea, Romania. The aim of SCDP is to improve the walnut assortment in Southwest Romania through selection of local walnut population and controlled hybridization. The walnut cultivar collection field is located north of Râmnicu Vâlcea city, at 45°8′22.35″N and 24°22′36.85″E. The climate of the area is temperate, of Cfb Köppen-Geiger type [32], the average annual temperature is 10.2 °C and rainfall 715 mm. The walnut collection was planted in 1997 on an alluvial soil with medium content of nitrogen, phosphorus, and potassium, and with a pH of 6.8. No irrigation was provided to the trees. Fruits were harvested at full maturity, in 2015 and/or 2016, from a number of 11 walnut representative cultivars originated from USA (seven from

California, three from Oregon and one from Idaho), four cultivars from France, and 14 from Romania. Additionally, five advanced selections and six hybrids from the breeding program were sampled. Another set of samples was represented by walnuts grown in other locations than SCDP: Two cultivars from both Greece (Tripoli City area) and China (Taigu county and Taiyuan City from Shanxi Province), and the other four local selections from Romania (three from Vâlcea county and one with red kernel from Dâmboviţa county). A total of 64 samples were investigated, from which 15 were taken both in 2015 and 2016 (see Table 4).

Table 4. Description of walnut samples collection.

Geographical Origin of the Cultivar	Description	Name	Genetic Origin	Harvest Year
Fruit Growing Location-Research Station (RO-VL1), Romania				
USA (Oregon)	Cultivar	Adams 10	Open pollinated seedling	2015 and 2016
	Cultivar	Chase D9	Open pollinated seedling	2015 and 2016
	Cultivar	Howe	Chance seedling	2016
USA (California)	Cultivar	Tehama	'Waterloo' × 'Payne'	2015
	Cultivar	Hartley	Open pollinated seedling	2015 and 2016
	Cultivar	Payne	Chance seedling	2015 and 2016
	Cultivar	Pedro	'Conway Mayette' × 'Payne'	2015 and 2016
	Cultivar	Vina	'Franquette' × 'Payne'	2015 and 2016
	Cultivar	Wilson Franquette	Selection of 'Franquette'	2015 and 2016
	Cultivar	Serr	'Payne' × PI 159568	2016
USA (Idaho)	Cultivar	Idaho	Selection from local populations	2016
France	Cultivar	Fernette	'Franquette' × 'Lara'	2015 and 2016
	Cultivar	Fernor	'Franquette' × 'Lara'	2015 and 2016
	Cultivar	Lara	Chance seedling of 'Payne'	2015 and 2016
	Cultivar	Ferjean	Grosvert' × 'Lara'	2016
Romania (Argeş)	Cultivar	Argesan	Selection from local populations	2015
	Cultivar	Jupâneşti	Selection from local populations	2015
	Cultivar	Mihaela	Selection from local populations	2015
	Selection	Ignat Tudor	Selection from local populations	2015
	Selection	Munteanu	Selection from local populations	2015
	Selection	Vladesti	Selection from local populations	2015
	Cultivar	Muscelean	Selection from local populations	2015 and 2016
	Cultivar	Roxana	Selection from local populations	2015 and 2016
Romania (Hunedoara)	Cultivar	Sibisel Precoce	Selection from local populations	2015
	Cultivar	Orastie	Selection from local populations	2015
	Cultivar	Germisara	Selection from local populations	2015 and 2016
	Cultivar	Sibişel 44	Selection from local populations	2015 and 2016
Romania (Vâlcea)	Cultivar	Valcris (syn. VL202 PO)	Selection from local populations	2015
	Selection	VL 206 S	Selection from local populations	2015
	Cultivar	Timval (syn. VL 54 B)	Selection from local populations	2015 and 2016
	Cultivar	Unival	Selection from local populations	2016
	Cultivar	Valcor	Selection from local populations	2016
	Cultivar	Valmit	Selection from local populations	2016
Romania (Craiova)	Hybrid	C2	Open pollinated seedling of 'Ideal'	2015
	Hybrid	HC 02	Open pollinated seedling of 'Ideal'	2015
	Hybrid	HC 3390	Open pollinated seedling of 'Ideal'	2015
	Hybrid	T2/3	Open pollinated seedling of 'Ideal'	2015
	Hybrid	UC 3/1	Open pollinated seedling of 'Ideal'	2015
	Hybrid	UC 4/12	Open pollinated seedling of 'Ideal'	2015
Romania (Bucureşti)	Selection	VL 301 B	Selection from local populations	2015
Romania (Iaşi)	Cultivar	Velniţa	Selection from local populations	2015
Fruit Growing Location-Vâlcea county (RO-VL2), Romania				
Romania (Vâlcea)	Selection	Brezoi 1	Selection from local populations	2016
	Selection	Fumureni 1	Selection from local populations	2016
	Selection	Fumureni 2	Selection from local populations	2016
Fruit Growing Location-Dâmboviţa county (RO-DB), Romania				
Romania (Dâmboviţa)	Selection	Târgovişte 1-red kernel	Selection from local populations	2016

Table 4. *Cont.*

Geographical Origin of the Cultivar	Description	Name	Genetic Origin	Harvest Year
		Fruit Growing Location-Tripoli (GR), Greece		
Greece (Tripoli)	Cultivar	Nut sample 1	Franquette' × 'Hartley' x 'Chandler'	2015
Greece (Tripoli)	Cultivar	Nut sample 2	× 'Meylannaise'	2016
		Fruit Growing Location-PRC		
China (Fruit Growing Institute in Taigu-Shanxi)	Cultivar	Jin Bo Feng no.1	Unknown	2016
China (Taiyuan market, Shanxi)	Cultivar	Zhong Lin no.1	Unknown	2016

3.2. Sample Preparation

After harvest, the in-shell walnuts were naturally dried. Nuts were cracked, and kernels were extracted from each cultivar sample. The mean sample consisted of 30 g of healthy kernels selected to be representative for the lot, grounded to a fine homogenous mix. From this, 1 g was dried by lyophilization and C/H/N composition and δ^{13}C-IRMS were performed. Another 5 g of mean grounded walnut kernels were used to extract oil for 16 h with 100 mL of petroleum ether using a Buchi B-811 Soxhlet extractor (Büchi Labortechnik AG, Flawil, Switzerland). The oil resulted after the removal of the solvent and filtration represented the fat content of the kernels and its composition was determined–fatty acids content by NMR, C/H/N and δ^{13}C-IRMS. The dried residue after oil extraction (consisting mainly of the proteins and carbohydrates of the walnuts) was also analyzed for the C/H/N composition and δ^{13}C-IRMS.

3.3. C/H/N Composition

The elemental composition (C/H/N) of the walnut, oil, and residue was determined in triplicate using a Flash2000 Thermo (Thermo Fisher Scientific, Leicestershire, UK) with a TCD detector (Thermo Fisher Scientific, UK) [33]. Analysis conditions: Chromatographic columns—(i) molecular sieve 5A for separating flue gases after combustion, gas type CO, and (ii) PoraPLOT Q (30 cm, 20 μm, 65 °C) for separating flue gases after combustion, gas type NO_2, CO_2, H_2O, SO_2, furnace temperature: 950 °C, TCD detector: 1000 mA, measuring time—12 min, sample mass 2.5 mg. Linear regression analysis, based on seven points created using certified reference material, traceable at ISO GUIDE 34/17025, was used to generate the calibration equation for each element. The condition of a correct measurement was to draw a calibration curve with an RSD < 0.5% and the correlation coefficients $r^2 > 0.99$. To verify the conformity of the results or if the calibration curve did not suffer changes, before every sample injection, a quality control (QC) gas mixture, rich in permanently gases and hydrocarbons, was used. The measured values were used to calculate the energy values for walnut, oil and residue, and the proximate composition, crude protein (N × 5.30) (AOAC 950.48) and carbohydrate content.

3.4. NMR Analysis

The walnut oil samples for the NMR analysis were prepared in duplicate by dissolving 70 mg of oil in 700 μL of CDCl3 (99.96% at.) with TMS (NMR grade) as internal standard. The reagents were purchased from Aldrich (St. Louis, MO, USA) and, respectively, Sigma-Aldrich (Darmstadt, Germany). All NMR experiments were recorded at 300 K using a Bruker AvanceIII400 spectrometer (Bruker France SAS, Wissembourg, France) operating at 9.4 T, equipped with a 5 mm BBO probe, observing ^{1}H at 400.2 MHz and ^{13}C at 100.6 MHz. ^{1}H-NMR spectra were acquired using spectral width of 8224 Hz; 65536 data points; pulse width of 10.5 μs; relaxation delay of 1.0 s; acquisition time of 4 s and 16 scans. To verify the spectra quality, the solvent peak was examined on the ^{1}H spectrum for each sample. Since the same quantity of deuterated chloroform was always added for the NMR analysis, the resolution of the peak at 7.3 ppm was used to attest the correct adjustment of the shims. Spectra were processed by applying an exponential line broadening of 0.3 Hz for sensitivity enhancement before Fourier transforms and were accurately phased and baseline adjusted. The determination of the fatty acids composition of the walnut oil was done according to Guillén and Ruiz [34].

3.4.1. δ^{13}C Measurement

The overall carbon-13 composition of the walnut kernel, oil, and residue was obtained using a Thermo DeltaVPlus Spectrometer (Thermo Fisher Scientific, Bremen, Germany) coupled with an 1112 Elemental Analyser equipped with an autosampler for solids. The isotopic composition is expressed as δ-values (‰), which refers the isotope ratio of the sample (S) to that of the international reference PDB:

$$\delta^{13}C = 1000 \cdot \left[\frac{(^{13}C/^{12}C)_S - (^{13}C/^{12}C)_{PDB}}{(^{13}C/^{12}C)_{PDB}} \right]$$

The δ^{13}C isotopic values were calibrated against an international reference material, sugar IAEA-CH-6 (IAEA—International Atomic Energy Agency, Vienna, Austria), and three other laboratory standards, namely, L-Alanine IA-R041 (Iso-Analytical Laboratory Standard), Beet Sugar IA-R005, and Cane Sugar IA-R006.

During measurements, a reference was injected at regular intervals as a working standard to check the accuracy of the results. Each sample was analyzed in triplicate and the results were validated if the difference was below 0.4‰.

3.5. *Data Analysis*

The data were evaluated by statistical methods. Descriptive statistical analysis (mean, standard deviation, etc.), agglomerative hierarchical clustering (AHC), and discriminant analysis (DA) were performed using commercial software packages as Microsoft Excel 2013 (Microsoft, Redmond, WA, USA) and XLSTAT Addinsoft 2014.5.03 version (Addinsoft Inc, Long Island, NY, USA).

4. Conclusions

The determined parameters (fat, protein, nutritional value, fatty acids profile by ^{1}H-NMR, and carbon-13 isotopic composition) for the walnuts and walnut oil were used to differentiate with certain degrees of success the cultivars grown in the same location or different locations and to show the composition differences for the same cultivar in different years of production. The study allowed the comparison of the Romanian cultivars, selections and hybrids with the internationally available ones. The data also showed that in order to enhance the discrimination of walnut and walnut oil samples a wider range of parameters is needed, for example to couple the isotopic fingerprint (of bulk, fractions, or individual compounds) with the composition data or vice versa.

Author Contributions: Conceptualization, R.P. and R.E.I.; methodology, R.P. and R.E.I.; software, R.P. and O.R.B.; validation, R.P., O.R.B. and D.C.; formal analysis, R.P., D.C., F.B. and E.I.G.; investigation, R.P.; resources, R.E.I.; data curation, Y.F.J.'A. and M.B.; writing—original draft preparation, R.P.; writing—review and editing, R.E.I.; visualization, M.B.; supervision, R.P. and R.E.I.; project administration, R.P.; funding acquisition, R.E.I.

References

1. Pereira, J.A.; Oliverira, I.; Sousa, A.; Ferreira, I.C.F.R.; Bento, A.; Estevihno, L. Bioactive properties and chemical composition of six walnut (*Juglans regia* L.) cultivars. *Food Chem. Toxicol.* **2008**, *46*, 2103–2111. [CrossRef]

2. Cosmulescu, S.; Baciu, A.; Achim, G.; Botu, M.; Trandafir, I. Mineral Composition of fruits in different walnut (*Juglans regia* L.) cultivars. *Notulae Botanicae Horti Agrobotanici Cluj-Napoca* **2009**, *37*, 156–160.
3. Li, L.; Tsao, R.; Yang, R.; Kramer, J.K.G.; Hernandez, M. Fatty acids profiles, tocopherol contents, and antioxidant activities of heartnut (Juglans ailanthiofolia Var. cordiformis) and Persian walnut (*Juglans regia* L.). *J. Agric. Food Chem.* **2007**, *55*, 1164–1168. [CrossRef]
4. Tapia, M.I.; Sárnchez-Morgado, J.R.; Garcia-Para, J.; Ramirez, R.; Hernández, T.; Gonzáles Gómez, D. Comparative study of the nutritional and bioactive compounds content of four walnut (*Juglas regia* L.) cultivars. *J. Food Compost. Anal.* **2013**, *31*, 232–237. [CrossRef]
5. Sabaté, J.; Fraser, G.E.; Burke, K.; Knutsen, S.F.; Bennett, H.; Lindsted, K.D. Effects of walnuts on serum lipid levels and blood pressure in normal men. *N. Engl. J. Med.* **1993**, *328*, 603–607. [CrossRef]
6. Lavedrine, F.; Zmirou, D.; Ravel, A.; Balducci, F.; Alary, J. Blood cholesterol and walnut consumption: A cross-sectional survey in France. *Prev. Med.* **1999**, *28*, 333–339. [CrossRef] [PubMed]
7. Hayes, D.; Angove, M.J.; Tucci, J.; Dennis, C. Walnuts (*Juglas regia* L.) chemical composition and research in human health. *Crit. Rev. Food Sci. Nutr.* **2016**, *56*, 1231–1241. [CrossRef]
8. Molnar, T.J.; Zaurov, D.E.; Capik, J.M.; Eisenman, S.W.; Ford, T.; Nikolyi, L.V.; Funk, C.R. Persian Walnut (*Juglas regia* L.) in central Asia. *Annu. Rep. North. Nut Growers Assoc.* **2011**, *101*, 56–69.
9. Mitra, S.K.; Rathore, D.S.; Bose, T.K. Walnut. In *Temperate Fruits*; Horticulture and Allied Publishers: Calcutta, India, 1991; pp. 377–414.
10. Food and Agriculture Organisation of United Nations. Available online: http://www.fao.org/faostat (accessed on 9 April 2018).
11. Marcone, M.F.; Wang, S.; Albabish, W.; Nie, S.; Somnarain, D.; Hill, A. Diverse food-based applications of nuclear magnetic resonance (NMR) technology. *Food Res. Int.* **2013**, *51*, 729–747. [CrossRef]
12. Popescu, R.; Costinel, D.; Dinca, O.R.; Marinescu, A.; Stefanescu, I.; Ionete, R.E. Discrimination of vegetable oils using NMR spectroscopy and chemometrics. *Food Control* **2015**, *48*, 84–90. [CrossRef]
13. Spangenberg, J.E.; Ogrinc, N. Authentication of vegetable oils by bulk and molecular carbon isotope analyses with emphasis on olive oil and pumpkin seed oil. *J. Agric. Food Chem.* **2001**, *49*, 1534–1540. [CrossRef]
14. Paolini, M.; Bontempo, L.; Camin, F. Compund-specific δ^{13}C and δ^{2}H analysis of olive oil fatty acids. *Talanta* **2017**, *174*, 38–43. [CrossRef]
15. Guyader, S.; Thomas, F.; Portaluri, V.; Jamin, E.; Akoka, S.; Silvestre, V.; Remaud, G. Authentication of edible fats and oils by non-targeted ^{13}C INEPT NMR spectroscopy. *Food Control* **2018**, *91*, 216–224. [CrossRef]
16. Esteki, M.; Vander Heyden, Y.; Farajmand, B.; Kolahderazi, Y. Qualitative and quantitative analysis of peanut adulteration in almond powder samples using multi-elemental fingerprinting combined with multivariate data analysis methods. *Food Control* **2017**, *82*, 31–41. [CrossRef]
17. Amaral, J.S.; Casal, S.; Pereira, J.A.; Seabra, R.M.; Oliveira, B.P.P. Determination of sterol and fatty acid compositions, oxidative stability, and nutritional value of six walnut (*Juglans regia* L.) cultivars grown in Portugal. *J. Agric. Food Chem.* **2003**, *51*, 7698–7702. [CrossRef] [PubMed]
18. Batun, P.; Bakkalbasi, E.; Kazankaya, A.; Cavioglu, I. Fatty acid profiles and mineral contents of walnuts from different provinces of Van Lake. *GIDA* **2017**, *42*, 155–162. [CrossRef]
19. Arranz, S.; Cert, R.; Perez-Jimenez, J.; Cert, A.; Saura-Calixto, F. Comparison between free radical scavenging capacity and oxidative stability of nut oils. *Food Chem.* **2008**, *110*, 985–990. [CrossRef]
20. Bujdoso, G.; Konya, E.; Berki, M.; Nagy-Gasztonyi, M.; Bartha-Szughi, K.; Marton, B.; Izsepi, F.; Adanyi, N. Fatty acid composition, oxidative stability, and antioxidant properties of some Hungarian and other Persian walnut cultivars. *Turk. J. Agric. For.* **2016**, *40*, 160–168. [CrossRef]
21. Al-Bachir, M. Effect of gamma irradiation on fungal load, chemical and sensory characteristics of walnuts (*Juglans regia* L.). *J. Stored Prod. Res.* **2004**, *40*, 355–362. [CrossRef]
22. Crews, C.; Hough, P.; Godward, J.; Brereton, P.; Lee, M.; Guiet, S.; Winkelmann, W. Study of the main constituents of some authentic walnut oils. *J. Agric. Food Chem.* **2005**, *53*, 4853–4860. [CrossRef]
23. Christopoulos, M.; Tsantili, E. Oil composition in stored walnut cultivars–quality and nutritional value. *Eur. J. Lipid Sci. Technol.* **2014**, *116*, 1–11. [CrossRef]
24. Spangenberg, J.E.; Macko, S.A.; Hunziker, J. Characterization of olive oil by carbon isotope analysis of individual fatty acids: implications for authentication. *J. Agric. Food Chem.* **1998**, *46*, 4179–4184. [CrossRef]
25. Morrison, D.J.; Dodson, B.; Slater, C.; Preston, T. ^{13}C natural abundance in the British diet: implications for ^{13}C breath tests. *Rapid Commun. Mass Spectrom.* **2000**, *14*, 1321–1324. [CrossRef]

26. Guo, L.X.; Xu, X.M.; Yuan, J.P.; Wu, C.F.; Wang, J.H. Characterization and authentication of significant Chinese edible oilseed oils by Stable Carbon Isotope Analysis. *J. Am. Oil Chem. Soc.* **2010**, *87*, 839–848. [CrossRef]
27. Martinez, M.L.; Labuckas, D.O.; Lamarque, A.L.; Maestri, D.M. Walnut (*Juglans regia* L.): Genetic resources, chemistry, by-products. *J. Sci. Food Agric.* **2010**, *90*, 1959–1967. [CrossRef] [PubMed]
28. National Meteorological Administration-Climate Monitoring. Available online: http://www.meteoromania.ro/anm2/clima/monitorizare-climatica/ (accessed on 9 April 2018).
29. Iacumin, P.; Bernini, L.; Boschetti, T. Climatic factors influencing the isotope composition of Italian olive oils and geographic characterization. *Rapid Commun. Mass Spectrom.* **2009**, *23*, 448–454. [CrossRef]
30. Costinel, D.; Tudorache, A.; Ionete, R.E.; Vremera, R. The impact of grape varieties to wine isotopic characterization. *Anal. Lett.* **2011**, *44*, 2856–2864. [CrossRef]
31. Hoefs, J. *Stable Isotope Geochemistry*; Springer: Berlin, Germany, 1980; Volume 9.
32. Rubel, F.; Kottek, M. Observed and projected climate shifts 1901-2100 depicted by world maps of the Köppen-Geiger climate classification. *Meteorol. Z.* **2010**, *19*, 135–141. [CrossRef]
33. Constantinescu, M.; Oancea, S.; Bucura, F.; Ciucure, C.; Ionete, R.E. Evaluation of the fuel potential of sewage sludge mixtures with beech sawdust and lignite. *J. Renew. Sustain. Ener.* **2018**, *10*, 053106. [CrossRef]
34. Guillén, M.D.; Ruiz, A. Rapid simultaneous determination by proton NMR of unsaturation and composition of acyl groups in vegetable oils. *Eur. J. Lipid Sci. Technol.* **2003**, *105*, 688–696. [CrossRef]

11

Development of an Impedimetric Aptasensor for Label Free Detection of Patulin in Apple Juice

Reem Khan [1,2,3], Sondes Ben Aissa [1,4], Tauqir A. Sherazi [2], Gaelle Catanante [1], Akhtar Hayat [3,*] and Jean Louis Marty [1,*]

1 BAE: Biocapteurs-Analyses-Environnement, Universite de Perpignan Via Domitia, 52 Avenue Paul Alduy, 66860 Perpignan CEDEX, France; Kreemjadoon@gmail.com (R.K.); sondes.benaissa@fst.utm.tn (S.B.A.); gaelle.catanante@univ-perp.fr (G.C.)

2 Department of chemistry, COMSATS University Islamabad, Abbottabad Campus, Abbottabad 22060, Pakistan; sherazi@cuiatd.edu.pk

3 Interdisciplinary Research Centre in Biomedical Materials (IRCBM), COMSATS University Islamabad, Lahore Campus, Lahore 54000, Pakistan

4 Université de Tunis El Manar, Faculté des Sciences de Tunis, Laboratoire de Chimie Analytique et Electrochimie (LR99ES15), Sensors and Biosensors Group, Campus Universitaire de Tunis El Manar, Tunis 2092, Tunisia

* Correspondence: akhtarhayat@cuilahore.edu.pk (A.H.); jlmarty@univ-perp.fr (J.L.M.)

Abstract: In the present work, an aptasensing platform was developed for the detection of a carcinogenic mycotoxin termed patulin (PAT) using a label-free approach. The detection was mainly based on a specific interaction of an aptamer immobilized on carbon-based electrode. A long linear spacer of carboxy-amine polyethylene glycol chain (PEG) was chemically grafted on screen-printed carbon electrodes (SPCEs) via diazonium salt in the aptasensor design. The NH_2-modified aptamer was then attached covalently to carboxylic acid groups of previously immobilized bifunctional PEG to build a diblock macromolecule. The immobilized diblocked molecules resulted in the formation of long tunnels on a carbon interface, while the aptamer was assumed as the gate of these tunnels. Upon target analyte binding, the gates were assumed to be closed due to conformational changes in the structure of the aptamer, increasing the resistance to the charge transfer. This increase in resistance was measured by electrochemical impedance spectroscopy, the main analytical technique for the quantitative detection of PAT. Encouragingly, a good linear range between 1 and 25 ng was obtained. The limit of detection and limit of quantification was 2.8 ng L^{-1} and 4.0 ng L^{-1}, respectively. Selectivity of the aptasensor was confirmed with mycotoxins commonly occurring in food. The developed apta-assay was also applied to a real sample, i.e., fresh apple juice spiked with PAT, and toxin recovery up to 99% was observed. The results obtained validated the suitability and selectivity of the developed apta-assay for the identification and quantification of PAT in real food samples.

Keywords: impedimetric aptasensor; screen-printed interface; bifunctional polymer arms; PAT detection; apple juice

1. Introduction

Effective detection of low molecular weight toxic molecules is vital in areas such as environmental monitoring and the food industry [1–3]. Mycotoxins are small agro-based food contaminants that pose serious health threats to humans and animals [4]. Patulin (PAT) is a low molecular weight mycotoxin produced by various fungal species. This small mycotoxin is commonly present in fruit- and vegetable-based products, especially apples [5,6]. The antibiotic properties of PAT were firstly reported in 1940. Some studies have also shown the adverse health effects of PAT on higher plants and animals [7]. Extensive studies have been carried out to highlight the health risks of PAT, and it has been concluded that PAT exhibits several chronic, acute, and cellular level toxic effects both on human and animals [8]. PAT has neurotoxic, immunosuppressive, and mutagenic effects on several animal species and cause severe damage to the intestinal epithelium, which leads to degeneration, inflammation, ulceration, and bleeding [8,9]. Due to such pronounced toxicities, different authorities have set regulatory limits for PAT in food. The European Union (EU) is the first organization that defined a maximum admissible limit, i.e., 50 $\mu g\ L^{-1}$ [10].

To monitor the level of PAT in food, several analytical techniques have been established including HPLC, thin layer chromatography with mass spectrometry and UV detection [11], gas chromatography coupled with mass spectrometry [12], micellar electrokinetic chromatography (MEKC), and colorimetry [13–16]. Despite their high reliability and accuracy, these techniques suffer many disadvantages such as complex instrumentation, high running cost, and complicated handling methods that make on-site analysis unsuitable. Owing to such drawbacks, there is an immense need of reliable, fast, and easy-to-operate screening methods.

Biosensors, being simple, reliable, and fast screening tools, have emerged as an attractive alternative to classical analytical methods. Among different types of biosensors, electrochemical aptasensors offer significant advantages over the others. [17]. Change in the unique three-dimensional structure of aptamers upon target binding provides great flexibility in developing electrochemical aptasensors. In most cases, an aptamer is tagged with a redox label such as ferrocene or methylene blue [18]. These types of biosensor designs have been successfully employed to detect different targets, including mycotoxins. However, their main problem is that they require large conformational changes in the aptamer structure to tune the distance of the electro-active label from the transducer surface. Moreover, the process of labeling aptamers is expensive, and labeling sometimes decreases the activity of the modified biomolecules, which results in a more complex, laborious, and time-consuming assay design. [19]. As an alternative, interest in the development of simpler and cheaper label-free aptasensors has recently increased. These platforms are generally based on electrochemical impedance spectroscopy (EIS) as a powerful technique, allowing sensitive detection of the smallest variations in electron transfer processes. Still, in the case of small analytes, target binding events do generate an easily measurable impedimetric signal, because physical hindrance induced by these low molecular weight analytes is often insufficient to create detectable variation. However, ultrasensitive detection of low weight targeted analytes by a simple aptasensor design is quite difficult. Different methods have been used for signal amplification, such as strand displacement amplification, rolling circle amplification, and the use of nanomaterials [20,21]. Although these approaches offer advantages in signal intensification, they are costly, are complex, and have difficult operating conditions. These problems have resulted in the exploration of novel aptasensor designs to ensure the sensitive detection of low molecular weight targets. Integration of a spacer in aptasensor design could be a simple and effective method of signal amplification. In this work, we present this strategy for the detection of a small mycotoxin, i.e., PAT.

Furthermore, considering the mycotoxin analysis, multiple efforts have been made to design and develop biosensors for aflatoxins and ochratoxins, but very few reports are available for PAT detection [22,23]. In the present work, a novel impedimetric aptasensor was designed for the label-free detection of PAT. A long spacer, i.e., carboxy-amine PEG, was used to facilitate the label-free detection and enhance the sensitivity of the developed aptasensor. Disposable screen-printed carbon electrodes (SPCEs) were used as transducers. The SPCEs were based on a conventional three-electrode system with a working electrode made up of graphite (with a circular disk 4 mm in diameter), a counter electrode (curved line: 16×1.5 mm), and Ag/AgCl (straight line: 16×1.5 mm) as a pseudo reference electrode. All parameters were carefully optimized. The developed aptasensor was then assessed for a real sample, i.e., apple juice. To the best of our knowledge, no such impedimetric aptasensor has yet been reported for the detection of PAT.

2. Materials and Methods

2.1. Design and Working Principle of Impedimetric Aptasensor

A schematic representation of the working principle and the stepwise design strategy of the developed impedimetric aptasensor for PAT detection is shown in Scheme 1. As indicated, three major steps were carried out for the fabrication process. In the proposed aptasensor, a diblock macromolecule (carboxy-amine PEG + anti-PAT-aptamer) was chemically grafted onto the surface of SPCEs to form clusters of long spacer arms on the working electrode. Aminobenzoic acid was first electrochemically grafted onto the surface of SPCEs through linear sweep voltammetry (LSV). This technique is based on the electrochemical reduction of diazonium salt, leading to aryl-centered radicals by the spontaneous release of nitrogen. The voltammogram in Figure S1 shows the successful reduction of diazonium cations via a one-electron process. An insulating layer of carboxyphenyl was covalently formed on the surface.

Diazonium deposition was then followed by the chemical grafting of carboxy-amine PEG through the amide bond. The long spacer arms of heterobifunctional PEG acted like tunnels for the electrons of the redox marker to reach the surface of the electrode. The second part of the diblock, i.e., the anti-PAT aptamer, acted as gates for these tunnels. In the absence of a targeted analyte, the aptamer remained unfolded, and it is assumed that the gates were opened, thereby permitting the electron transfer from the redox probe to the surface of the electrode. Meanwhile, in the presence of the targeted analyte, 3D conformational changes in the aptamer structure resulted in the formation of an aptamer-PAT quadruplex complex that locked the gates of tunnels, resulting in blockage of the electron flow toward the electrode surface [24]. Furthermore, the conformational changes in the aptamer structure also resulted in the exposition of the negatively charged backbone of the aptamer. Consequently, steric hindrance after the formation of the quadruplex complex and the electrostatic repulsion between the anionic ferri/ferrocyanide redox system and the negatively charged aptamer collectively increased the impedance of the system. The higher the concentration of PAT is, the higher the impedance will be. This concept makes the quantitative detection of PAT feasible.

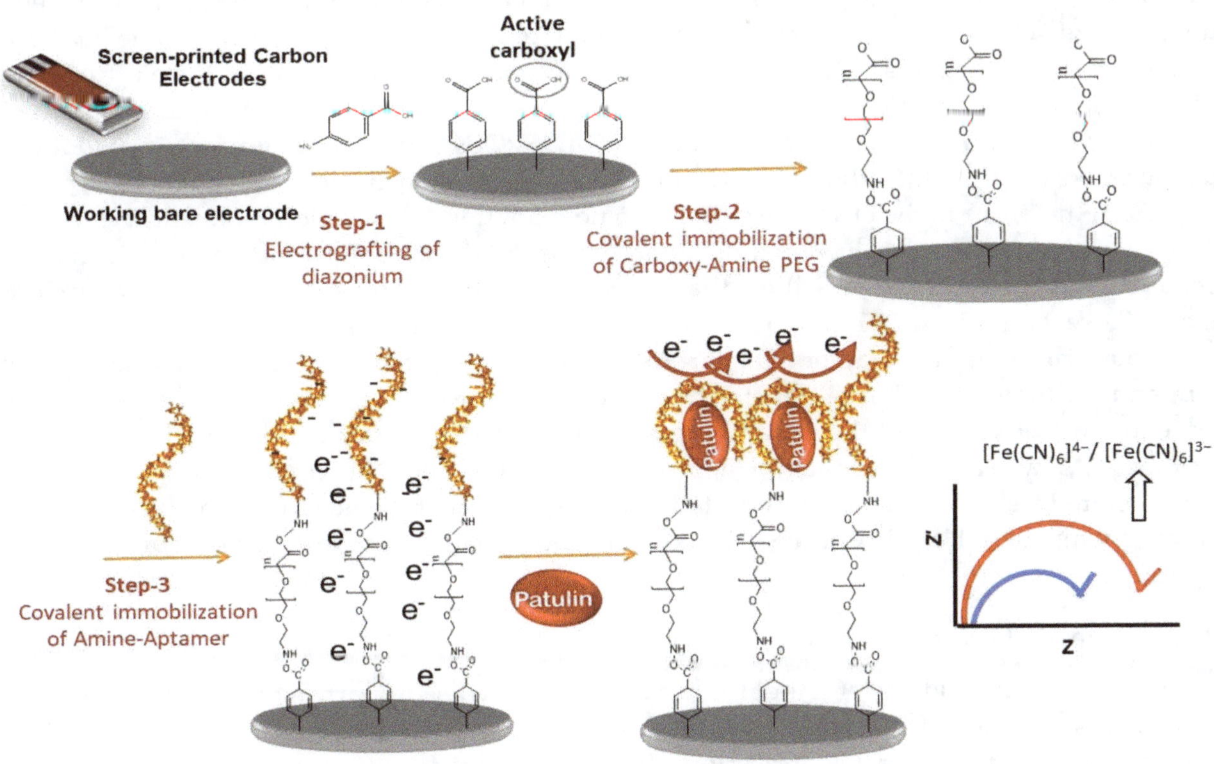

Scheme 1. Design and working principle of the proposed aptasensor.

2.2. Electrochemical Characterization of the Aptasensors

Cyclic voltammetry (CV) and EIS are widely used techniques to characterize the electrode/electrolyte interfacial properties at different modification steps of aptasensor fabrication. Thus, the different preparation stages were investigated by recording the impedance spectra and cyclic voltammograms of the modified electrode in the presence of the reversible $[Fe\ (CN)_6]^{4-/3-}$ redox system, shown in Figure 1.

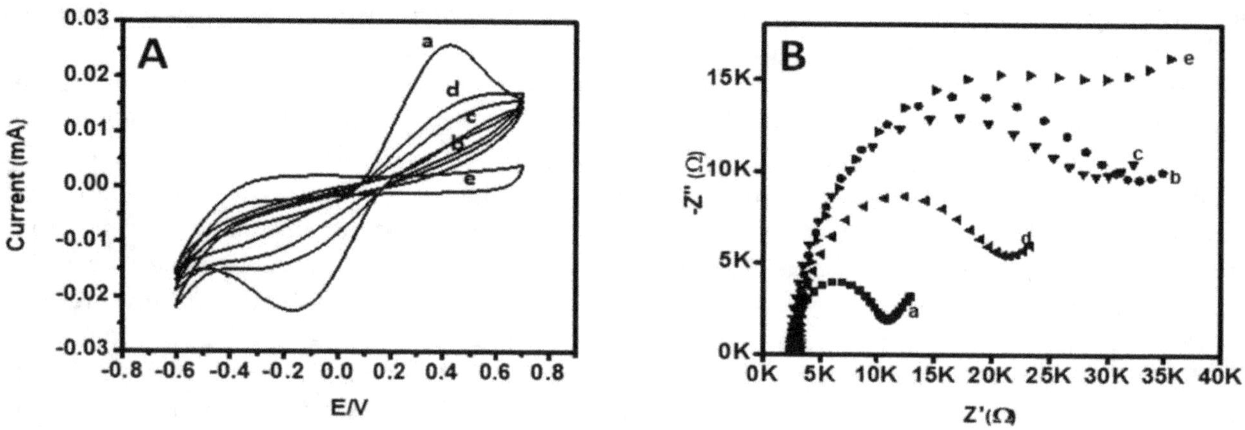

Figure 1. Characterization of modification steps during aptasensor fabrication. (**A**) Cyclic voltammogram obtained in 1 mM Ferri/Ferrocyanide after each modification step during aptasensor fabrication: (a) bare SPCE; (b) carboxyphenyl modified SPCE; (c) after immobilization of carboxy-amine PEG; (d) anti-patulin (anti-PAT) aptamer modified electrode; (e) after incubation of 50 ng L^{-1} patulin (PAT). (**B**) Nyquist plots of a 1 mM $[Fe(CN)_6]^{3-/4-}$ probe in PBS, pH 7.4.

2.2.1. Cyclic Voltammetry Characterization

CV in the ferri/ferrocyanide solution is a useful tool for examining the electrochemical behavior of functionalized electrodes, as electron transfer is more or less feasible depending on the hindrance of steric barriers. Therefore, this technique was used to investigate and characterize the electrode surface modifications. Figure 1A shows the cyclic voltammograms obtained after each modification of the electrode surface during aptasensor fabrication. Bare SPCEs exhibited a symmetric oxidation and reduction peaks (Figure 1A(a)), but the insulating layer of 4-aminobenzoic acid (ABA) blocked electron transfer from the probe to the electrode surface, so no oxidation or reduction took place, as we can see in Figure 1A(b). After that, bifunctional PEG was deposited through the amide bond. The long linear spacer arms of carboxy-amine PEG allow the electron to move from the probe solution to the electrode surface. Hence, an increase in redox peaks of cyclic voltammogram can be observed in Figure 1A(c). The immobilization of the aptamer over carboxy-amine PEG further decreased the resistance of the system and facilitated the charge transfer from the solution to the electrode surface, resulting in an increase in the redox shape of the voltammogram (Figure 1A(d)). After the completion of aptasensor fabrication, PAT was incubated over the aptasensor. Changes in the 3D conformation of the aptamer structure offered steric hindrance and resulted in the exposition of the negative charges of the aptamer that strongly repelled the redox probe. No redox reaction occurred at the electrode surface, as shown in Figure 1A(e).

2.2.2. Impedance Characterizations

EIS is a powerful and a useful tool for depicting the barrier properties of the modified electrode. The R_{et} values obtained after fitting the curves to a Randles equivalent circuit are shown in Table S1 for various stages of aptasensor fabrication. Faradic impedance spectra presented in the form of

Nyquist plots are shown in Figure 1B for the stepwise modification process. Bare SPCEs exhibited a small semicircle, as shown in Figure 1B(a), indicating a very low resistance in electron transfer from the redox probe to the electrode surface. After the electro-grafting of ABA via the diazonium ion, the resistance of the surface significantly increased (Figure 1B(b)). This increase in resistance was due to the presence of the COO^- surface, which acted as a barrier layer and hindered the electron transfer from the anionic ferri/ferrocyanide redox probe to the electrode due to strong electrostatic repulsion. The second modification was the covalent attachment of carboxy-amine PEG. For this purpose, the COOH group was first activated with EDC-NHS chemistry. After the activation of COOH carboxy-amine PEG was covalently attached via amide coupling between the terminal NH_2 group of carboxy-amine PEG and the COO^- group of the carboxyphenyl layer, there was a significant increase in R_{et} (Figure 1B(c)).

Contrary to the expected results, a small decrease in impedance was observed, as compared to that obtained after ABA electrodeposition, although in both cases $COO^-$was exposed on the surface. This comparative decrease in R_{et} can be attributed to the formation of tunnels on the electrode surface after the deposition of heterobifunctional PEG, which facilitates the electron flow from the probe solution to the electrode. In the case of a carboxyphenyl layer, these tunnels were absent, so the substrate was completely and uniformly covered with negative charges. The electrostatic repulsion between COO^- did not allow any electrons to transfer from the probe solution to the electrode. Such behavior of modified electrodes demonstrates that the formation of tunnels on the electrode surface is due to the long linear spacer arms of carboxy-amine PEG [24]. After successful immobilization of bifunctional PEG, the exposed COOH group was again activated by EDC-NHS. Subsequently, amine-terminated anti-PAT aptamer was immobilized via amide bond formation. A decrease in R_{et} in Figure 1B(d), as compared to 1B(c), proved the effective immobilization of the aptamer, as the electrostatic repulsion between COO^- and the redox probe was reduced after aptamer immobilization. The incubation of PAT increased in R_{et} values, which can be seen in Figure 1B(e). The quadruplex formed upon target recognition, blocked the molecular gates and exposed the negatively charged backbone of the aptamer toward the anionic probe that in-turn increased the impedance of system.

2.3. *Optimization of Experimental Parameters*

The factors that can affect the electrochemical properties of the proposed aptasensor are pH, temperature, analyte incubation time, and the concentration of the aptamer immobilized on the electrode surface. Therefore, these parameters were optimized to obtain better analytical performance. Results are shown in Figure 2. The optimal response of the aptasensor was obtained at room temperature and pH 7.4. The concentration of carboxy-amine PEG was optimized to be 6 mg mL^{-1}.

Furthermore, the time required to immobilize the maximum concentration of aptamers on the active surface of the SPCEs was optimized. Variation in the Δ_{ratio} was recorded in the time range of 5–90 min. It was observed that maximum immobilization of the aptamer was achieved after 30 min (Figure 2A).

The concentration of the aptamer is another important parameter. We recorded the Δ_{ratio} for aptamer concentration ranging between 0.5 and 5.0 μM. Results in Figure 2B indicated that the maximum Δ_{ratio} was observed at a 2.0 μM concentration. Therefore, 2.0 μM was considered as the optimum concentration and was used in the following experiments.

The analyte incubation time is also an important parameter optimized during the aptasensor fabrication. Optimum incubation time for the analyte was found to be 45 min, as shown in Figure 2C.

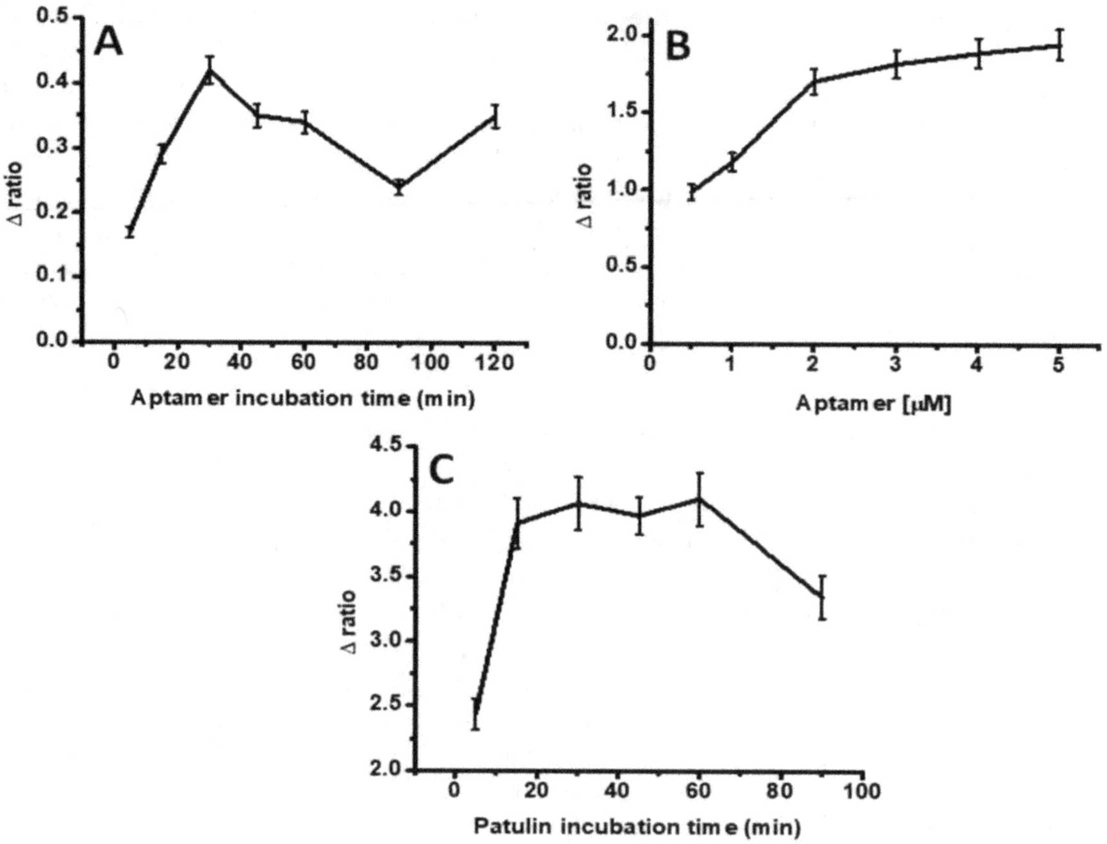

Figure 2. Optimization of experimental parameters. (**A**) Variation in Δ_{ratio} with increasing time after incubation with aptamer solution. (**B**) Variation in Δ_{ratio} with varying aptamer concentration. (**C**) Variation in Δ_{ratio} obtained at different intervals of time after the incubation of PAT.

2.4. Analytical Performance of the Aptasensors

To evaluate the analytical performance of the developed impedimetric aptasensor, different concentrations of PAT were incubated on the proposed aptasensing platform. A significant increase in the R_{et} was observed with increasing PAT concentration. A Nyquist plot for the electrochemical aptasensor in response to different concentrations of PAT is presented in Figure 3. The results indicated that the more PAT was bound to the electrode surface, the higher the R_{et} of the system was. This proportional increase in R_{et} indicated that the developed aptasensor can quantitatively detect PAT. In order to calculate a reliable estimation of increase or decrease in impedance, Δ_{ratio} was calculated for all aptasensors (different electrodes developed in triplicate).

Figure 3B shows that the Δ_{ratio} of the aptasensor gradually increases with an increasing concentration of PAT.

Limit of detection and limit of quantification were determined by calculating the signal-to-noise ratio in the blank. The formulae are

$$LOD = 3 * \text{standard deviation of blank} / S$$

$$LOQ = 10 * \text{standard deviation of blank} / S$$

where "S" is the slope of calibration curve.

The limit of detection obtained from the impedimetric aptasensor is 2.8 ng L^{-1}, while the limit of quantification is 4.0 ng L^{-1}.

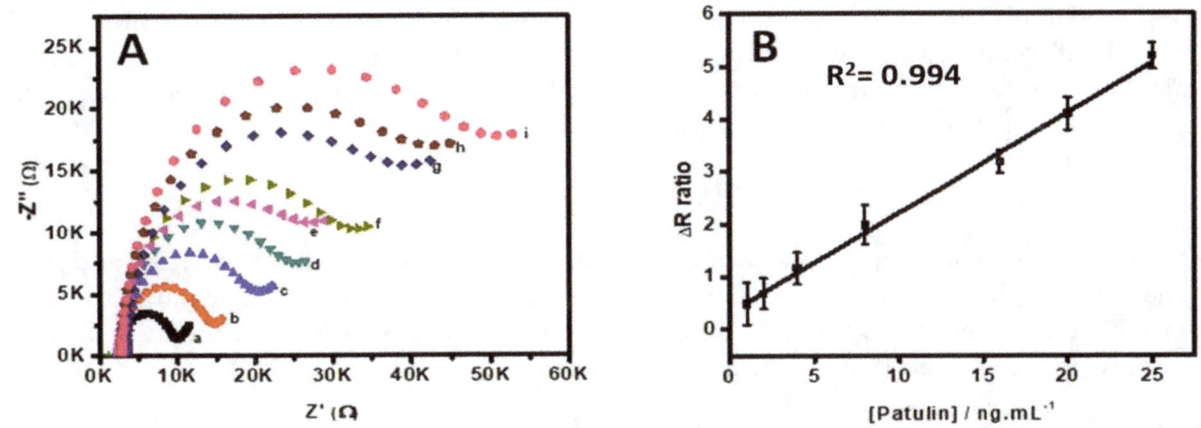

Figure 3. (**A**) Nyquist plot of the aptasensor after incubation with different PAT concentrations: (a) bare, (b) 1 ng mL^{-1}, (c) 2 ngmL^{-1}, (d) 4 ng mL^{-1}, (e) 8 ng mL^{-1}, (f) 16 ng mL^{-1}, (g) 20 ng mL^{-1}, (h) 25 ng mL^{-1}, and (i) 50 ng mL^{-1}. (**B**) Calibration plot of Δ_{ratio} with increasing PAT concentration.

2.5. Selectivity of the Aptasensors

Selectivity of the aptasensor against PAT was checked by incubating the aptasensor with different interfering analytes such as ochratoxin A, ochratoxin B, and aflatoxins. Obtained results are presented in Figure 4. No significant increase in R_{et} values can be seen in the case of mycotoxins other than PAT, which confirms the high selectivity and specificity of the aptasensor for PAT. In the same context, a control single-strand DNA was immobilized on the transducer surface in a fashion similar to that described for immobilization of the aptamer. No significant change in R_{et} was observed in the presence of PAT. These results also prove that there were no non-specific binding sites on the transducer surface.

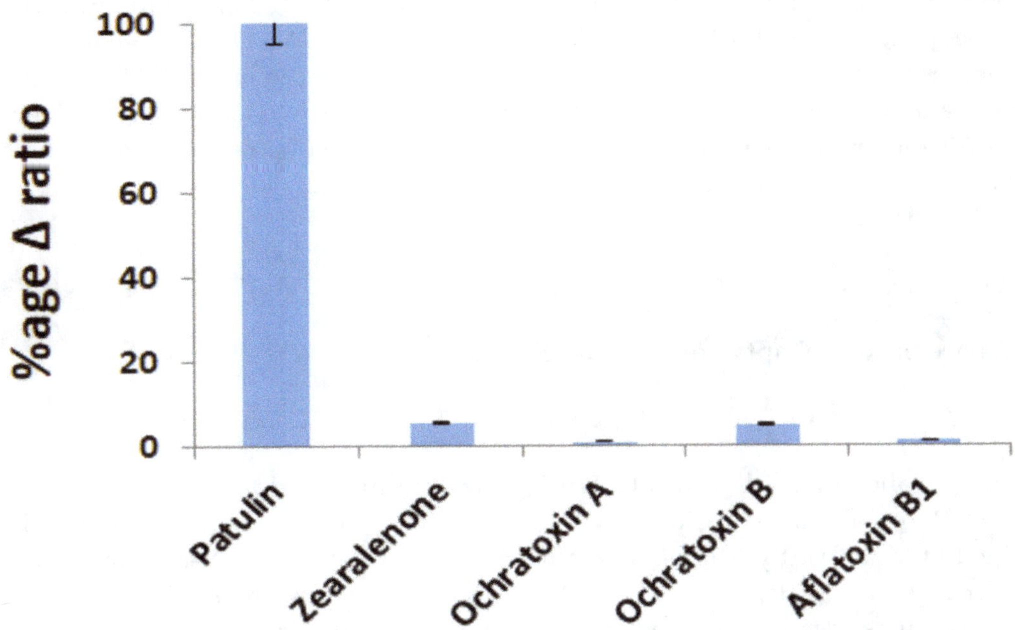

Figure 4. Selectivity test for the developed aptasensor.

2.6. Application to Spiked Apple Juice Sample

The developed impedimetric aptasensor was evaluated for real sample analysis. For a real sample application, PAT was detected in apple juice. Apple juice samples were spiked with three different concentrations of PAT, i.e., 4, 10, and 20 ng mL^{-1}. Controlled experiments were performed by detecting the same concentrations of PAT in phosphate binding buffer. The electrochemical response in apple juice exhibits the same phenomenon as observed in buffer. Results are presented in Table 1. This experiment confirms the feasibility of the aptasensor for field applications. The recovery values were calculated based on the integration of the Δ_{ratio} value (electrochemical impedimetric response) of the spiked value in the equation of the calibration curve.

$$Y = 0.3128 + 0.1901 * X$$

The precision and accuracy of the proposed aptasensor was confirmed by interday and intraday assay analysis. A relative standard deviation (RSD) of 2.8–4.0% ($n = 3$) was calculated for interday analysis, which indicated very good reproducibility of the assay. Similarly, for intraday analysis, RSD was 1.34%, which confirmed that results are reproducible.

Table 1. Recovery percentages obtained with designed electrochemical aptasensor for PAT monitoring in apple juice.

PAT Added (ng/mL)	PAT Found (ng/mL)	RSD %	RE %	R %
4	3.7	3.7	7.5	92.5
10	9.6	4.1	4	96
20	18.8	4.8	6	94

RSD % = relative standard deviation percentage; RE % = relative error percentage; R % = recovery percentage.

3. Experimental

3.1. Chemicals

Amine-terminated aptamer sequence for PAT was purchased from Microsynth (Balgach, Switzerland) The aptamer sequence was developed by Shijia Wu et al. by a graphene oxide-assisted SELEX (systematic evolution of ligands by exponential enrichment) process. Initially, eight aptamer sequences were shortlisted, and an aptamer named PAT-11 was then selected as the final recognition element based on its low K_d (dissociation constant) value, i.e., 21.83 nM, and the highest binding affinity and selectivity for PAT. The potential applicability of the selected aptamer sequence was further confirmed by developing an enzyme-chromogenic substrate system-based colorimetric aptasensor for the selective and sensitive detection of PAT in a real matrix [25]:

Sequence of PAT aptamer:

5′GGC CCG CCA ACC CGC ATC ATC TAC ACT GAT ATT TTA CCT T-NH_2-3′.

The sequence of control aptamers was as follows:

(5′ GCA GTT GAT CCT TTG GAT ACC CTG G3′)-NH_2.

PAT, sodium phosphate dibasic Na_2HPO_4, potassium chloride KCl, potassium phosphate monobasic KH_2PO_4, calcium chloride $CaCl_2$, magnesium chloride $MgCl_2$, 4-aminobenzoic acid, carboxy-amine PEG (3400) ethyl-dimethyl-amino-propyl carbodiimide, *N*-hydroxy succinimide and ferricyanide, ferrocyanide, sodium nitrite, and 4-aminobenzoic acid were obtained from Sigma-Aldrich (Lyon, France). Sterilized water was used for aptamer solution preparation, while distilled water was used throughout the experimental procedure for reagent preparation. For real sample analysis, fresh apple juice was bought from a supermarket in Perpignan, France.

Home-made SPCEs were used for the electrochemical experiments.

3.2. Instrumentation

Pre-treatment of the aptamer (heating at 90 °C for 8 min followed by immediate cooling to 4 °C for 4 min then 15 min at RT) was done on thermocycler mastercycler (Eppendorf, Le pecq, France).

All the electrochemical measurements were performed on an Electrochemical Workstation "Biologics" equipped with EC-Lab software.

Lyophilized aptamer was diluted in phosphate binding buffer (PBB 50 mM). PBB was composed of 2 mM KH_2PO_4, 8mM Na_2HPO_4, 135 mM KCl, 60 mM NaCl, and 5 mM $MgCl_2$ in deionized water. The pH of the buffer was adjusted to 7.4.

3.3. Experimental Protocol

3.3.1. Covalent Immobilization of Aptamer onto the SPCEs

Before use, all electrodes were electrochemically washed with washing buffer (0.5 M H_2SO_4 diluted in 0.1 M KCl) with 8–10 CV scans between 1 and −1 V at 100 $mV.s^{-1}$

Diazonium salt was generated by mixing 2 mM ABA in 0.5 M HCl with 2 mM sodium nitrite. The solution was left to react for 5 min for maximum generation of diazonium salt. After 5 min, 150–200 μL of this salt solution was electrodeposited on SPCEs via LSV with a potential ranging between 0.6 and −0.8 V and a scan rate of 50 $mV.s^{-1}$ [26]. Furthermore, the exposed COOH group of ABA was activated through EDC/NHS chemistry (by immersing the electrode for 60 min in solution of 100 mM EDC and 25 mM NHS solution in 100 mM MES buffer pH = 5.5). After activation of COOH, 6 mM carboxy-amine PEG solution was then incubated on a working electrode for 40 min to attach the spacer covalently with aminobenzoic acid via an amide bond. The terminal COOH group of carboxy amine PEG was again activated through carbodiimide chemistry. Afterward, 2 μM NH2-modified PAT-aptamer was cast (20 μL) on the activated surface of the working electrode for an optimized immobilization time of 30 min. The NH_2 group of the aptamer made a covalent bond with a terminal COOH group of PEG. Finally, the working area was treated with 1% BSA solution to block non-reactive sites, therefore avoiding the possibility of non-specific adsorption.

All the modification steps were characterized by CV and EIS. The CV was performed within a potential range from 0.7 to −0.6 V at a 50 $mV.s^{-1}$ scan rate. EIS measurements were performed using 1 mM $[Fe (CN)_6]^{3-/4-}$ solution as a free redox probe containing 10 mM KCl.

The prepared electrode was stored at 4 °C and employed as a ready-to-use device during the present work. The ready-to-use electrodes did not show any variation in the electrochemical response over the entire period, suggesting the stability of the aptasensor.

3.3.2. Detection of Patulin

A stock solution of PAT was prepared in acetonitrile, and further dilutions were prepared in PBB. For the detection of PAT, selected concentrations of PAT were incubated for 1 h with a fabricated aptasensor, followed by rinsing with binding buffer, and electrochemical measurements were then carried out.

3.3.3. Apple Juice Sample Preparation

In a 5 mL falcon tube, apple juice was vortex-mixed with ethyl acetate solution (1:1) and centrifuged at 10m000 rpm for 5 min. The supernatant was dried in a nitrogen environment. Afterward, the obtained dry powder was dissolved in PBS buffer after filtering with a nylon syringe filter (0.22 mm). Different PAT concentrations (a 4, 10, and 20 ng mL^{-1} final concentration) were then added to 50 μL of the pre-treated apple juice samples in order to perform standard additions and recovery experiments. The pH of the apple juice sample was adjusted to 7.4.

3.4. Impedimetric Measurements

Impedance measurements were obtained at an applied potential of 100 mV (vs. Ag/AgCl reference electrode), within a frequency range from 10 KHz to 0.5 Hz, an AC amplitude of 10 mV, and a sampling rate of 50 points. All measurements were performed in PBB containing a redox probe (1:1 mixture of 2 mM $K_3[Fe(CN)_6]/K_4[Fe(CN)_6]$). Impedance data was registered after each modification step of electrode surface in the following order: (i) bare electrode; (ii) diazonium salt deposition; (iii) carboxy-amine PEG immobilization; (iv) aptamer immobilization steps; and (v) PAT detection. The impedance spectra were presented as Nyquist plots, ($-Z_{im}$ vs. Z_{re}) and fitted to the Randles equivalent circuit model, as shown in Figure S1, with EC Lab software. The parameter R_s, corresponds to the solution resistance. R_{et} is the resistance in electron transfer between the electrode surface and the redox marker, whereas C is the double-layer capacitance and W (Warburg parameter) is associated with the diffusion of the redox probe. For all fittings, performed on EC-Lab software, the chi-square goodness-of-fit test was carefully checked to verify calculations (with 80.73 at a 92% confidence level). In this work, we focused on the variation in electron transfer resistance (R_{et}). In order to minimize the electrode to electrode variations and to obtain reproducible and independent results, relative and normalized signals were required. Thus, the Δ_{ratio} parameter was calculated for all the electrodes. Δ_{ratio} is the ratio of R_{et} of the bare aptasensor to the R_{et} of the aptasensor with a particular amount of analyte. The R_{et} of the bare aptasensor was subtracted from the final values to ignore the electrode-to-electrode variation.

$$\Delta_{ratio} = R_{et\ (after\ PAT)} - R_{et\ (bare)} \ / \ R_{et\ (before\ PAT)} - R_{et\ (bare)}$$

where $R_{et\ (after\ PAT)}$ is the value obtained incubating the aptasensor with the PAT, $R_{et\ (before\ PAT)}$ is the electron transfer resistance value of the bare electrode, and $R_{et\ (bare)}$ is the resistance in electron transfer of the blank electrode and buffer.

4. Conclusions

In summary, we successfully developed a field-portable impedimetric aptasensor for selective, sensitive, and label-free detection of the carcinogenic mycotoxin known as PAT. Electrochemical impedimetric aptasensing for label-free detection of PAT was based on diazonium chemistry and bifunctional polyethylene glycol as spacer. The carboxy-amine PEG proved to be an effective spacer that resulted in tunnel formation for the electron transfer from the redox probe to the electrode surface. The LOD and LOQ obtained were 2.8 ng L^{-1} and 4.0 ng L^{-1}, respectively. Real sample analysis and interference study confirms that the developed aptasensor has the potential for PAT detection in the real matrix.

Supplementary Materials:
Figure S1: Linear sweep voltammogram for electrodeposition of the in situ generated 4-carboxyphenyl diazonium salt in the diazotization mixture at SPCE, Figure S2: Randles Circuit for fitting of Nyquist Plot, Table S1: R_{et} Values obtained after each modification step of Aptasensor fabrication Electrode.

Author Contributions: Conceptualization, A.H. and J.L.M.; methodology, R.K. and S.B.A.; software, S.B.A.; validation and formal analysis, R.K., T.A.S. and S.B.A.; investigation, R.K.; resources, G.C.; data curation, R.K.; writing—original draft preparation, R.K. and S.B.A.; writing—review and editing, A.H.; visualization, T.A.S.; supervision, A.H.; project administration, A.H.; funding acquisition, A.H.

Acknowledgments: Akhtar Hayat would like to acknowledge the project; HEC (Higher Education Commission of Pakistan)-Technology Development Program (TDF) project; No. 028, 2017. S.B.A. acknowledges the University of Tunis El Manar (Tunisia) for the mobility grant "Bourse d'alternance", as well as the Tunisian Ministry of Higher Education and Scientific Research for LR99ES15 Lab funding.

References

1. Xia, X.; He, Q.; Dong, Y.; Deng, R.; Li, J. Aptamer-based Homogeneous Analysis for Food Control. *Curr. Anal. Chem.* **2018**, *14*, 1–9. [CrossRef]
2. Patra, J.K.; Mahato, D.K.; Kumar, P. Biosensor Technology—Advanced Scientific Tools, With Special Reference to Nanobiosensors and Plant- and Food-Based Biosensors. In *Nanomaterials in Plants, Algae and Microorganisms*; Elsevier Academic Press: Cambridge, MA, USA, 2019; pp. 287–303.

3. Vasilescu, A.; Marty, J.L. Electrochemical aptasensors for the assessment of food quality and safety. *Trends Anal. Chem.* **2016**, *79*, 60–70. [CrossRef]
4. Laganà, A. *Introduction to the Toxins Special Issue on LC-MS/MS Methods for Mycotoxin Analysis*; Multidisciplinary Digital Publishing Institute: Basel, Switzerland, 2017.
5. Moss, M. Fungi, quality and safety issues in fresh fruits and vegetables. *J. Appl. Microbiol.* **2008**, *104*, 1239–1243. [CrossRef] [PubMed]
6. Ioi, J.D.; Zhou, T.; Tsao, R.; Marcone, M.F. Mitigation of patulin in fresh and processed foods and beverages. *Toxins* **2017**, *9*, 157. [CrossRef]
7. Iwahashi, Y.; Hosoda, H.; Park, J.-H.; Lee, J.-H.; Suzuki, Y.; Kitagawa, E.; Murata, S.M.; Jwa, N.-S.; Gu, M.-B.; Iwahashi, H. Mechanisms of patulin toxicity under conditions that inhibit yeast growth. *J. Agric. Food Chem.* **2006**, *54*, 1936–1942. [CrossRef] [PubMed]
8. Glaser, N.; Stopper, H. Patulin: Mechanism of genotoxicity. *Food Chem. Toxicol.* **2012**, *50*, 1796–1801. [CrossRef] [PubMed]
9. Mahfoud, R.; Maresca, M.; Garmy, N.; Fantini, J. The mycotoxin patulin alters the barrier function of the intestinal epithelium: Mechanism of action of the toxin and protective effects of glutathione. *Toxicol. Appl. Pharmacol.* **2002**, *181*, 209–218. [PubMed]
10. Commission, E. Commission Regulation (EC) No 1881/2006 of 19 December 2006 setting maximum levels for certain contaminants in foodstuffs. *Off. J. Eur. Union* **2006**, *364*, 5–24.
11. Zheng, Y.; Wang, X.; He, S.; Gao, Z.; Di, Y.; Lu, K.; Li, K.; Wang, J. Aptamer-DNA Concatamer-Quantum Dots Based Electrochemical Biosensing Strategy for Green and Ultrasensitive Detection of Tumor Cells via Mercury-Free Anodic Stripping Voltammetry. *Biosens. Bioelectron.* **2018**, *126*, 261–268. [CrossRef] [PubMed]
12. Sadok, I.; Szmagara, A.; Staniszewska, M.M. The validated and sensitive HPLC-DAD method for determination of patulin in strawberries. *Food Chem.* **2018**, *245*, 364–370. [CrossRef]
13. Shephard, G.S.; Leggott, N.L. Chromatographic determination of the mycotoxin patulin in fruit and fruit juices. *J. Chromatogr. A* **2000**, *882*, 17–22. [CrossRef]
14. Takino, M.; Daishima, S.; Nakahara, T. Liquid chromatography/mass spectrometric determination of patulin in apple juice using atmospheric pressure photoionization. *Rapid Commun. Mass Spectrom.* **2003**, *17*, 1965–1972. [CrossRef] [PubMed]
15. Gimeno, A. Thin layer chromatographic determination of aflatoxins, ochratoxins, sterigmatocystin, zearalenone, citrinin, T-2 toxin, diacetoxyscirpenol, penicillic acid, patulin, penitrem A. *J. Assoc.* **1979**, *62*, 579–585.
16. Beltrán, E.; Ibáñez, M.; Sancho, J.V.; Hernández, F. Determination of patulin in apple and derived products by UHPLC–MS/MS. Study of matrix effects with atmospheric pressure ionisation sources. *Food Chem.* **2014**, *142*, 400–407. [CrossRef]
17. Grabowska, I.; Sharma, N.; Vasilescu, A.; Iancu, M.; Badea, G.; Boukherroub, R.; Ogale, S.; Szunerits, S. Electrochemical Aptamer-Based Biosensors for the Detection of Cardiac Biomarkers. *ACS Omega* **2018**, *3*, 12010–12018. [CrossRef]
18. Goud, K.Y.; Hayat, A.; Catanante, G.; Satyanarayana, M.; Gobi, K.V.; Marty, J.L. An electrochemical aptasensor based on functionalized graphene oxide assisted electrocatalytic signal amplification of methylene blue for aflatoxin B1 detection. *Electrochim. Acta* **2017**, *244*, 96–103. [CrossRef]
19. Rhouati, A.; Catanante, G.; Nunes, G.; Hayat, A.; Marty, J.-L. Label-free aptasensors for the detection of mycotoxins. *Sensors* **2016**, *16*, 2178. [CrossRef]
20. Wu, L.; Xiong, E.; Zhang, X.; Zhang, X.; Chen, J. Nanomaterials as signal amplification elements in DNA-based electrochemical sensing. *Nano Today* **2014**, *9*, 197–211. [CrossRef]
21. Huang, L.; Wu, J.; Zheng, L.; Qian, H.; Xue, F.; Wu, Y.; Pan, D.; Adeloju, S.B.; Chen, W. Rolling chain amplification based signal-enhanced electrochemical aptasensor for ultrasensitive detection of ochratoxin A. *Anal. Chem.* **2013**, *85*, 10842–10849. [CrossRef]
22. He, B.; Dong, X. Aptamer based voltammetric patulin assay based on the use of ZnO nanorods. *Microchim. Acta* **2018**, *185*, 462. [CrossRef] [PubMed]
23. Wu, Z.; Xu, E.; Jin, Z.; Irudayaraj, J. An ultrasensitive aptasensor based on fluorescent resonant energy transfer and exonuclease-assisted target recycling for patulin detection. *Food Chem.* **2018**, *249*, 136–142. [CrossRef] [PubMed]

24. Hayat, A.; Andreescu, S.; Marty, J.-L. Design of PEG-aptamer two piece macromolecules as convenient and integrated sensing platform: Application to the label free detection of small size molecules. *Biosens. Bioelectron.* **2013**, *45*, 168–173. [CrossRef] [PubMed]
25. Wu, S.; Duan, N.; Zhang, W.; Zhao, S.; Wang, Z. Screening and development of DNA aptamers as capture probes for colorimetric detection of patulin. *Anal. Biochem.* **2016**, *508*, 58–64. [CrossRef] [PubMed]
26. Mahouche-Chergui, S.; Gam-Derouich, S.; Mangeney, C.; Chehimi, M.M. Aryl diazonium salts: A new class of coupling agents for bonding polymers, biomacromolecules and nanoparticles to surfaces. *Chem. Soc. Rev.* **2011**, *40*, 4143–4166. [CrossRef] [PubMed]

12

Fatty Acid, Lipid Classes and Phospholipid Molecular Species Composition of the Marine Clam *Meretrix lyrata* (Sowerby 1851) from Cua Lo Beach, Nghe an Province, Vietnam

Quoc Toan Tran [1,2], Thi Thanh Tra Le [2,3], Minh Quan Pham [1,2], Tien Lam Do [1], Manh Hung Vu [4], Duy Chinh Nguyen [5], Long Giang Bach [5,6], Le Minh Bui [5,*] and Quoc Long Pham [1,2,*]

1 Institute of Natural Products Chemistry, Vietnam Academy of Science and Technology, Hanoi 122100, Vietnam; tranquoctoan2010@gmail.com (Q.T.T.); minhquanaries@gmail.com (M.Q.P.); dotienlam198@gmail.com (T.L.D.)

2 Graduate University of Science and Technology, Vietnam Academy of Science and Technology, Hanoi 122100, Vietnam; traltt@wru.vn

3 Department of Chemical Engineering, Faculty of Environment, Thuy loi University, Hanoi 122100, Vietnam

4 Institute of Marine Environment and Resources, Vietnam Academy of Science and Technology, Hanoi 122100, Vietnam; hungvm@imer.vast.vn

5 NTT Hi-Tech Institute, Nguyen Tat Thanh University, Ho Chi Minh City 700000, Vietnam; ndchinh@ntt.edu.vn (D.C.N.); blgiang@ntt.edu.vn (L.G.B.)

6 Center of Excellence for Biochemistry and Natural Products, Nguyen Tat Thanh University, Ho Chi Minh City 700000, Vietnam

* Correspondence: blminh@ntt.edu.vn (L.M.B.); mar.biochem@fpt.vn (Q.L.P.)

Abstract: This study aims to analyze compositions of fatty acids and phospholipid molecular species in the hard clams *Meretrix lyrata* (Sowerby, 1851) harvested from Cua Lo beach, Nghe An province, Viet Nam. Total lipid of hard clams *Meretrix lyrata* occupied 1.7 ± 0.2% of wet weight and contained six classes: hydrocarbon and wax (HW), triacylglycerol (TAG), free fatty acids (FFA), sterol (ST), polar lipid (PoL), and monoalkyl diacylglycerol (MADAG). Among the constituents, the proportion of PoL accounted was highest, at 45.7%. In contrast, the figures for MADAG were lowest, at 1.3%. Twenty-six fatty acids were identified with the ratios of USAFA/SAFA was 2. The percentage of n-3 PUFA (ω-3) and n-6 PUFA (ω-6) was high, occupying 38.4% of total FA. Among PUFAs, arachidonic acid (AA, 20:4n-6), eicosapentaenoic acid (EPA, 20:5n-3), docosapentaenoic acid (DPA, 22:5n-3), and docosahexaenoic acid (DHA, 22:6n-3) accounted for 3.8%, 7.8%, 2.2% and 12.0% of total lipid of the clam respectively. Phospholipid molecular species were identified in polar lipids of the clams consisting six types: phosphatidylethalnolamine (PE, with 28 molecular species), phosphatidylcholine (PC, with 26 molecular species), phosphatidylserine (PS, with 18 molecular species), phosphatidylinositol (PI, with 10 molecular species), phosphatidylglycerol (PG, with only one molecular species), and ceramide aminoethylphosphonate (CAEP, with 15 molecular species). This is the first time that the molecular species of sphingophospholipid were determined, in *Meretrix lyrata* in particular, and for clams in general. Phospholipid formula species of PE and PS were revealed to comprise two kinds: Alkenyl acyl glycerophosphoethanolamine and Alkenyl acyl glycerophosphoserine occupy 80.3% and 81.0% of total PE and PS species, respectively. In contrast, the percentage of diacyl glycero phosphatidylcholine was twice as high as that of PakCho in total PC, at 69.3, in comparison with 30.7%. In addition, phospholipid formula species of PI and PG comprised only diacyl glycoro phospholipids. PE 36:1 (p18:0/18:1), PC 38:6 (16:0/22:6), PS 38:1 (p18:0/20:1), PI 40:5 (20:1/20:4), PG 32:0 (16:0/16:0) and CAEP 34:2 (16:2/d18:0) were the major molecular species.

Keywords: hard clams; *Meretrix lyrata*; lipid classes; fatty acids; phospholipids; molecular species of phospholipid; high resolution mass spectrometry

---

1. Introduction

Investigation on marine and freshwater two-part shell mollusks has been growing rapidly due to their rich nutritional value, variety of biofunctions, potent activities, and most importantly, ease of exploitation [1–4]. Poly-unsaturated fatty acids, mainly omega-3 fatty acids, have been a valuable nutrient group and the nutraceutical of interest. In almonds (*Prunus dulcis*), it is shown that the content of linoleic acid, a poly-unsaturated fatty acid, ranged from 10 to 31% depending on the origin of the sample [5]. The high content of PUFA is suggested to be associated with the cardio protective effect. The mollusk, especially bivalve mollusks, is widely consumed in Asian countries, providing high protein content, sugars, lipids, especially omega-3-fatty acids and essential amino acids, vitamin B12 and essential elements including iron, zinc and copper. Among these constituents, phospholipids and fatty acids play a major role in the functions of the immune system and the maintenance all hormonal systems of the organism. These important functionalities promote the development of new and more efficient extraction methods for mass production and commercialization of clams.

Clams, cockles, and arkshells, occupying the largest share in worldwide production of shellfish (about 6.16 million tons in 2016), are widely used as an important nutriment source [6]. Investigations of several clam species such as *Meretrix lusoria* (Asian hard clam), *Meretrix meretrix*, *Cyclina sinensis*, and *Chamelea gallina* (venus clam) have revealed that phospholipid (PL) and ω-3 long-chain polyunsaturated fatty acids (LC-PUFA) are abundantly found in clams [7–10]. Beneficial fatty acids in ω-3 LC-PUFAs include eicosapentaenoic acid (EPA, 20:5n-3), docosahexaenoic acid (DHA, 22:6n-3) and docosapentaenoic acid (DPA, 22:5n-3) [11–13]. Such dietary fatty acids have been studied extensively, showing a wide range of positive clinical effects including improvement in treatment of heart diseases, effective intervention of insulin sensitivity, inhibition of tumors, inflammation and metastasis. The FDA (Food and Drug Administration) has approved a qualified health claim for conventional foods and dietary supplements that contain EPA and DHA [14]. It states, "Supportive but not conclusive research shows that consumption of EPA and DHA omega-3 fatty acids may reduce the risk of coronary heart disease." The FDA also specifies that the labels of dietary supplements should not recommend a daily intake of EPA and DHA higher than 2 g per day for health benefits [14]. For patients who need to lower their triglyceride levels, the American Heart Association recommends 2–4 g/day of EPA plus DHA under the care of a physician [15]. Several prescription omega-3 preparations are also available to treat hypertriglyceridemia [16].

Clams could be potentially a good source of PL and PUFAs. Much evidence regarding lipid composition in various clam species has been presented. For example, in the striped Venus Clam (*Chamelea gallina*), it was shown that n-3 LC-PUFA constituted a large quantity in total fatty acids of, ranging from 33.7 to 41.9%. In addition, EPA (8.2–20.0% of total fatty acids) and DHA (12.5–20.3% of total fatty acids) were also found [8]. This is also similar to the Asian hard clam (*Meretrix lusoria*), where PUFA (46.8–49.2% of total fatty acids), DHA (13.3–16.5% of total fatty acids) and EPA (4.8–7.1% of total fatty acids) were found in high levels [9]. In the *Ruditapes philippinarum* clam, PLs were the principal lipids (57–75% of total lipids) [7]. In the lipid composition of the *Calyptogena phaseoliformis* clam, major lipids are the n-4 family non-methylene interrupted polyunsaturated fatty acids (NMI-PUFA) including 20:3n-4,7,15, 20:4n-1,4,7,15, and 21:3n-4,7,16, with significant levels of 20:2n-7,15 and 21:2n-7,16 as non-methylene interrupted n-7 dienes [17]. The major fatty acids of lipids in *Meretrix lamarckii* and *Ruditapes philippinarum* were 14:0, 16:0, 18:0, 16:1n-7, 18:1n-9, 18:1n-7, 20:4n-6, 20:5n-3, and 22:6n-3, while those of *Mesolinga soliditesta* were 16:0, 18:0, 16:1n-7, 18:1n-7, 20:1n-7, 20:1n-13, 20:2n-7, 15 (Δ5,13-20:2), and 22:2n-7,15 (Δ7,15-22:2) [18].

Due to the immensity of the sea aquaculture reservoir, the Vietnamese fishery industry was ranked very highly. In terms of fish landings from marine fishing areas, Vietnam occupied the 8th position in 2015 and 2016, reaching a volume of 2.607 and 2.678 million tons respectively [19]. Among Vietnam's highly valued aquatic organisms, the *Meretrix lyrata* (Sowerby 1851) clam, belonging to the *Veneridae* genus, also known as the hard clam, could be found on seashores and in estuarine areas. Apart from being an export product of high economic value, *Meretrix lyrata* also acts as a protein source for domestic consumption [20,21]. Considering the limited data on plasmalogens of bivalve species and the prevalence of the hard clam in Vietnamese diet, our study, for the first time, aims to report the fatty acids, lipid classes and phospholipid molecular species of the hard clams *Meretrix lyrata* S. The sample was collected in Cua Lo Beach, Nghe An Province, located in the North Central Coast region of Vietnam and populated with diversified aquaculture and ample clam species, in particular. We showed that the lipids from clams contained high percentages of PL. In addition, most of the predominant glycerophospholipid (GP) molecular species, such as phosphatidylethalnolamine (PE), phosphatidylcholine (PC), phosphatidylserine (PS), phosphatidylinositol (PI) and phosphatidylglycerol (PG) and sphingophospholipid (ceramide aminoethylphosphonate—CAEP), were identified.

2. Results and Discussion

2.1. Total Lipid

Total lipid (TL) constituted 1.7 ± 0.2% of wet weight of the clams. The composition and content of TL of *Meretrix lyrata* (Figure 1 and Table 1) was similar to that of other clams investigated previously [18,22–24]. Overall, hard clam species contained common lipids hydrocarbon and wax (HW), triacylglycerol (TAG), free fatty acids (FFA), sterol (ST), polar lipid (PoL) and monoalkyl diacylglycerol (MADAG). In contrast with lipid compositions of cnidarians and coral in which MADAG accounts for a significant proportion, MADAG component of the hard clam only represents a marginal content of 1.3 ± 0.2 of TL.

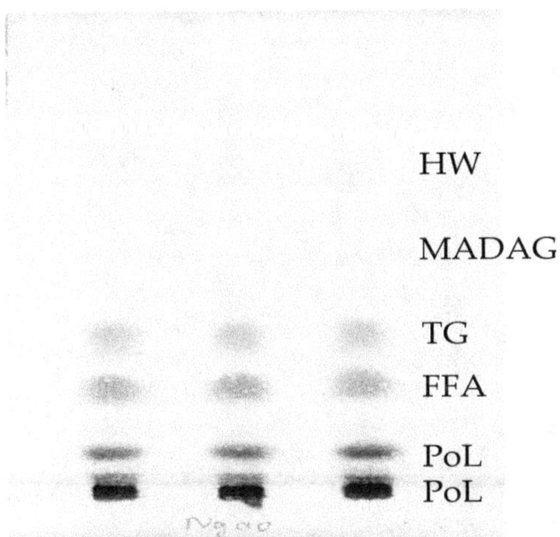

Figure 1. TLC determination of lipid classes of the clams.

Table 1. Composition of main lipid classes (% of total lipid) of the clams.

No.	Lipid Class	Content (%)
1	Hydrocarbon and wax (HW)	2.6 ± 0.1
2	Monoalkyl diacylglycerol (MADAG)	1.3 ± 0.2
3	Triacylglycerol (TG)	11.7 ± 0.3
4	Free fatty acids (FFA)	18.6 ± 0.8
5	Sterols (ST)	20.2 ± 0.5
6	Polar lipid (PoL)	45.7 ± 0.6

The findings for PoL levels in the total lipids of the hard clam are similar to those of other studies which have investigated lipid profiles and phospholipids. In addition, this result also highlights the seasonal variation of PoL content in clam. Specifically, the analyzed hard clams, which were collected in summer, exhibited a lower PoL content in comparison to those of other clam species collected in other seasons [25,26].

Regarding other lipid classes, TAG, FFA and ST were the major lipid classes in the non-polar lipid, representing 11.7%, 18.6% and 20.2% of TL respectively. This observation was in good agreement with previous publications of other clam species [22–24]. Notably, the content of FFA was found to be high in hard clam, suggesting possible involvement of lipase enzymes in the intestinal organs in the hydrolysis of TAG into FFA. The presence of sterols in hard clam is also consistent with the ubiquity of the lipid class in most marine organisms, which plays many key roles, such as presenting in membrane composition [26,27].

2.2. Fatty Acids in Total Lipids

Twenty-six fatty acids in the lipid sample of hard clam were found with the number of carbon atoms ranging from 14 to 22 (Table 2). Major FA were Acid 16:0, 16:1n-7, 18:0, 18:1n-9, 20:4n-6 (arachidonic acid AA), 20:5n-3 (EPA) and 22:6n-3 (DHA), respectively accounting for 16.0, 6.3, 5.0, 4.7, 3.8, 7.8, and 12.0% of the total FA content. Saturated fatty acids occupied 26.3% of total FA content.

Table 2. Fatty acid composition (% of total) of clams.

	R.t (min)	Area	Content (%)
Σ SAFAs		1,971,006	26.3
14:0 *	3.748	291,352.4	3.9
16:0 *	5.491	1,197,827	16.0
17:0 *	6.812	105,581.2	1.4
18:0 *	8.686	376,245.5	5.0
Σ MUFAs		1,191,078	15.9
16:1n-7 *	5.863	469,554.2	6.3
18:1n-9 *	9.228	355,129.6	4.7
18:1n-7 *	9.378	147,985.6	2.0
20:1n-11	15.324	99,722.7	1.3
20:1n-9 *	15.511	42,995.4	0.6
20:1n-7 *	15.875	75,690.3	1.0
Σ PUFAs		2,831,606	37.7
16:3n-3	7.292	56,506.2	0.8
18:2n-6 *	10.411	63,571.7	0.8
18:3n-3	12.383	77,148.8	1.0
18:4n-3 *	13.518	80,259.3	1.1
20:2n-6 *	17.775	108,771.3	1.4
20:3n-6	19.233	31,176.5	0.4
20:4n-6 *	20.617	283,411	3.8
20:4n-3 *	23.19	36,235.8	0.5
20:5n-3 *	24.929	588,733	7.8
21:3n-3	28.661	85,752.1	1.1
21:5n-6	29.156	144,854.4	1.9
21:5n-3	33.552	53,585.8	0.7
22:4n-6 *	36.426	115,438.2	1.5
22:5n-6	39.584	96,337.1	1.3
22:5n-3 *	44.249	161,865.8	2.2
22:6n-3 *	48.476	901,545.1	12.0
Octadecanal [a]	7.882	808,630.9	10.8
Other [b]		702,316.4	9.4

* Fatty acids were identified by GCMS and GC. [a] Fatty aldehyde was determined as its dimethylacetal derivatives by GC and GCMS. [b] 15:0, i-16:0, 16:1n-5, i-17:0, a-17:0, i-18:0, 17:3n-6, a-19:0, 19:0, 18:3n-6, 19:1n-9, 19:1n-7, 20:0, i-21:0, a-21:0, 20:3n-3, hexadecanal, hexadecenal, octedecenal. SAFAs: saturated fatty acids. MUFAs: monounsaturated fatty acids. PUFAs: poliunsaturated fatty acids.

Monounsaturated fatty acids (MUFA) accounted for 15.9% of total FA content and 29.6% of unsaturated fatty acids. Two major MUFA in the composition were C16 and C18 MUFA. Regarding polyunsaturated fatty acids (PUFA), content of PUFA represented 37.7% and 70.7% of total FA content and total unsaturated fatty acids respectively. Among PUFA, three C18 PUFA were found including18:2n-6, 18:3n-3 and 18:4n-3. However, C18 PUFA only accounted for 2.9% of the total FA content. In addition, the percentages of n-6 PUFA (ω-6) were 11.2% of total FA. The figure for n-3 PUFA (ω-3) was approximately 26.5%. Highly unsaturated fatty acids (HUFA) accounted for 32.1% in total and 59.9% in unsaturated fatty acids.

These results were similar to those of various reports [23,28,29]. In the composition, several odd-chain acids were found in small quantities and methyl-branched FA including i-16:0, i-17:0, a-17:0, i-18:0 and i-21:0 were detected as trace amounts. Among many long-chain DMA produced by plasmalogen PL under acid methanolysis of clam lipids, octadecanal dimethylacetal was the major DMA and amounted to 10.8% of FA composition. Three FAs with 21 carbon atoms were identified, including 21:3n-3, 21:5n-6 and 21:5n-3 and accounted for 1.1, 1.9, and 0.7% of FA, respectively. According to a previous study [26], 6,9,12,15,18-Heneicosapentaenoic acid (21:5n-3) (HPA), was shown to exhibit stronger inhibition for arachidonic acid synthesis from alpha-linoleic acid and dihomo-gamma-linolenic acid in hepatoma cells.

2.3. *Fatty Acids in Fractions of Phospholipid*

As shown in Table 3, GC analysis indicated that the fractions of phospholipid of the clams contained a high proportion of SAFAs. Saturated fatty acids exist dominantly in CAEP, at 75.8%. This figure was more than twice as much as of PS + PI. The figures for PE + PG and PC were lower, at 40.5, and 52.2% respectively. The abundance of SAFAs in the FA composition of CAEP was consistent with the molecular species content of 16:0, 17:0, and 18:0 acid, which were reported later in this study and totaled at 67.1% of CAEP from hard clam (Table S6).

Compared with the lipids recovered from the other clam species [17], the FA composition results of hard clam were also similar regarding DMA in PE and PC fraction. To be specific, both studies suggested that DMAs were non-existent in FA composition of PC in investigated clam species. In addition, present results also showed that CAEP fraction contains no DMA and PE is the only fraction which contains all three DMAs including of 16:0, 17:0 and 18:0 DMA, with corresponding contents of 0.4, 1.2, and 11.3%. In fractions of PS + PI, only 17:0 DMA and 18:0 DMA exists at 1.5% and 14.6% respectively. Thus, two kinds of phospholipid formula species contained in PE and PS were plasmenyl and diacyl phospholipids and two phospholipid formula species of PC includes plasmanyl and diacyl glycero phospholipids.

Table 3. Fatty acid composition (% of each type phospholipid) of clams.

Fraction	PE + PG	PC	CAEP	PS + PI
Σ SAFAs	40.5	52.2	75.8	32.5
14:0	4.3	nd	nd	0.7
16:0	22.0	36.9	57.0	12.9
17:0	3.2	4.1	4.3	2.1
18:0	11.0	9.9	13.0	16.2
20:00	nd	1.3	1.5	0.4
Σ MUFAs	25.3	14.4	11.3	22.5
16:1n-7	10.9	3.6	nd	1.2
18:1n-9	8.4	7.9	8.5	9.5
18:1n-7	3.5	1.5	0.4	0.7
20:1n-11	1.1	0.5	0.3	7.4
20:1n-9	0.6	0.5	0.7	1.2
20:1n-7	0.8	0.4	1.4	2.5

Table 3. *Cont.*

Fraction	PE + PG	PC	CAEP	PS + PI
Σ PUFAs	14.9	29.9	9.1	22.7
16:3n-3	0.9	0.8	nd	nd
18:2n-6	0.9	1.0	2.1	1.5
18:3n-3	1.2	0.5	0.1	nd
18:4n-3	0.8	0.6	nd	nd
20:2n-6	nd	1.2	0.2	nd
20:3n-6	nd	0.7	0.3	nd
20:4n-6	1.7	2.4	1.7	5.8
20:4n-3	nd	0.3	nd	nd
20:5n-3	2.9	3.8	1.2	1.4
21:3n-6	nd	2.6	0.8	nd
21:3n-3	2.2	0.4	0.2	4.9
21:5n-6	nd	0.8	0.3	nd
21:5n-3	nd	0.1	1.1	nd
22:4n-6	nd	1.0	nd	0.7
22:5n-6	1.1	0.9	nd	1.5
22:5n-3	nd	1.8	nd	1.3
22:6n-3	3.2	11.0	1.0	5.5
Σ DMAs	12.9			16.0
16:0	0.4	nd	nd	nd
17:0	1.2	nd	nd	1.5
18:0	11.3	nd	nd	14.6
Other	6.3	3.5	3.8	6.3

nd: not detected or under 0.1% of fatty acid composition in each type phospholipid. SAFAs: saturated fatty acids. MUFAs: monounsaturated fatty acids. PUFAs: poliunsaturated fatty acids. DMAs: dimethylaxetal. PE: Phosphatidylethanolamine. PC: Phosphatidylcholine. PS: Phosphatidylserine. PI: Phosphatidylinositol. PG: Phosphatidylglycerol. CAEP Ceramide aminoethylphosphonate.

2.4. Molecular Species of Phospholipid

Molecular species of polar lipids from the hard clam were determined following the previously described HRMS fragmentations of PL standards [30]. TL of Clams contained 48.96% of polar lipids (PoL) (Table 1). PoL have got glycolipid and phospholipid. Especially in the hard clams, phospholipid dominated, with six types of phospholipids, including phosphatidylethanolamine (PE), phosphatidylcholine (PC), phosphatidylserine (PS), phosphatidylinositol (PI) and phosphatidylglycerol (PG) and Ceramide aminoethylphosphonate (CAEP). The molecular species and qualification of each phospholipid were performed by HPLC-HRMS [31].

Among phosphatidylethanolamine (PE) of clams, we determined 28 constituents. Alkenyl acyl glycerophosphoethanolamine (ethanolamine plasmalogen PlsEtn) was the major component in PE, with 80.3% of total PE species (Table S1), followed by PE 40:6 (m/z (M+) 776.5556), at 17.13%, Plasmalogen PE 36:1 (m/z (M+) 730.5794), at 15.4%, and PE 40:2 (m/z (M+) 784.6161), at 10.3%.

PlsEtn can be characterized according to their MS and MS/MS data. The signals of positive quasi-molecular ions $[M + H]^+$, cluster ions $[M + H + C_6H_{15}N]^+$ and negative quasi-molecular ions $[M - H]^-$ were observed in the HRMS spectra of all components of formula species of PE. For instance, we detected signals of negative quasi-molecular ions $[M - H]^-$ at m/z 728.5544 ($[C_{41}H_{79}NO_7P]^-$, calculated 728.5594, different 0.00556), positive quasi-molecular ions $[M + H]^+$ at m/z 730.5794 ($[C_{41}H_{81}NO_7P]^+$, calculated 730.5751, different 0.00488), and positive cluster ions $[M + C_6H_{15}N]^+$ at m/z 831.6822 ($[C_{47}H_{96}N_2O_7P]^+$, calculated 831.6955, different 0.01277) of PE 36:1 (Figure 2).

From the the MS^{2-} spectrum of the ions $[M - H]^-$ of PE 36:1, one signal corresponding to one carboxylate anion of 18:1 was detected at m/z 281.2466. For the plasmalogen, the fatty acid (FA) anion ($[RCOO]^-$) can only be liberated from the *sn*-2 position due to the ankenyl linkages at the *sn*-1 position [32]. Therefore, three possible molecular species including plasmenyl p18:0/18:1. Component PE 36:1 was identified as alkenyl acyl glycerophosphoethanolamine, p18:0/18:1 PlsEtn (Figure S1).

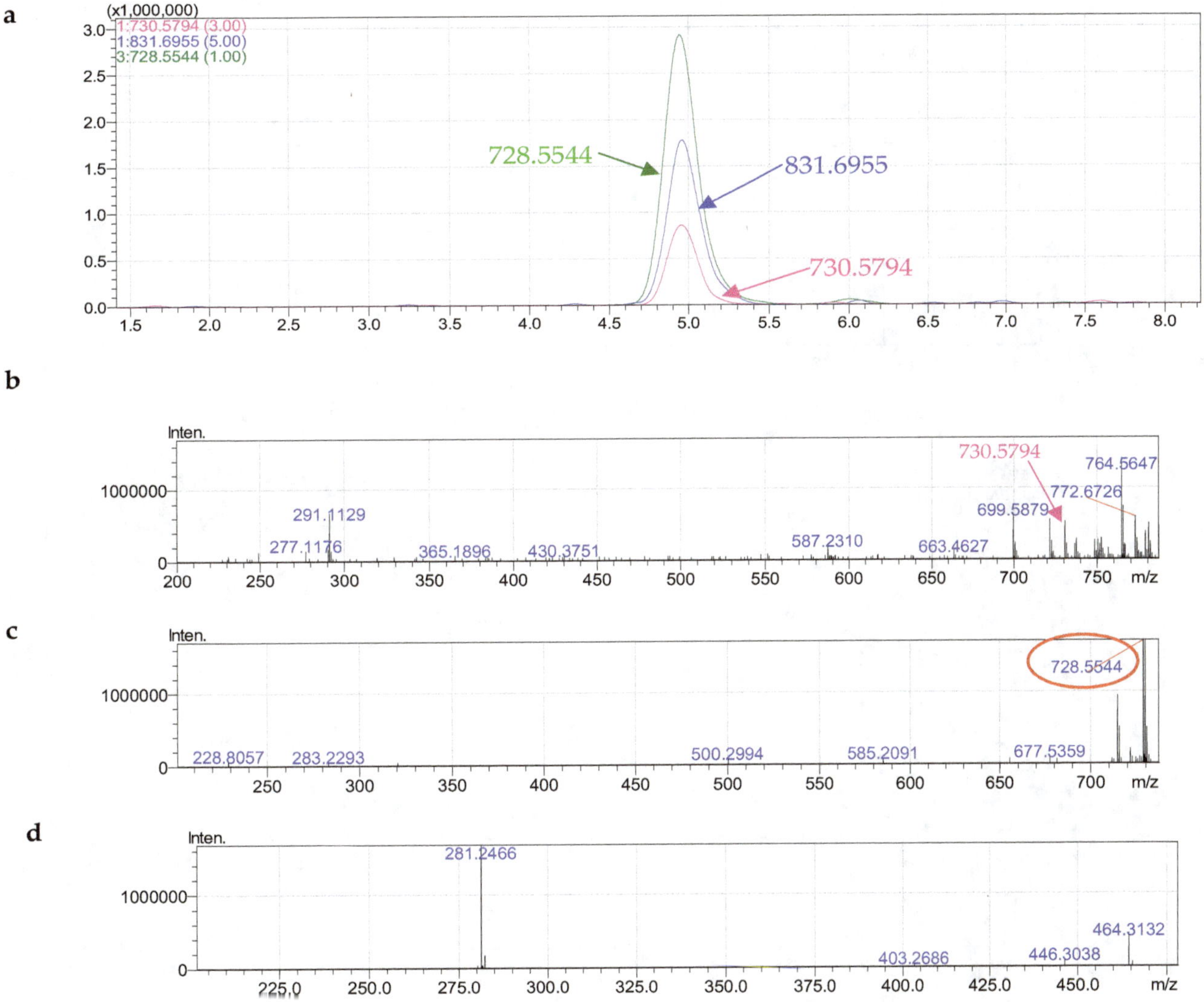

Figure 2. HPLC–HRMS (**a**) and fragmentation of PE 36:1(**b**—MS^+, **c**—MS^-, **d**—MS^{2-}).

Diacyl glycerophosphoethanolamine only accounted for 19.7% of total PE species. For an unknown PL containing two FAs, MS and MS/MS data also are needed to determine the two individual FAs esterified at the *sn*-1 and *sn*-2 positions of the glycerol backbone.

For example, for the unknown PE with measured *m/z* of 730.5411 (quasi-molecular ion ($[M + H]^+$)), PE 35:2 could tentatively be deduced according to the previously described formulae [30,31]. The MS^{2-} spectrum of the ions $[M - H]^-$ of component PE 35:2 contained the signal of two carboxylate anion of 18:1 at *m/z* 281.2466 and 17:1 at *m/z* 267.2315 (Figure 3 and Figure S2). The component PE 35:2 was determined as diacy glycerophosphoethanolamine, 18:1/17:1.

The choline glycerophospholipids (ChoGpl) of clams is summarized in Table S2. Among 26 components, 9 of which constituted more than 51% of total ChoGpl. The percentage of diacyl glycero phosphatidylcholine was twice as abundant as PakCho in total PC, at 69.3% in comparison with 30.7%.

The signals of positive quasi-molecular ions $[M + H]^+$, negative formate molecular ions $[M + HCOO]^-$ and cluster ions $[M - CH_3]^-$ were observed in the HRMS spectra of all components of formula species of PC. Formate molecular ions ($[M + HCOO]^-$) of each component lost methyl formate at the MS^{2-} stage (Figure 4). For instance, with the highest percentage at 16.0%, PC 38:6 formed negative acetylated molecular ions $[M + HCOO]^-$ at *m/z* 850.5571, positive quasi-molecular ions $[M + H]^+$ at *m/z* 806.5637 and cluster ions $[M - CH_3]^-$ at 790.5373 corresponding to composition $[C_{47}H_{81}NO_{10}P]^-$ (calculated850.5598, different 0.00326), $[C_{46}H_{81}NO_8P]^+$ (calculated806.5700, different 0.00573) and $[C_{45}H_{77}NO_8P]^-$ (calculated 790.5387, different 0.00193), respectively (Figure 4).

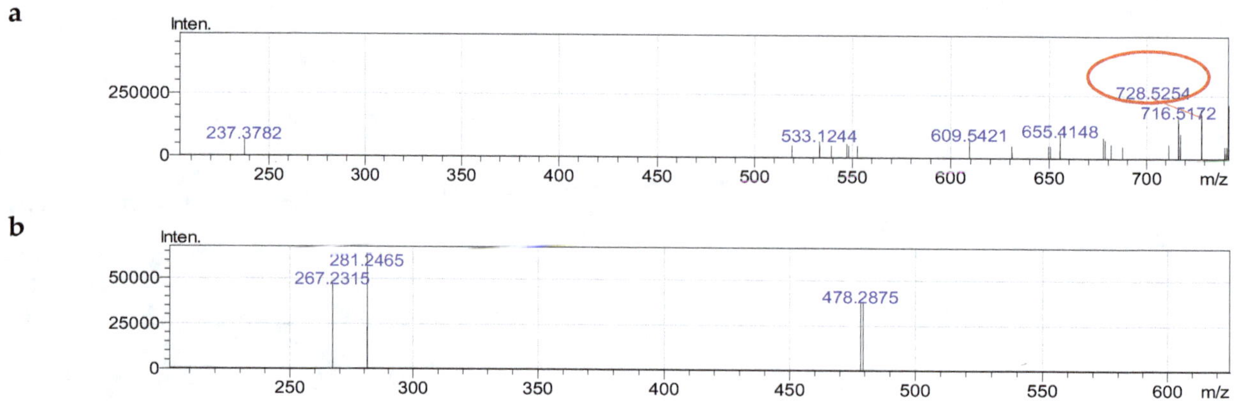

Figure 3. Fragmentation of PE (35:2): **a**—MS$^-$, **b**—MS^{2-}.

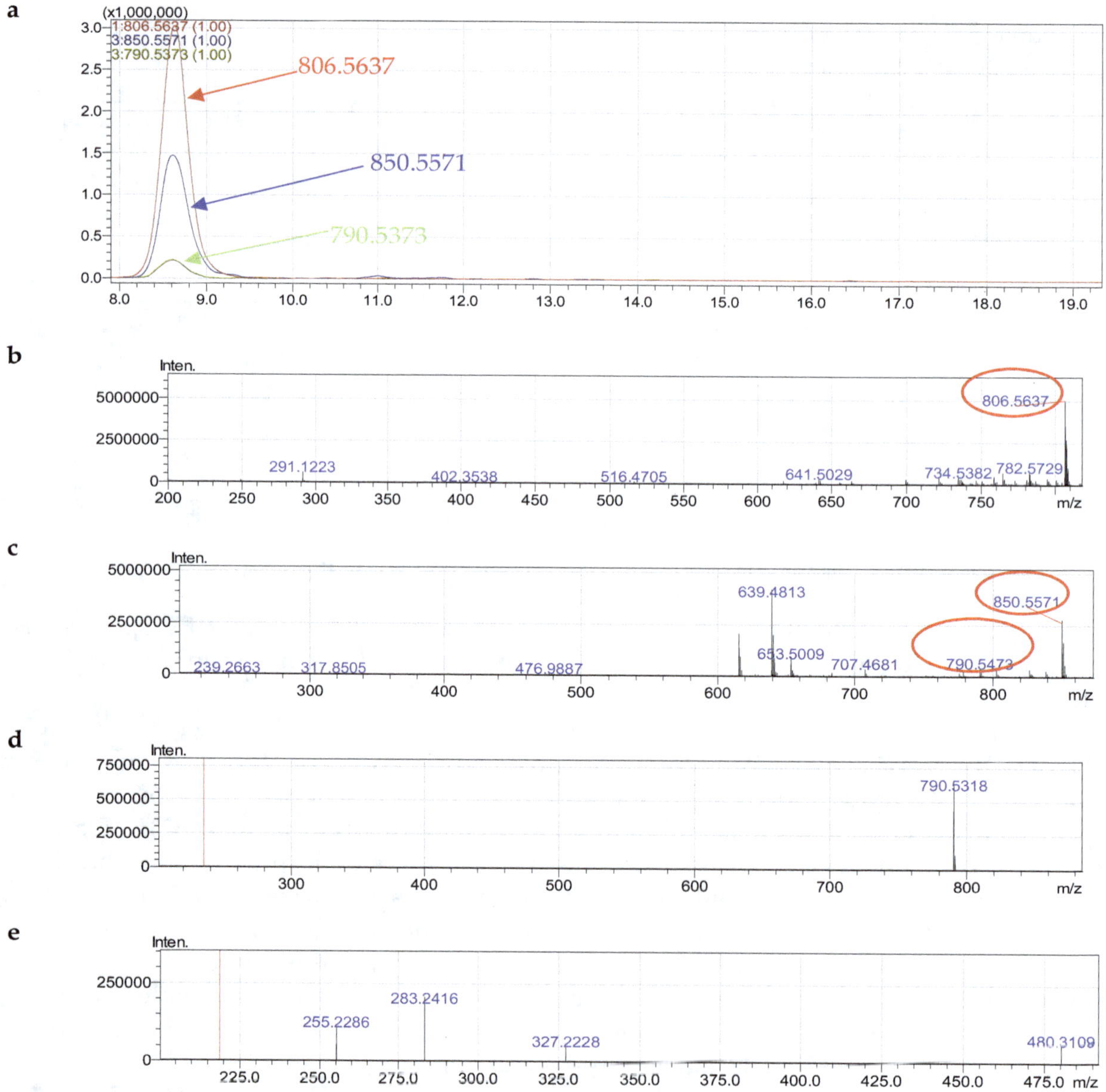

Figure 4. Fragmentation of PC (38:6): **a**—HPLC-HRMS, **b**—MS$^+$; **c**—MS$^-$ of fragmentation 850.5490 and 790.5301; **d**, **e**—MS^{2-} of fragmentation 850.5490 and 790.5301.

At the MS^{2-} stage, the ions at m/z 850.5571 eliminated a molecule of $C_2H_4O_2$ (methyl formate) and formed ions at m/z 790.5303, suggesting the formation of the quasi-molecular negative ions by the addition of formate ion to the lipid molecule. On MS^{2-} of component lost methyl formate, two carboxylate anion of 22:6 at m/z 327.2228 (calculated 327.2324, different 0.0105) and 16:0 at m/z 255.2286 (calculated 255.2324, different 0.00435) were observed (Figure 4 and Table S3). Normally, the *sn*-2 of PL is the preferred position for PUFAs [32]. Therefore, PC 38:6 was characterized as diacy glycerophosphocholine 16:0/22:6.

Among phosphotidylserine PS of clams, we determined 18 constituent components (Table S3). Alkenyl acyl glycerophosphoserine was the major component in PS, with 11 components occupying 81.0% of total PS species.

All components of formula species of PS had a signal of negative quasi-molecular ions $[M-H]^-$. On the other hand, positive quasi-molecular ions did not form. The MS^{2-} spectrum of $[M-H]^-$ of each component contained a signal of characteristic ion $[M–H–C_3H_5NO_2]^-$ corresponding to the loss of serine group (Figure 5). This was a specific fragmentation difference from the fragmentations of negative quasi-molecular ions of PE.

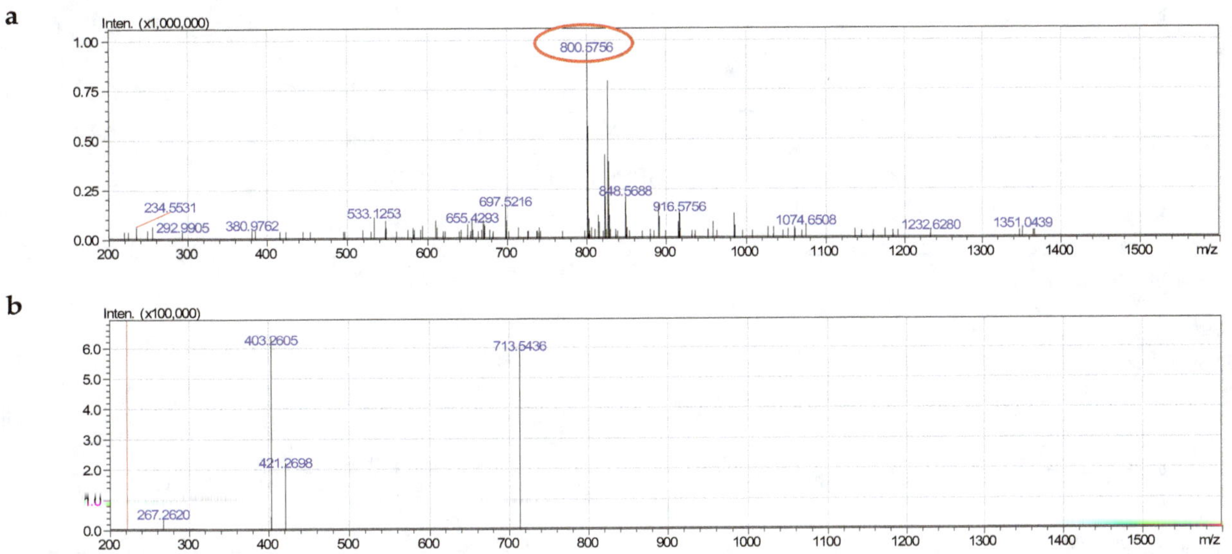

Figure 5. Fragmentation of PS (38:1): **a**—MS$^-$, **b**—MS^{2-}.

For example, PS 38:1 formed negative quasi-molecular ion $[M-H]^-$ at 800.5756 ($[C_{44}H_{83}NO_9P]^-$, calculated 800.5805, different 0.00549). On MS^{2-}, the absence of serine (713.6436) and serine and acyl groups (403.2605) was observed. However, signal of fatty acid was not detected (Figure S4). The component PS 38:1 was determined as alkenyl acyl glycerophosphoserine, p18:0/20:1.

Diacyl glycerophosphoserines only accounted for 19% of total PS species. On the MS^{2-} of signal negative quasi-molecular ion [M-H]$^-$ at 836.5402 ([C46H79NO10P]-, mass 836.5402, different 0.00451), the signal lost serine (749.4945) and lost simultaneously serine and acyl groups (419.2543) and one carboxylate anion of fatty acid 18:0 at m/z 283.2521 (Figure 6 and Figure S5). The component PS 40:5 was determined as diacy glycerophosphoserine, 18:0/22:5.

Among phosphotidylinositol PI of clams, we determined 10 components (Table S4). In addition, Alkenyl acyl glycerophosphoinositol was absent in PI. All ten components PI were diacyl glycerophosphoinositol with fatty acids 16:0. 17:0, 18:0, 19:0, 20:1, 20:4, 20:5.

The molecular species of clam PtdIns indicated the formation of both negative and positive quasi-molecular ions $[M-H]^-$. In addition, MS^{2-} fragmentation of the ions $[M-H]^-$ of PtdIns also reveals the presence of several characteristic ions such as PI 40:5 (Figure 7 and Figure S6). The fragmentation 911.5593 had formula $[C_{49}H_{84}O_{13}P]^-$, calculated 911.5650 (different 0.00620). The MS^{2-} spectra of component PI 40:5 also contained signals corresponding to the quasi-molecular ion, whose inositol, acyl groups and carboxylate anion of fatty acid have been simultaneously lost. In addition, fragmentation 297.0467 was quasi-molecular ion that lost diacyl groups. This fragmentation was important in determining the molecular species of PI.

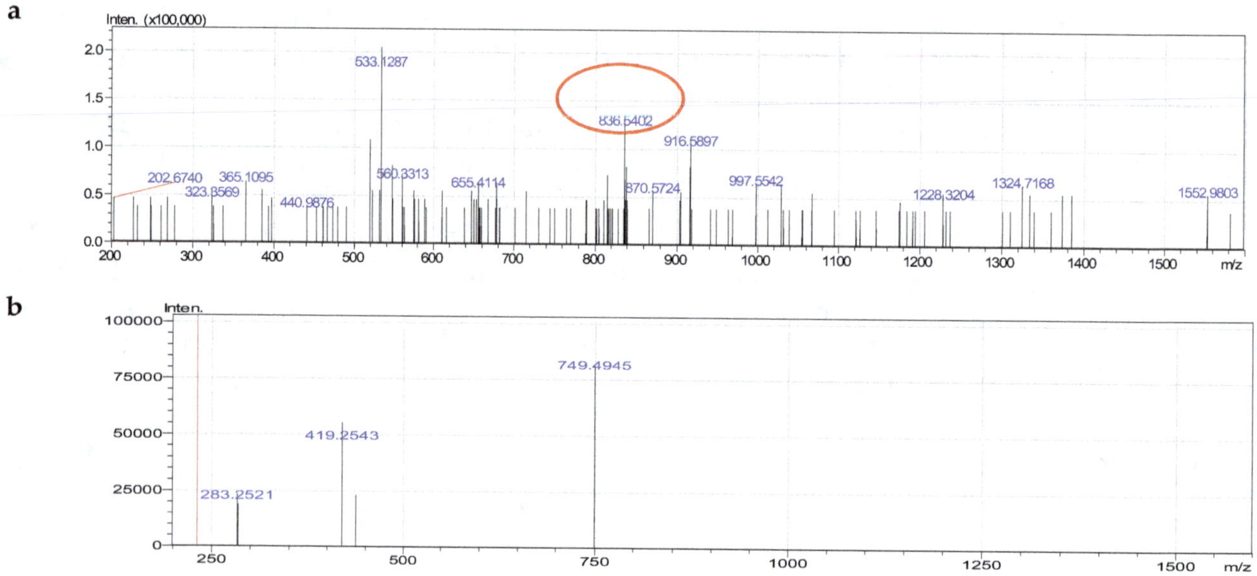

Figure 6. Fragmentation of PS (40:5): **a**—MS$^-$, **b**—MS^{2-}.

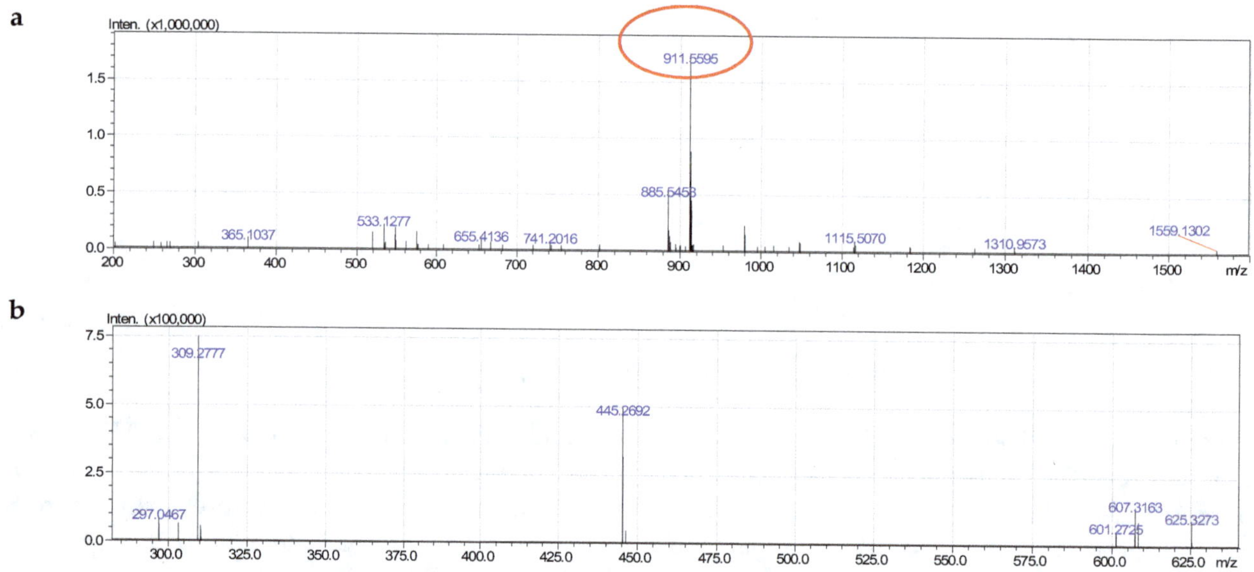

Figure 7. Fragmentation of PI (40:5): **a**—MS$^-$, **b**—MS^{2-}.

The component PI 40:5 (Figure S6) was determined as diacy glycerophosphoinositol 20:1/20:4.

We determined a single component, which was isomers of PG 32:0 constituting phosphatidylglycerol (PG) (Table S5).

PG 32:0 was described as follows (Figure 8 and Figure S7). The MS^{2-} spectra of component PG 32:0 contained signals corresponding to the negative quasi-molecular ion 721.4975 ([$C_{38}H_{74}O_{10}P]^-$, calculated 721.5020, different 0.00501). On MS^{2-}, we obtained fragmentations of carboxylate anion of fatty acid 16:0 (255.2304) with the simultaneous absence of glycerol and acyl groups (391.2255).

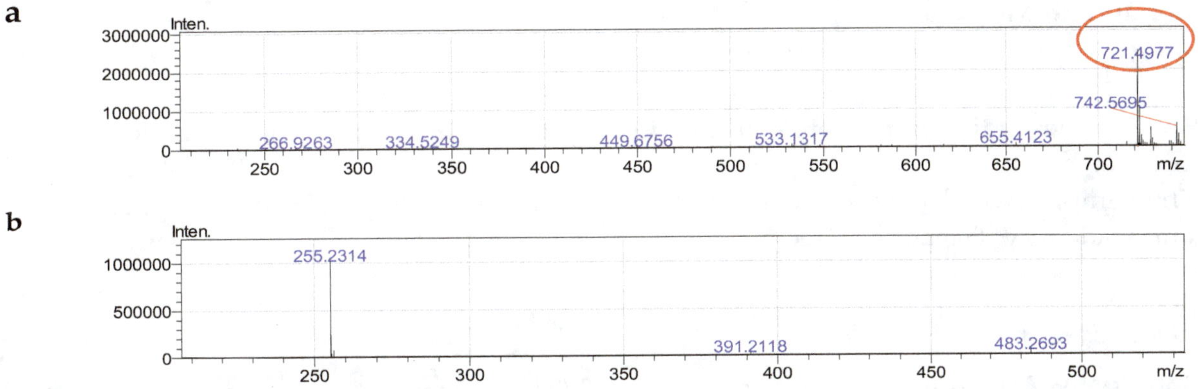

Figure 8. Fragmentation of PG (32:0): **a**—MS$^-$, **b**—MS^{2-}.

The component PG 32:0 (Figure S7) was determined as diacy glycerophosphoglycerol 16:0/16:0. Fifteen molecular species of CAEP were determined (Table S6). In mass spectra of all molecular species, we observed signals of negative quasi-molecular ions [M − H]$^-$, positive quasi-molecular ions [M + H]$^+$ and positive quasi-molecular ions [M + Na]$^+$.

For example, we had signals of negative quasi-molecular ions [M − H]$^-$, at m/z 685.5565 ([$C_{39}H_{78}N_2O_5P$]$^-$, calculated 685.5648, different 0.00888), positive quasi-molecular ions [M + H]$^+$ at m/z 687.5830 ([$C_{39}H_{80}N_2O_5P$]$^+$, calculated 687.5805, different 0.00306), and positive adductions [M + Na]$^+$ at m/z 709.5627 ([$C_{39}H_{79}N_2O_5NaP$]$^+$, calculated 709.5619, different 0.00082) of CAEP 37:1 (Figure 9).

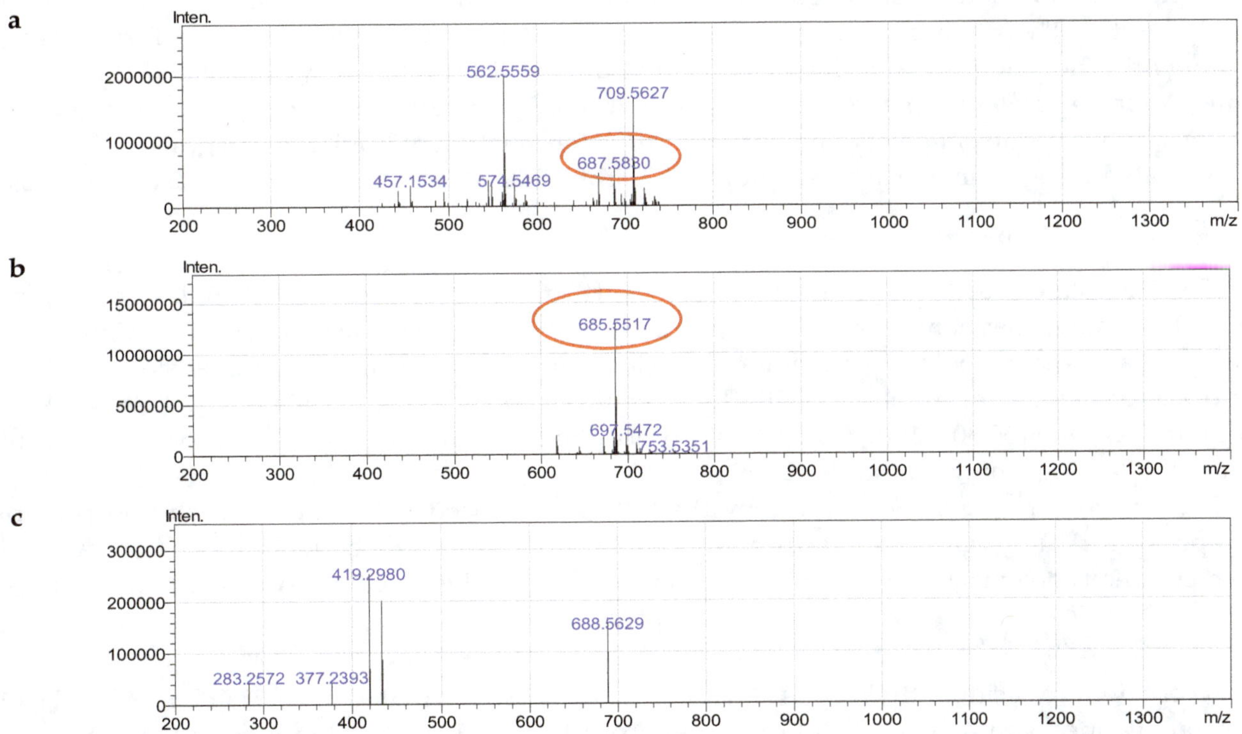

Figure 9. Fragmentation of CAEP (37:1): **a**—MS$^+$, **b**—MS$^-$, **c**—MS^{2-}.

The MS^{2-} spectrum of [M − H]$^-$ of CAEP which was sphingolipid signified a specific fragmentation similar to the fragmentations of negative quasi-molecular ions of glycerophospholipids (Figure S8). On MS^{2-}, we obtained carboxylate anion fragmentation of fatty acid 18:0 (283.2572). In addition, [M − H]$^-$ lost a neutral fragment amounted to 266.2567 and formed a single ion at 419.2980. Other fragmentations are also explained in Figure S8.

3. Materials and Methods

3.1. Material

The colonies of the clams were collected in May 2017 in Cua Lo beach, Cua Lo town, Nghe An provine, Vietnam.

The samples were transported immediately to the Institute of Natural Products Chemistry, Vietnam Academy of Science and Technology. Then, the shells and the meat were separated and stored at −5 °C.

3.2. Total Lipid Extraction

Soft tissue of clams was crushed and total lipid (TL) was extracted using modified Blight–Dyer extraction technique [31]. To be specific, lipids were extracted by homogenizing with the $CHCl_3/CH_3OH$ solution (1:2, *v*/*v*) (30 mL per 10 g of clams wet weight) in 6 h, at 30 °C. The obtained homogenate was filtered, and the residue was repeatedly extracted (6 h, 30 °C) in $CHCl_3$ (20 mL) a second time. After extraction, the homogenates were then mixed and separated into layers by adding 20 mL of H_2O. The lower layer was evaporated and the obtained TL obtained was dissolved in $CHCl_3$ and stored at −5 °C.

3.3. Analyses of Lipid Classes

To determine lipid class compositions, the extracted lipids were first dissolved in chloroform (10 mg/mL) and then spotted onto the one-dimensional thin-layer chromatography (TLC) using the pre-coated silica gel plates (6 cm × 6 cm) Sorbfil PTLC-AF-V (Sorbfil, Krasnodar, Russia). The first development of the plate was carried out with n-hexane/diethyl ether/acetic acid (85:15:1, *v*/*v*/*v*) for full length. Then, the plate was redeveloped with $CHCl_3/CH_3OH$ (2:1, *v*/*v*) for 5% length. Subsequently, the plates were air-dried, sprayed with 10% H_2SO_4/CH_3OH and heated at 240 °C for 10 min. An image scanner (Epson Perfection 2400 Photo, Epson, Suwa, Japan) operating in grayscale mode was employed to obtain chromatograms. For determination of lipid percentages, band intensity was evaluated using an image analysis program (Sorbfil TLC Video densitometer, Krasnodar, Russia).

3.4. Separation of Phospholipid Classes

The TLC plates used in this study were 10 × 10 cm glass-backed HPTLC Silica gel 60 plates (Merck, Darmstadt, Germany). Prior to sample loading, the plates were activated at 110 °C for at least 90 min and allowed to cool to room temperature in a vacuum desiccator. One dimensional TLC was employed to analyze phospholipids with the solvent system of chloroform–methanol–25% aqueous ammonia–benzene (65:30:6:10, *v*/*v*/*v*/*v*) [33]. Phosphomolybdate spray reagent for detection was prepared following a previous study [34].

Spots from TLC plates was eluted by the following procedure. First, for each phospholipid, corresponding band was scraped from the plate developed twice in 1st direction with the aforementioned solvent system. Second, elution was performed with chloroform–methanol (1:1, *v*/*v*).

3.5. Analyses of Fatty Acids

To obtain fatty acid methyl esters (FAME), the lipids were treated with 2% H_2SO_4 in CH_3OH in a screw-capped vial (2 h, 80 °C) under air and purified by TLC development in hexane–diethyl ether (95:5, *v*/*v*). FAME were analyzed with GC at 210 °C and identified with reference to authentic standards and a table of equivalent chain lengths [35]. Injector and detector temperatures were 240 °C.

To confirm structures of FA, corresponding FAME were analyzed with GC–MS and resulted spectra were matched with the NIST library and FA mass spectra archive [36]. Temperature of the GC–MS analysis initiated at 160 °C and then progressed at 2 °C/min to 240 °C which was kept constant for 20 min. Injector and detector temperatures were 250 °C.

3.6. Analysis of Molecular Species of Phospholipids

Phospholipids were analyzed with high performance liquid chromatography–high resolution mass spectrometry (HPLC–HRMS) to identify and quantify chemical structures of molecular species of phospholipids.

The high performance liquid chromatography (HPLC) separation of PL was performed at the constant content of $(C_2H_5)_3N$/acid formic (0.08:1, *v*/*v*) in the solvent system [37] that allowed carrying out efficient ionization in ESI conditions and obtaining a stable ion signal by the simultaneous registration of positive and negative ions. For polar lipids, HPLC separation was performed using the binary solvent gradient consisted of solvent mixture A: *n*-hexane/2-propanol/acid formic/$(C_2H_5)_3N$ (82:17:1:0.08, *v*/*v*/*v*/*v*) and mixture B: 2-propanol/H_2O/acid formic/$(C_2H_5)_3N$ (85:14:1:0.08, *v*/*v*/*v*/*v*). The gradient started at 5% of mixture B, and its percentage was increased to 80% over 25 min. This composition was maintained for 1 min before being returned to 5% of mixture B over 10 min and maintained at 5% for another 4 min (the total run time was 40 min). The flow rate was 0.2 mL/min. Polar lipids were detected by high resolution mass spectrometry (HRMS) and identified by a comparison with authentic standards using a Shimadzu LCMS Solution control and processing software (v.3.60.361, Shimadzu, Kyoto, Japan). The quantification of individual molecular species within each polar lipid class was carried out by calculating the peak areas for the individual extracted ion chromatograms [38].

3.7. Instrumental Equipment

The GC analysis was carried out on a Shimadzu GC-2010 chromatography (Kyoto, Japan) with a flame ionization detector on a SUPELCOWAX 10 (Supelco, Bellefonte, PA, USA) capillary column (30 m × 0.25 mm × 0.25 μm). Carrier gas was He at 30 cm/s. The GC–MS analysis was performed with a Shimadzu CMS-QP5050A instrument (Kyoto, Japan) (electron impact at 70 eV) with a MDN-5s (Supelco, Bellefonte, PA, USA) capillary column (30 m × 0.25 mm ID). Carrier gas was He at 30 cm/s.

The HPLC–HRMS analysis of polar lipids was performed with a Shimadzu Prominence liquid chromatograph equipped with two LC-20AD pump units, a high pressure gradient forming module, CTO-20A column oven, SIL-20A auto sampler, CBM-20A communications bus module, DGU-20A3 degasser, and a Shim-Pack diol column (50 mm × 4.6 mm ID, 5 μm particle size) (Shimadzu, Kyoto, Japan). Lipids were detected by a high resolution tandem ion trap–time of flight mass spectrometry with a Shimadzu LCMS-IT-TOF instrument (Kyoto, Japan) operating both at positive and negative ion mode during each analysis at electrospray ionization (ESI) conditions. Ion source temperature was 200 °C, the range of detection was m/z 200–1600, and potential in the ion source was −3.5 and 4.5 kV for negative and positive modes, respectively. The drying gas (N_2) pressure was 200 kPa. The nebulizer gas (N_2) flow was 1.5 L/min.

4. Conclusions

Lipid analysis of the present study revealed that the total lipid in hard clam (*Meretrix lyrata*) occupies 1.7 ± 0.2% of wet weight and that six lipid classes consisting of hydrocarbon and wax (HW), triacylglycerol (TAG), free fatty acids (FFA), sterol (ST), polar lipid (PoL) and monoalkyl diacylglycerol (MADAG) were detected. Among these classes, the proportion of PoL and MADAG accounted for the highest, at 45.7%, and the lowest, at 1.3% of total lipid, respectively. In addition, the ratios of PUFA/SAFA was 1.4. The total percentage of n-3 PUFA (ω-3) and n-6 PUFA (ω-6) was relatively high, at 38.4% of total FA.

To our knowledge, this is the first report to determine chemical structures and quantities of molecular species of phospholipid, in general, and phosphoethanolamines in particular, on the hard clams *Meretrix lyrata* from Cua Lo beach, Nghe An province, Viet Nam. Five types glycerophospholipid were identified: PE, PC, PS, PI and PG. One type sphingophospholipid was identified to be CAEP. Ninety-eight molecular species were identified in polar lipids of the clams. Alkenyl acyl forms of glycerophospholipids predominated in the molecular species determined. PE 36:1 (p18:0/18:1), PC 38:6 (16:0/22:6), PS 38:1 (p18:0/20:1), PI 40:5 (20:1/20:4), PG 32:0 (16:0/16:0) and CAEP 34:2 (16:2/d18:0) were the major molecular species.

Author Contributions: Investigation, T.T.T.L., M.Q.P., T.L.D., M.H.V. and L.G.B.; Supervision, L.M.B. and Q.L.P.; Writing—original draft, Q.T.T.; Writing—review & editing, D.C.N.

Acknowledgments: We are grateful to Andrey B. Imbs and his group at Laboratory of Comparative Biochemistry A.V. Zhirmunsky Institute of Marine Biology, Far-Eastern Branch of the Russian Academy of Sciences, 17 Palchevskogo str., Vladivostok 690041, Russian Federation.

Abbreviations

AA	Arachidonic acid
CAEP	Ceramide aminoethylphosphonate
ChoGpl	Choline glycerophospholipids
DHA	Docosahexaenoic acid
DMA	Dimethylacetal
DPA	Docosapentaenoic acid
EPA	Eicosapentaenoic acid
FFA	Free fatty acids
GC	Gas chromatography
GCMS	Gas chromatography—mass spectrometry
HPA	Heneicosapentaenoic acid
HPLC-HRMS	High performance liquid chromatography–high resolution mass spectrometry
HRMS	High resolution mass spectrometry
HUFA	Highly unsaturated fatty acid
HW	Hydrocarbon and wax
LC-PUFA	Long-chain polyunsaturated fatty acid
MADAG	Monoalkyl diacylglycerol
MS	Mass spectrometry
MS/MS	Mass spectrometry / Mass spectrometry (or tandem mass spectrometry)
MUFA	Monounsaturated fatty acid
NMI-PUFA	Non-methylene interrupted polyunsaturated fatty acid
PakCho	1-*O*-alkyl-2-acyl-*sn*-glycero-3-phosphocholine
PC	Phosphatidylcholine
PE	Phosphatidylethalnolamine
PG	Phosphatidylglycerol
PI	Phosphatidylinositol
PL	Phospholipid
PlsEtn	Ethanolamine plasmalogen
PoL	Polar lipid
PS	Phosphatidylserine
PtdIns	Phosphatidylinositol
PUFA	Poly-unsaturated fatty acid
SAFA	Saturated fatty acid
ST	Sterol
TAG	Triacylglycerol
TL	Total lipid
TLC	Thin-layer chromatography
USAFA	Unsaturated fatty acid

References

1. Loureiro, D.R.P.; Soares, J.X.; Costa, J.C.; Magalhães, A.F.; Azevedo, C.M.G.; Pinto, M.M.M.; Afonso, C.M.M. Structures, Activities and Drug-Likeness of Anti-Infective Xanthone Derivatives Isolated from the Marine Environment: A Review. *Molecules* **2019** *24*, 243. [CrossRef] [PubMed]

2. Huang, K.C.; Wu, W.T.; Yang, F.L.; Chiu, Y.H.; Peng, T.C.; Hsu, B.G.; Liao, K.W.; Lee, R.P. Effects of Freshwater Clam Extract Supplementation on Time to Exhaustion, Muscle Damage, Pro/Anti-Inflammatory Cytokines, and Liver Injury in Rats after Exhaustive Exercise. *Molecules* **2013**, *18*, 3825–3838. [CrossRef] [PubMed]
3. Greetham, D.; Zaky, A.; Makanjuola, O.; Du, C. A brief review on bioethanol production using marine biomass, marine microorganism and seawater. *Curr. Opin. Green Sustainable Chem.* **2018**, *14*, 53–59. [CrossRef]
4. Patil, M.P.; Kim, G.D. Marine microorganisms for synthesis of metallic nanoparticles and their biomedical applications. *Colloids Surf. B* **2018**, *172*, 487–495. [CrossRef] [PubMed]
5. Amorello, D.; Orecchio, S.; Pace, A.; Barreca, S. Discrimination of almonds (Prunus dulcis) geographical origin by minerals and fatty acids profiling. *Nat. Prod. Res.* **2016**, *30*, 2107–2110. [CrossRef] [PubMed]
6. FAO Yearbook of Fishery and Aquaculture Statistics 2018. Available online: http://www.fao.org/fishery/statistics/en (accessed on 20 February 2019).
7. Fernández, M.J.; Labarta, U.; Albentosa, M.; Pérez-Camacho, A. Lipid composition of Ruditapes philipponarum spat: Effect of ration and diet quality. *Comp. Biochem. Physiol.* **2006**, *144*, 229–237.
8. Orban, E.; Di, L.G.; Nevigato, T.; Casini, I.; Caproni, R.; Santaroni, G.; Giulini, G. Nutritional and commercial quality of the striped venus clam, Chamelea gallina, from the Adriatic sea. *Food Chem.* **2007**, *101*, 1063–1070. [CrossRef]
9. Karnjanapratum, S.; Benjakul, S.; Kishimura, H.; Tsai, Y. Chemical compositions and nutritional value of Asian hard clam (Meretrix lusoria) from the coast of Andaman Sea. *Food Chem.* **2013**, *141*, 4138–4145. [CrossRef] [PubMed]
10. Boselli, E.; Pacetti, D.; Lucci, P.; Frega, N. Characterization of phospholipid molecular species in the edible parts of bony fish and shellfish. *J. Agric. Food Chem.* **2012**, *60*, 3234–3245. [CrossRef] [PubMed]
11. Janssen, C.I.; Kiliaan, A.J. Long chain polyunsaturated fatty acids (LC-PUFA) from genesis to senecscence: The influence of LC-PUFA on neural development, aging and neurodegeneration. *Prog. Lipid Res.* **2014**, *5*, 1–7. [CrossRef] [PubMed]
12. Lorente-Cebrian, S.; Costa, A.; Navas-Carretero, S.; Zabala, M.; Martinez, J.; Moreno-Aliaga, M. Role of omega-3 fatty acids in obesity, metabolic syndrome and cardiovascular diseases: A review of the evidence. *J. Physiol. Biochem.* **2013**, *69*, 633–651. [CrossRef] [PubMed]
13. Cottin, S.; Sanders, T.A.; Hall, W.L. The differential effects of EPA and DHA on cardiovascular risk factors. *Proc. Nutr. Soc.* **2011**, *70*, 215–231. [CrossRef] [PubMed]
14. Edelstein, S. *Food Science: An Ecological Approach*; Jones & Bartlett Publishers: Burlington, MA, USA, 2014; pp. 70–71.
15. Miller, M.; Stone, N.J.; Ballantyne, C.; Bittner, V.; Criqui, M.H., Ginsberg, H.N. Triglycerides and cardiovascular disease: A scientific statement from the American Heart Association. *Circulation* **2011**, *123*, 2292–2333. [CrossRef] [PubMed]
16. Weintraub, H.S. Overview of prescription omega-3 fatty acid products for hypertriglyceridemia. *Postgrad Med.* **2014**, *126*, 7–18. [CrossRef] [PubMed]
17. Saito, H. Identification of novel n-4 series polyunsaturated fatty acids in a deep-sea clam, Calyptogena phaseoliformis. *J. Chromatogr. A* **2007**, *1163*, 247–259. [CrossRef] [PubMed]
18. Saito, H.; Murata, M.; Hashimoto, J. Lipid characteristics of a seep clam, Mesolinga soliditesta: Comparison with those of two coastal clams, Meretrix lamarckii and Ruditapes philippinarum. *Deep Sea Res. Part I* **2014**, *94*, 150–158. [CrossRef]
19. FAO Yearbook. Fishery and Aquaculture Statistics 2016. Available online: http://www.fao.org/fishery/static/Yearbook/YB2016_USBcard/navigation/index_intro_e.htm (accessed on 20 February 2019).
20. Hamli, H.; Idris, M.H.; Rajaee, A.H.; Kamal, A.H.M. Reproductive Cycle of Hard Clam, Meretrix lyrata Sowerby, 1851 (Bivalvia: Veneridae) from Sarawak, Malaysia. *Trop Life Sci. Res.* **2015**, *26*, 59–72. [PubMed]
21. Thao, N.T.; Nha le, Q. Effects of glucose and probiotic supplementation in nursing juvenile clam, Meretrix lyrata. *Commun. Agric. Appl. Biol. Sci.* **2013**, *78*, 304–307. [PubMed]
22. Klingensmith, J.S.; Stillway, L.W. Lipid composition of selected tissues of the hardsell clam, Mercenaria mercenaria. *Comp. Biochem. Physiol.* **1981**, *71B*, 111–112.
23. Caers, M.; Coutteau, P.; Sorgeloos, P. Impact of starvation and of feeding algal and artificial diets on the lipid content and composition of juvenile oysters (Crassostrea gigas) and clams (Tapes philippinarum). *J. Mar. Biol.* **2000**, *136*, 891–899. [CrossRef]
24. Aouini, F.; Ghribi, F.; Boussoufa, D.; Bejaoui, S.; Navarro, J.; Cafsi, M.E. Changes in lipid classes, fatty acid

composition and lipid peroxidation in the gills of the clam Donax Trunculus after permethrin exposure. *Research Gate* **2016**, *186*.

25. Beninger, P.G. Seasonal variation of the mayor lipid classes in relation to the reproductive activity of two species of clams raised in a common habitat: Tapes decssatus L. (Jeffreys, 1863) and Philippinraum (Adams and Reeve), 1850. *J. Exp. Mar. Bio. Ecol.* **1984**, *79*, 79–90. [CrossRef]
26. Prato, E.; Danieli, A.; Michele, M.; Francesca, B. Lipid and Fatty Acid Compositions of Mytilus galloprovincialis Cultured in the Mar Grande of Taranto (Southern Italy): Feeding Strategies and Trophic Relationships. *Zool. Stud.* **2010**, *49*, 211–219.
27. Bernsdorff, C.; Winter, R. Differential properties of the sterols cholesterol, ergosterol, β-sitosterol, trans-7-dehydrocholesterol, stigmasterol and lanosterol on DPPC bilayer order. *J. Phys. Chem. A* **2003**, *107*, 10658–10664. [CrossRef]
28. Chen, T.C.C. A comparison of the behavior of cholesterol, 7-dehydrocholesterol and ergosterol in phospholipid membranes. *Biochim. Biophys. Acta* **2012**, *1818*, 1673–1681. [CrossRef] [PubMed]
29. Dennis, F.; Riccardo, G.; Agnese, P.; Deborah, P.; Barrara, Z.; Luciiano, B.; Enrico, M. Comparison of eleven extraction methods for quantitative evalution of total lipids and fatty acids in the clam Anadara Inaequivalvis (Bruguiere). *J. Shellfish Res.* **2013**, *32*, 285–290.
30. Andray, B.I.; Ly, P.D.; Viacheslav, G.R.; Vasily, I.S. Fatty acid, Lipid class, and Phospholipid Molecular Species Composition of the Soft Coral Xenia sp. (Nha Trang Bay, the South China Sea, Vietnam). *Springer* **2015**, *50*, 575–589.
31. Liu, Z.Y.; Zhou, D.-Y.; Wu, Z.-X.; Yin, F.-W.; Zhao, Q.; Xie, H.-K.; Zhang, J.-R.; Qin, L.; Shahidi, F. Extraction and detailed characterization of phospholipid-enriched oils from six species of edible clams. *Food Chem.* **2017**, *239*, 1175–1181. [CrossRef] [PubMed]
32. Peterson, B.L.; Cummings, B.S. A review of chromatographic methods for the assessment of phospholipids in biological samples. *Biomed. Chromatogr.* **2006**, *20*, 227–243. [CrossRef] [PubMed]
33. Bligh, E.G.; Dyer, W.J. A rapid method of total lipid extraction and purification. *Can. J. Biochem. Phys.* **1959**, *37*, 911–917. [CrossRef]
34. Vaskovsky, V.E.; Terekhova, T.A. HPTLC of phospholipid mixtures containing phosphatidylglycerol. *J. High. Resolut. Chromatogr. Chromatogr. Commun.* **1979**, 671–672. [CrossRef]
35. Vaskovsky, V.E.; Kostetsky, E.Y.; Vasendin, I.M. A universal reagent for phospholipid analysis. *J. Chromatogr.* **1975**, *114*, 129–141. [CrossRef]
36. Christie, W. Equivalent chain lengths of methyl ester derivatives of fatty acids on gas chromatography a reappraisal. *J. Chromatogr.* **1988**, *447*, 305–314. [CrossRef]
37. Mass Spectrometry of Fatty Acid Derivatives. Available online: http://lipidlibrary.aocs.org/ms/masspec.html (accessed on 10 August 2014).
38. Harrabi, S.; Herchi, W.; Kallel, H.; Mayer, P.; Boukhchina, S. Liquid chromatographic-mass spectrometric analysis of glycerophospholipids in corn oil. *Food Chem.* **2009**, *114*, 712–716. [CrossRef]

13

Novel Electrochemical Sensors Based on Cuprous Oxide-Electrochemically Reduced Graphene Oxide Nanocomposites Modified Electrode toward Sensitive Detection of Sunset Yellow

Quanguo He [1,2,†], Jun Liu [1,2,†], Xiaopeng Liu [2], Yonghui Xia [3], Guangli Li [1,2,*], Peihong Deng [4] and Dongchu Chen [1,*]

1 School of Materials Science and Energy Engineering, Foshan University, Foshan 528000, China; hequanguo@126.com (Q.H.); liu.jun.1015@163.com (J.L.)
2 College of Life Sciences and Chemistry, Hunan University of Technology, Zhuzhou 412007, China; amituo321@163.com
3 Zhuzhou Institute for Food and Drug Control, Zhuzhou 412000, China; sunnyxia0710@163.com
4 Department of Chemistry and Material Science, Hengyang Normal University, Hengyang 421008, China; dph1975@163.com
* Correspondence: guangli010@hut.edu.cn (G.L.); chendc@fosu.edu.cn (D.C.)

† These authors contributed equally to this work.

Abstract: Control and detection of sunset yellow is an utmost demanding issue, due to the presence of potential risks for human health if excessively consumed or added. Herein, cuprous oxide-electrochemically reduced graphene nanocomposite modified glassy carbon electrode (Cu_2O-ErGO/GCE) was developed for the determination of sunset yellow. The Cu_2O-ErGO/GCE was fabricated by drop-casting Cu_2O-GO dispersion on the GCE surface following a potentiostatic reduction of graphene oxide (GO). Scanning electron microscope and X-ray powder diffractometer was used to characterize the morphology and microstructure of the modification materials, such as Cu_2O nanoparticles and Cu_2O-ErGO nanocomposites. The electrochemical behavior of sunset yellow on the bare GCE, ErGO/GCE, and Cu_2O-ErGO/GCE were investigated by cyclic voltammetry and second-derivative linear sweep voltammetry, respectively. The analytical parameters (including pH value, sweep rate, and accumulation parameters) were explored systematically. The results show that the anodic peak currents of Cu_2O-ErGO /GCE are 25-fold higher than that of the bare GCE, due to the synergistic enhancement effect between Cu_2O nanoparticles and ErGO sheets. Under the optimum detection conditions, the anodic peak currents are well linear to the concentrations of sunset yellow, ranging from 2.0×10^{-8} mol/L to 2.0×10^{-5} mol/L and from 2.0×10^{-5} mol/L to 1.0×10^{-4} mol/L with a low limit of detection (S/N = 3, 6.0×10^{-9} mol/L). Moreover, Cu_2O-ErGO/GCE was successfully used for the determination of sunset yellow in beverages and food with good recovery. This proposed Cu_2O-ErGO/GCE has an attractive prospect applications on the determination of sunset yellow in diverse real samples.

Keywords: cuprous oxide nanoparticles; reduced graphene oxide; modified electrode; sunset yellow; second-derivative linear sweep voltammetry

1. Introduction

Sunset yellow, as a common azo colorant, has been widely added in several beverages (such as carbonated beverage, orange juice, and Fanta drink) and food (i.e. candies, cakes, cheese) to improve the appearance and appetite [1,2]. Sunset yellow contains an azo group (-N=N-) and aromatic ring structure, which may cause mutagenic and carcinogenic risk for human [3,4]. As a result, it can bring many serious health problems, such as hepatocellular damage, kidney failures, headache, cancers, and attention deficit hyperactivity disorder (ADHD), when it is excessively consumed [5]. Hence, the additive amount of sunset yellow in beverages and food demands strict control and regulation. European Food Safety Authority (EFSA) recommends an acceptable daily intake (ADI) for sunset yellow of 1.0 mg kg^{-1} body weight [6]. In some countries, it was explicitly stated that the permitted maximum content of sunset yellow in nonalcoholic beverages is 100 μg mL^{-1} [7]. Moreover, Finland and Norway even has already banned the use of sunset yellow in foods [6]. While considering the food quality and safety, it is urgent to develop reliable analytical techniques for the quick detection of this food dye.

At present, various techniques have been developed for sunset yellow, including high performance liquid chromatography (HPLC) [8], thin layer chromatography [9], spectrophotometry [10], fluorometry [11], and capillary electrophoresis [12]. Although these methods have been proved to be reliable, they still have some disadvantages, such as time-consuming, complicated preprocessing, and expensive equipment. Recently, electrochemical analysis has been widely used for the detection of bioactive molecules, nutrients, food additives, as well as contaminants, due to its considerable merits such as low cost, simple operation, rapid response, high sensitivity and sensitivity. As we all know, the key issue for electrochemical detection toward sunset yellow is to develop ultrasensitive modified electrodes.

Nanostructure precious metals or alloys have become the preferred electrode modification materials for the sensitive detection of sunset yellow, due to their superior electrocatalytic activity [7,13–17]. For example, Wang and coworkers developed a promising electrochemical sensor based on Au-Pd and reduced graphene oxide nanocomposite decorated electrode (Au-Pd-RGO/GCE) [17]. The Au-Pd-RGO/GCE exhibited good stability, superior electrocatalytic performance, low detection limit (1.5 nmol/L), and wide response range (0.686–331.686 μmol/L). Pd-Ru nanoparticles incorporated carbon aerogel nanocomposites (Pd-Ru/CA) have been successfully used for electrochemical detection and catalytic degradation of sunset yellow [16]. The Pd-Ru/CA nanocomposite decorated screen printed carbon electrode (Pd-Ru/CA/SPCE) showed a low detection limit (7.1 nmol/L) and high sensitivity (3.571 μA/(μmol/L cm^2). Electrochemical sensing platform based on Au NPs and reduced graphene modified GCE (Au NPs/RGO/GCE) was constructed for the quantitative analysis of sunset yellow, and the Au NPs/RGO/GCE had excellent catalytic activity toward the oxidation of sunset yellow [15]. The proposed electrode showed wide linear response range of 0.002–109.14 μmol/L and low detection limit of 2 nmo/L (S/N = 3). Au NPs decorated carbon-paste electrode (Au NPs/CPE) was also fabricated for the simultaneous detection sunset yellow and Tartrazine [7]. The Au NPs/CPE displayed low detection limits of 3.0×10^{-8} and 2.0×10^{-9} mol/L for sunset yellow and Tartrazine, respectively. These precious metals-based modified electrodes can detect sunset yellow at the nanomole level; however, the scarcity and high cost of precious metals or alloys have seriously hindered the broad practical applications.

When compared to precious metals or alloys, transition metals and metal oxides have outstanding advantages in terms of abundance and cost. What is more, transition metals and metal oxides also have excellent catalytic activity. In our previous work, Cu_2O-RGO nanocomposite [18], NH_2-Fe_3O_4-RGO nanocomposite [19], MnO_2-RGO nanocomposite [20], TiO_2-RGO nanocomposite [21], and α-MnO_2/N-doped ketjenblack carbon composite [22,23] demonstrated excellent electrochemical sensing performance toward the detection of dopamine and Tartrazine and superior electrocatalytic activity in oxygen reduction reaction (ORR) & oxygen evolution reaction (OER). Recently, few researchers have devoted to developing transition metal oxide modified electrodes for the determination of sunset yellow. For example, Dorraji et al. [24] developed ZnO/cysteic acid nanocomposite modified electrode (ZnO/CA/GCE), and successfully used for simultaneous determination of sunset yellow and Tartrazine. The ZnO/CA/GCE exhibited two linear response ranges in the concentration ranges of 0.1–3.0 μmol/L, and 0.07–1.86 μmol/L, and detection limits of 0.03 μmol/L and 0.01 μmol/L for sunset yellow and Tartrazine, respectively. However, the linear dynamic response range is limited for trace detection of sunset yellow. The detection capacity of sunset yellow has been improved with graphene and mesoporous TiO_2 composite [25] and ZnO/RGO/ZnO@Zn [26] modified electrodes, and they showed superior sensing performance (i.e., linear ranges, detection of limit) that was comparable to precious metal modified electrode. Although some progress has been made, there are only few related reports concerning the transition metal oxide modified electrodes. Therefore, it is still worthwhile to develop novel transition metal oxide modified electrodes for sensitive detection of sunset yellow.

Among transition oxides, cuprous oxide (Cu_2O) is an environmentally friendly p-type semiconductor material, which has been widely used in solar cells and photo catalysis [27,28], due to its unique electronic structure and excellent catalytic performances. However, its electrical conductivity is poor due to the nature of semiconductor. To resolve this problem, Cu_2O nanoparticles are often composited or hybrided with conductive materials [29–33], to decrease the charge transfer resistance and eventually enhance the electrochemical performance. Graphene, as an emerging two-dimensional (2D) carbon material, has been usually used as conductive materials in modified electrodes, owing to its high specific surface area, excellent electrical conductivity, superior electrochemical performance, and fast heterogeneous electron transfer rate. It has been reported that graphene-based modified electrodes have been widely employed for the determination of azo dyes, such as sunset yellow, Tartrazine, and Amaranth [34–36]. However, to our best knowledge, Cu_2O/reduced graphene oxide nanocomposite modified electrode toward sensitive detection of sunset yellow has not been reported.

Graphene usually prepared from graphene oxide (GO) by electrochemical reduction method in the field of electrochemical analysis. The chemically reduced graphene oxide is hydrophobic, due to the removal of most oxygen-containing functional groups. As a result, the chemically graphene oxide tends to agglomerate resulting in a degradation on sensing performance. The agglomeration issue can be overcomed by introducing the surfactants [37], which can effectively improve the dispensability. However, the electrical conductivity also declined due to the use of surfactants. Electrochemically reduction method is a green and efficient method to obtain reduced graphene oxide that not require any reductants. Moreover, the residual oxygen-containing functional groups can be tuned by facial adjusting the electrochemical parameter, such as reduction potential, reduction time, and scanning cycles [18,19,21]. In other words, the property of reduced graphene oxide can be tailored by electrochemical parameters. For these reasons, the electrochemically reduced graphene oxide (ErGO) have been widely used for constructing diverse sensors.

Inspired by the foregoing reports, herein ErGO was composited with low cost and excellent electrocatalytic activity Cu_2O nanoparticles, aiming to develop a cost-effective, high sensitive, and good selective modification materials to substitute the precious metal-based materials. Meanwhile, the Cu_2O-ErGO nanocomposites are expected to exert their synergistic sensitizing effects to improve the sensing performance. Then, Cu_2O-ErGO nanocomposites was modified on the surface of the glassy carbon electrode (GCE) to construct a novel sensor toward sunset yellow. The Cu_2O-ErGO modified glassy carbon electrode (Cu_2O-ErGO/GCE) was prepared while using a facile drop-casting technique in combination with electrochemical reduction. The electrochemical behavior of sunset yellow on the Cu_2O-ErGO/GCE were investigated by cyclic voltammetry and second-derivative linear sweep voltammetry. The effect of detection conditions (such as pH value, sweep rate, and accumulation parameters) on the electrochemical response were also explored. Finally, the proposed Cu_2O-ErGO/GCE was used to detect the content of sunset yellow in soda drinks, orange juice, and candies samples while using second-derivative linear sweep voltammetry.

2. Results and Discussion

2.1. Morphology and Microstructural Characterization

The surface morphologies of Cu_2O nanoparticles (Cu_2O NPs) and Cu_2O-ErGO nanocomposites were characterized by scanning electron microscope (SEM, Hitachi S-3000N, Tokyo, Japan). The SEM images are shown in Figure 1A,B, respectively. The Cu_2O NPs exhibit cubic-like structure with uniform size, and the particle size is estimated to about 150 nm. Obviously, the thin layer ErGO sheets were successfully coated on the surface of Cu_2O nanoparticles. Moreover, the particle size of Cu_2O-ErGO nanocomposite increases slightly, which facilitates the adsorption of sunset yellow. The Cu_2O NPs were further characterized by X-ray diffraction (XRD, JEOL JEM-2010 (HT, Tokyo, Japan) and the XRD pattern of Cu_2O NPs is plotted in Figure 1C. The diffraction peaks of Cu_2O nanoparticles are clearly indexed into the pure cubic phase of Cu_2O (JSPDS78-2076), suggesting that the cubic phase of Cu_2O nanoparticles was prepared.

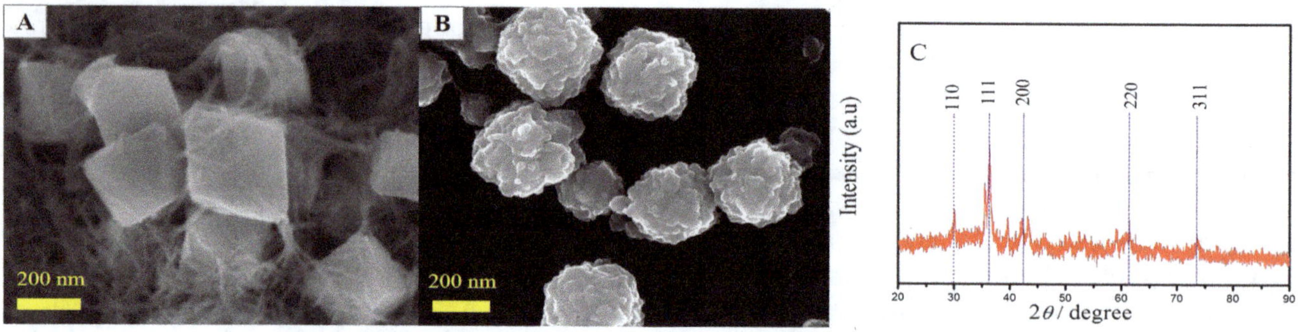

Figure 1. The scanning electron microscope (SEM) photos of cuprous oxide (Cu_2O) nanoparticles (**A**) and cuprous oxide-electrochemically reduced graphene oxide (Cu_2O-ErGO) nanocomposites (**B**); and, (**C**) The X-ray diffraction (XRD) pattern of Cu_2O nanoparticles.

2.2. Electrochemical Behavior of Sunset Yellow on Modified Electrodes

Second-derivative linear sweep voltammetric responses of 1.0×10^{-5} mol/L sunset yellow on different electrodes are presented in Figure 2. On the bare GCE, a weak anodic peak of sunset yellow appears at 798 mV with the anodic peak current (i_{pa}) of 0.725 μA. On the GO/GCE, the oxidation peak current of sunset yellow decreases to 0.497 μA, which is mainly due to the present of the poor electrical conductivity of GO. On the Cu_2O-GO/GCE, an apparent anodic peak of occurred at 770 mV, and the i_{pa} increases to 0.925 μA, probably owing to the electrocatalytic activity of Cu_2O nanoparticles. When the GO was electrochemically reduced to ErGO, the anodic peak appears at 792 mV and the i_{pa} increases to 16.93 μA. This phenomenon may be related to the high electrical conductivity, large specific surface area, and rapid heterogeneous electron transfer rate of ErGO. Moreover, the adsorption capacity of sunset yellow on the electrode surface is improved greatly by the π-π interaction, because the conductive carbon-conjugated networks are restored after the reduction process. When the GCE was modified with Cu_2O-ErGO nanocomposites, the i_{pa} is the largest (18.08 μA), which is about 25 fold greater than that of bare GCE. It mainly due to the synergistic enhancement effect between Cu_2O nanoparticles and ErGO sheets, which significantly improves the sensitivity of sunset yellow detection.

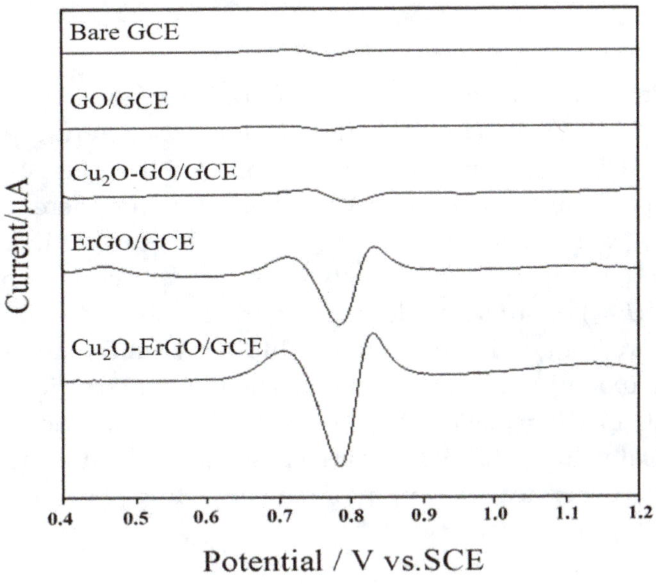

Figure 2. Second-derivative linear sweep voltammograms of 1.0×10^{-5} mol/L sunset yellow on the different electrodes.

The electrochemical behavior of 1.0×10^{-5} mol/L sunset yellow on the GCE (a), GO/GCE (b), Cu_2O-GO/GCE (c), ErGO/GCE (d), and Cu_2O-ErGO/GCE (e) were further investigated by cyclic voltammetry (Figure 3). All of the electrodes appear a pair of redox peaks, meaning that sunset yellow undergoes a quasi-reversible process. Obviously, a pair of sharp redox peaks occurs on the ErGO/GCE and Cu_2O-ErGO/GCE. Furthermore, the order of anodic peak currents obtained from cyclic voltammograms is consistent with the second-derivative linear sweep voltammograms, which further confirms that Cu_2O-ErGO nanocomposites can significantly enhance the electrochemical response toward sunset yellow.

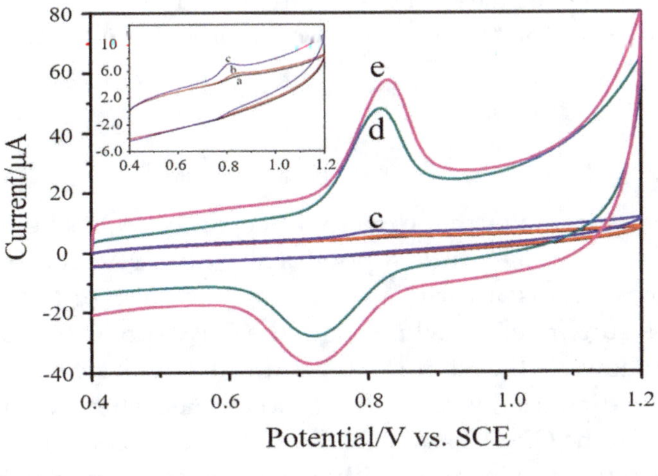

Figure 3. Cyclic voltammograms of 1.0×10^{-5} mol/L sunset yellow recorded on the glassy carbon electrode (GCE) (a), graphene oxide/glassy carbon electrode (GO/GCE) (b), Cu_2O-graphene oxide nanocomposite modified glass carbon electrode (Cu_2O-GO/GCE) (c), electrochemically reduced graphene oxide modified glass carbon electrode (ErGO/GCE) (d) and Cu_2O-ErGO modified glassy carbon electrode (Cu_2O-ErGO/GCE) (e). The inset is the magnification of the cyclic voltammograms recorded on the GCE (a), GO/GCE (b) and Cu_2O-GO/GCE (c).

2.3. Effect of pH Value

Since proton (H^+) plays an important role on the redox of sunset yellow, so it is worthwhile investigating the influence of pH value on the response peak current of sunset yellow. The i_{pa} of sunset yellow recorded in various pH PBS solution are depicted in Figure 4A. The i_{pa} of sunset yellow increases gradually as the pH value increases. When the pH value increases to 3.8, the largest i_{pa} is obtained. Afterwards the i_{pa} decreases slowly with the pH value further increasing. Hence, the pH 3.8 PBS solution was employed as supporting electrolytes on the subsequent experiments. Furthermore, the anodic peak potential (E_{pa}) of sunset yellow is negatively shifted with the increase of pH value. As plotted in Figure 4B, there is a good linear relationship between E_{pa} and pH value, confirming that protons are involved in the oxidation of sunset yellow. The linear regression equation can expressed as E_{pa} (V) = -0.0570 pH + 1.0167 (R^2 = 0.998). According to Nernst equation, its slope (-0.0570 V/pH) approaches to the theoretical value (-0.0590 V/pH), suggesting that the equal amounts of proton and electron involve in the electrochemical oxidation of sunset yellow [38].

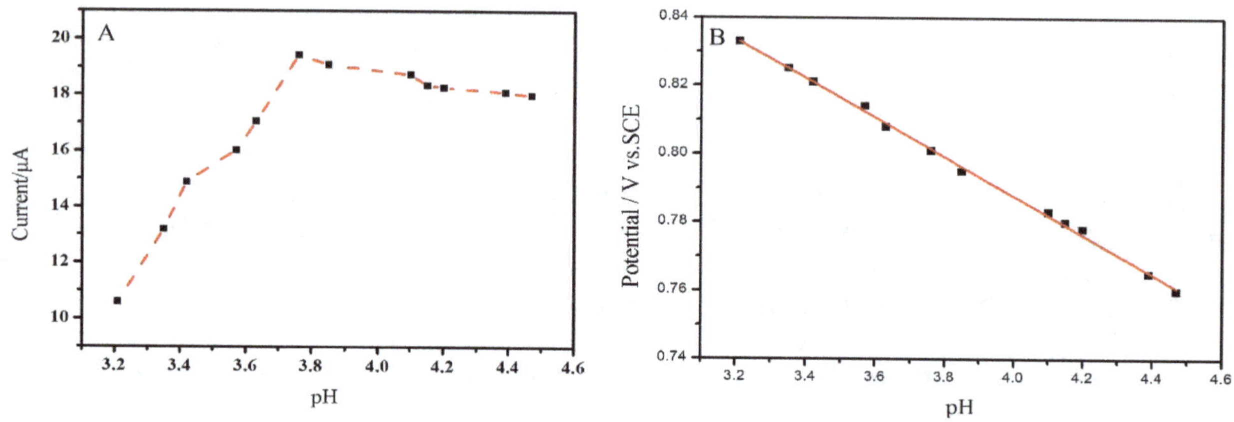

Figure 4. (**A**) Effects of pH value on the anodic peak currents of 1.0×10^{-5} mol/L sunset yellow on the Cu_2O-ErGO/GCE; and, (**B**) The plot of the anodic peak potential of 1.0×10^{-5} mol/L sunset yellow against pH value.

2.4. Effect of Sweep Rates

Sweep rate is a crucial parameter that directly affects the electrochemical response of analysts on the modified electrodes. Moreover, it is a powerful tool to reveal the electrochemical reaction mechanism. Cyclic voltammograms at various sweep rates were recorded at 0.1 mol/L PBS solution containing 1×10^{-5} mol/L sunset yellow while using the Cu_2O-ErGO/GCE, and their corresponding cyclic voltammograms are shown in the Figure 5A. As expected, with sweep rates increasing, the anodic peaks shift toward positive direction, while the cathodic peaks shift negatively, indicating that the oxidation of sunset yellow on the Cu_2O-ErGO/GCE is quasi-reversible. Both the anodic peak currents (i_{pa}) and cathodic peak currents (i_{pc}) increase with the sweep speeding up, however, the background currents also increase. To purist high signal-to-noise (S/N), a suitable sweep rate is recommended as 100 mV/s. It can be clearly seen from Figure 5B that both the anodic peak currents (i_{pa}) and cathodic peak currents (i_{pc}) of sunset yellow is nearly linear with the sweep rates (v). Their linear equations are expressed as: i_{pa} (μA) = 0.2968 v (mV/s) − 4.791 (R^2 = 0.998) and i_{pc} (μA) = -0.1471 v (mV/s) + 1.659 (R^2 = 0.997), suggesting that the oxidation of sunset yellow on the Cu_2O-ErGO/GCE is controlled by the adsorption process [39].

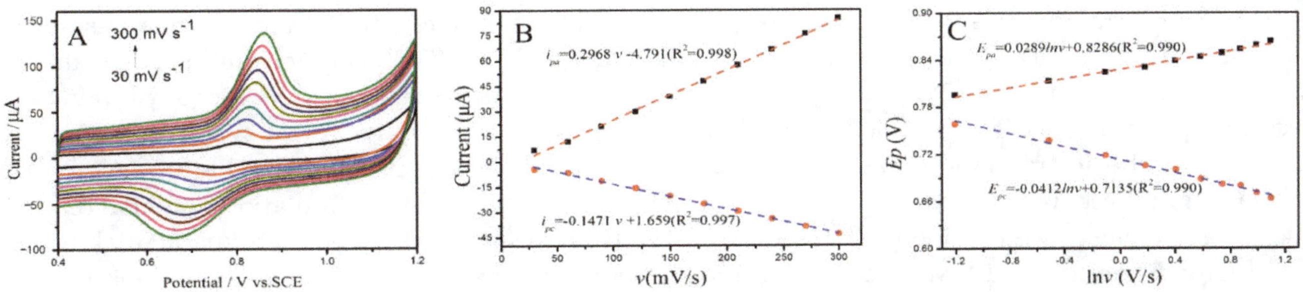

Figure 5. (**A**) Cyclic voltammograms of 1.0×10^{-5} mol/L sunset yellow on the Cu_2O-ErGO/GCE recorded at various sweep rates (30 mV/s–300 mV/s); (**B**) Linear plots of anodic and cathodic peak currents (i_{pa} and i_{pc}) of sunset yellow against sweep rates (v); and (**C**) Linear plots of andic and cathodic peak potential (E_{pa} and E_{pc}) of sunset yellow against Napierian Logarithm of sweep rates (ln$v$). Supporting electrolytes: 0.1 mol/L PBS.

It is observed that both the anodic peak potential (E_{pa}) and the cathodic peak potential (E_{pc}) is well linear to the Napierian Logarithm of sweep rates (ln$v$). Their corresponding linear equations are E_{pa} (V) = 0.0289ln$v$ (V/s) + 0.8286 (R^2 = 0.990) and E_{pc} (V) = − 0.0412ln$v$ (V/s) + 0.7135 (R^2 = 0.990). As for an adsorption-controlled and quasi-reversible process, according to the Lavrion equation [40], the peak potential and the sweep rate follows the following relationship:

$$E_{pa} = E^{0'} + \frac{RT}{(1-\alpha)nF}[0.78 + \ln(\frac{D^{1/2}}{K_s}) - \frac{1}{2}\ln\frac{RT}{(1-\alpha)nF}] + \frac{RT}{2(1-\alpha)nF}\ln v \quad (1)$$

$$E_{pc} = E^{0'} + \frac{RT}{\alpha nF}[0.78 + \ln(\frac{D^{1/2}}{K_s}) - \frac{1}{2}\ln\frac{RT}{\alpha nF}] - \frac{RT}{2\alpha nF}\ln v \quad (2)$$

where E_{pa} (V) and E_{pc} (V) represents the anodic peak potential and the cathodic peak potential, respectively; v (V/s) denotes the sweep rate; α is the charge transfer coefficient; k_s is the heterogeneous electron transfer rate; n is the electron transferred number; T is Kelvin temperature; F is Faraday constant (96,480 C/mol); and, R is molar gas constant (8.314 J/(mol·K)). Combining the slopes of Equations (1) and (2) with the E_{pa}/E_{pc} vs. lnv equations, the charge transfer coefficient α is estimated to be 0.45 and the electron transferred number n is around 1. Since the equal amount of proton and electron participates in the oxidation process, the electrochemical oxidation of sunset yellow is 1 electron and 1 proton process, which is in accordance with the previous studies [39,41]. Hence, the electrochemical oxidation mechanism of sunset yellow on the Cu_2O-ErGO/GCE can be inferred in Figure 6.

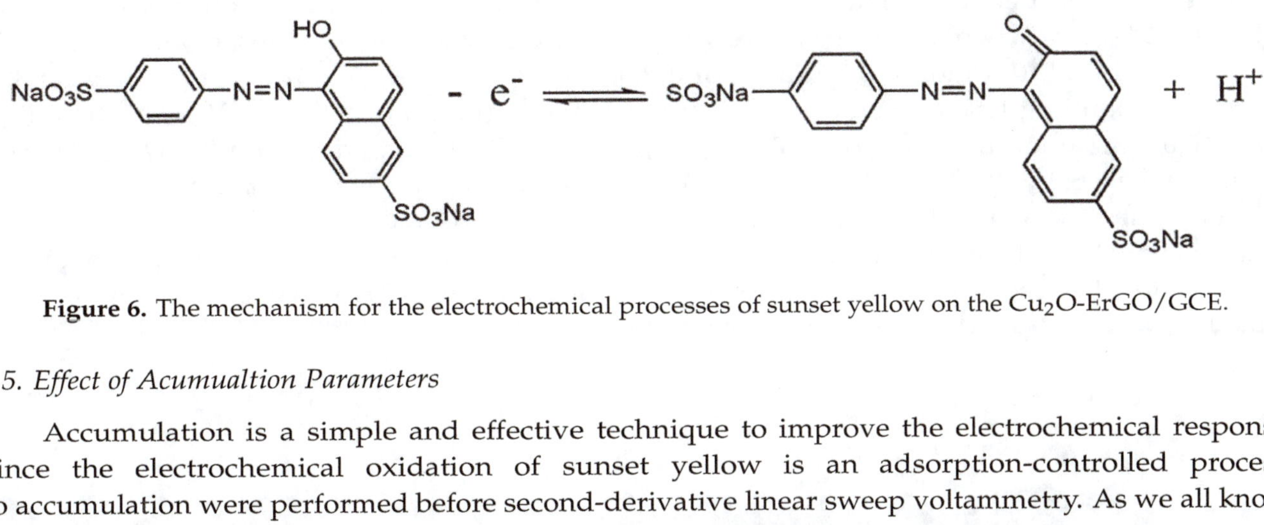

Figure 6. The mechanism for the electrochemical processes of sunset yellow on the Cu_2O-ErGO/GCE.

2.5. *Effect of Acumualtion Parameters*

Accumulation is a simple and effective technique to improve the electrochemical response. Since the electrochemical oxidation of sunset yellow is an adsorption-controlled process, so accumulation were performed before second-derivative linear sweep voltammetry. As we all know, accumulation potential as well as time are two important parameters that affect the response peak current greatly, so it is a worthwhile optimization. The Cu_2O-ErGO/GCE was accumulated at different accumulation potential for 240 s firstly. Then, their anodic peak currents (i_{pa}) of sunset yellow were recorded in 0.1 mol/L PBS solution (pH 3.8) while using second-derivative linear sweep voltammetry. The effect of the accumulation potential on the i_{pa} of sunset yellow is presented in Figure 7A. The i_{pa} increases gradually with the rising of accumulation potential. When the accumulation potential reaches 0.4 V, the strongest i_{pa} is obtained. Afterwards, the i_{pa} decreases with the accumulation potential further increasing. Therefore, 0.4 V was selected in the subsequent experiments. Furthermore, the influence of accumulation time was also explored. Similarly, the Cu_2O-ErGO/GCE was accumulated at an optimized accumulation potential for various time. Then, their i_{pa} of sunset yellow were recorded and compared. As shown in Figure 7B, the i_{pa} increases with the prolong of the accumulation during the first 180 s; then i_{pa} keep stable with the accumulation time further prolonging, demonstrating that the adsorption of sunset yellow achieved saturated. Hence, 180 s is recommended as the optimum accumulation time.

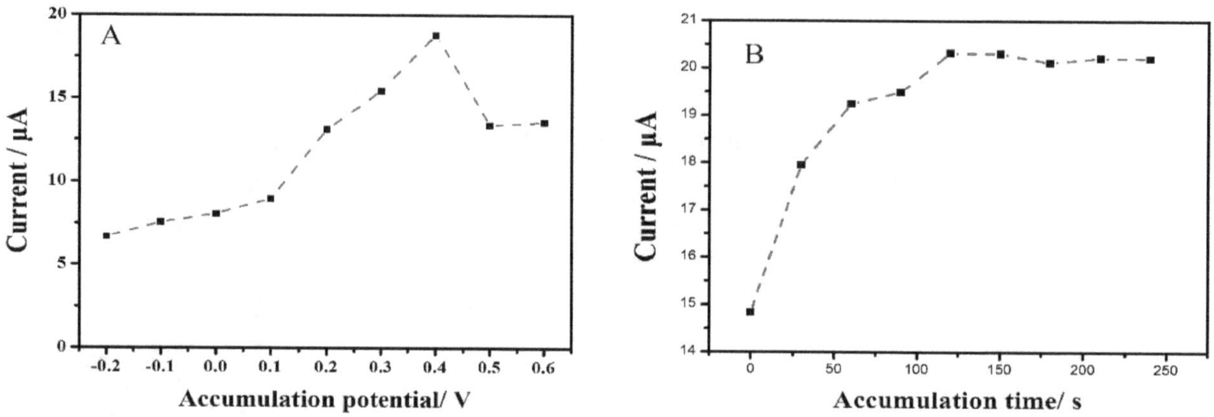

Figure 7. Influences of accumulation potential (**A**), and accumulation time (**B**) on the anodic peak currents of 1.0×10^{-5} mol/L sunset yellow on the Cu_2O-ErGO/GCE.

2.6. Standard Curves, Linear Range and Limit of Detection

With the optimal analytical parameters, the i_{pa} of different concentrations of sunset yellow standard solution was determined by second-derivative linear sweep voltammetry. Figure 8A shows the second-derivative linear sweep voltammograms of various concentrations of sunset yellow. There are two linear response ranges for the detection of sunset yellow, namely 2.0×10^{-8} ~2.0×10^{-5} mol/L (Figure 8B) and 2.0×10^{-5} mol/L ~1.0×10^{-4} mol/L (Figure 8C). Their corresponding linear equations are i_{pa} (μA) = 1.597c (μmol/L) + 2.628 (R^2 = 0.973) and i_{pa} (μA) = 0.0775c (μmol/L) + 30.36 (R^2 = 0.992), respectively. The limit of detection (LOD, S/N = 3) is estimated to be 6.0×10^{-9} mol/L. The linear response range is lower than the permitted maximum content of sunset yellow (2.2×10^{-4} mol/L), so that the Cu_2O-ErGO/GCE can be applied to detection sunset yellow by a direct or the dilution method. A comparison on sensing performances toward sunset yellow between the existing modified electrodes and Cu_2O-ErGO/GCE is summarized on Table 1.

Table 1. Comparison the sensing performances toward the detection of sunset yellow between the existing modified electrodes and the proposed Cu_2O-ErGO/GCE.

Modified Electrodes	Method	Metrological Parameters	Linear Range (μmol/L)	LOD (μmol/L)	Reference
PLPA/GCE	DPV	0.1 M phosphate-citrate buffer solution (pH 7.0); accumulation for 60 s	0.04–14	0.040	[42]
MIP/f-MWCNTs/GCE	DPV	0.1 M CBS solution (pH 5.0); accumulation for 30 min; scanned at 10 mV/s	0.05–100	0.005	[43]
Au-Pd-RGO/GCE	DPV	0.1 M PBS (pH 4.0); scanned at 50 mV/s	0.69–332	0.0015	[17]
CTAB-Gr-Pt/GCE	DPV	0.1 M PBS (pH 3.0); accumulation for 3 min	0.0085–1.0; 1.0–30	0.0042	[44]
GO/AgNPs-MIPs/GCE	LSV	0.1 M PBS (pH 5.5); accumulation for 7 min; scanned at 50 mV/s	0.1–0.6; 0.6–12	0.02	[45]
ILRGO-Au/GCE	SWV	0.1 M BR buffer solution (pH 7.0); accumulation for 300 s	0.004–1.0	0.00052	[13]
Au NPs/CPE	DPV	0.1 M PBS (pH 4.0); accumulation for 1 min; modulation amplitude = 60 mV and scan rate = 60 mV/s	0.1–2.0	0.03	[7]
MWCNT/GCE	DPV	0.1 M PBS (pH 7.0); accumulation at open circuit potential for 2 min; potential increment of 0.004 V, pulse amplitude of 0.05 V, and pulse period of 0.2 s	0.55–7.0	0.12	[46]
PDDA-Gr-Pd/GCE	DPV	0.1 M PBS (pH 3.0); accumulation for 5 min	0.01–10	0.002	[47]
Fe_3O_4@SiO_2-NPs@MIP/Gr/GCE	DPV	0.1 M PBS (pH 8.0); pulse amplitude = 0.05 V; pulse interval time = 0.05 s, and scan rate = 0.02 V/s for differential pulse voltammetry	0.02–20	0.0055	[48]
Cu_2O-ErGO/GCE	SDLSV	0.1 M PBS (pH 3.8); accumulation at 0.4 V for 180 s; scanned at 100 mV/s	0.02–20; 20–100	0.006	This work

The linear ranges and LOD of the proposed Cu_2O-ErGO/GCE are at least comparable to and even better than most of the previous reports. Moreover, Cu_2O-ErGO/GCE have outstanding advantages over noble metal modified electrodes (such as Au NPs/CPE [7], Au-Pd-RGO/GCE [17], CTAB-Gr-Pt/GCE [44], GO/AgNPs-MIPs/GCE [45] and PDDA-Gr-Pd/GCE [47] in terms of the cost and electrode fabrication.

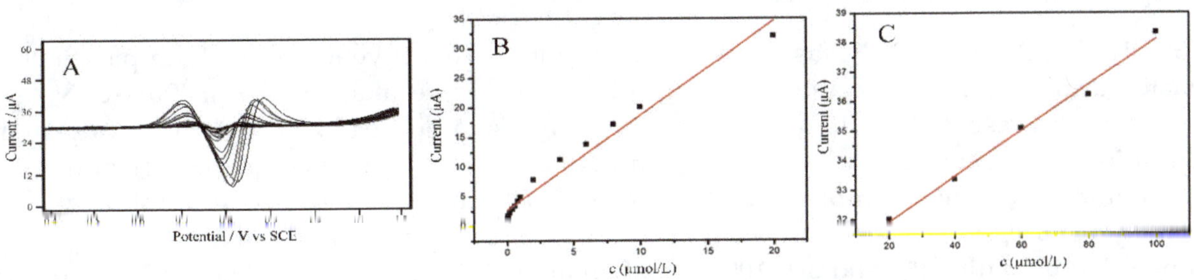

Figure 8. (**A**) Second-derivative linear sweep voltammograms of sunset yellow with various concentrations on the Cu_2O-ErGO/GCE; Calibration curves between the anodic peak current and the concentrations of sunset yellow ranging from 2.0×10^{-8} to 2.0×10^{-5} mol/L (**B**) and from 2.0×10^{-5} to 1.0×10^{-4} mol/L(**C**).

2.7. Interference and Reproducibility Investigation

Prior to the detection of real samples, the anti-interference and reproducibility was also investigated to validate the practicability of the proposed Cu_2O-ErGO/GCE. The response peak current of pure sunset yellow solution and potential interfering compounds mixture solution were recorded and compared. It is observed that the change of i_{pa} of 10 μmol/L sunset yellow is less than 5% in the presence of a 100-fold concentration of glucose, benzoic acid, citric acid, Na^+, K^+, Fe^{3+}, and 10-fold concentration of Tartrazine, quinoline yellow (Figure 9). It is demonstrating that our proposed Cu_2O-ErGO/GCE exhibits good selectivity toward sunset yellow. The reproducibility of Cu_2O-ErGO/GCEs were examined by continuous measurement the response peak currents of 10 μmol/L sunset yellow seven times. The result shows that the i_{pa} remain stable with relative standard deviation (RSD) of 2.78% (Table 2), indicating that the Cu_2O-ErGO/GCE has good reproducibility.

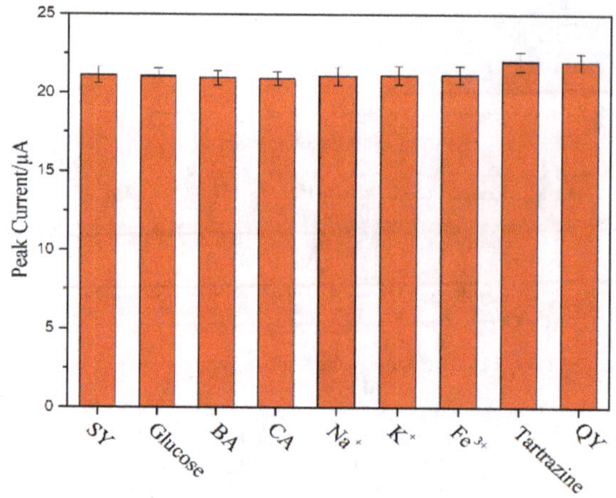

Figure 9. The response peak current of Cu_2O-ErGO/GCE for 1×10^{-5} mol/L sunset yellow (SY) in the presence of interfering substances ($n = 3$). The concentration of glucose, benzoic acid (BA), citric acid (CA), Na^+, K^+, Fe^{3+}, were 1×10^{-3} mol/L. The concentrations of Tartrazine, quinoline yellow (QY) were 1×10^{-4} mol/L.

Table 2. The reproducibility of Cu_2O-ErGO/GCE for detection of sunset yellow.

No.	1	2	3	4	5	6	7
i_{pa} (μA)	21.07	22.34	21.36	22.04	22.29	22.85	22.24
Average value (μA)				22.02			
RSD (%)				2.78			

2.8. Detection Sunset Yellow in Real Samples

Finally, Cu_2O-ErGO/GCE was applied to determinate sunset yellow in real samples, including carbonated drinks, orange juice, and candy samples. The detect results are listed in Table 3. No anodic peak current is presented in the carbonated beverage sample near 798 mV, indicating that the concentration of sunset yellow is low than the limit of detection. The concentrations of sunset yellow in orange juice and candy samples are detected by 0.085 μmol/L and 0.162 μmol/L, respectively. The content of sunset yellow in these samples is lower than the permitted maximum content that is recommended by national standard (100 μg/mL, namely 2.2×10^{-4} mol/L). Then, an appropriate amount of sunset yellow standard solution was added to the above three samples. Standard addition test results suggests that the recoveries of sunset yellow were 98.75~102.0%, with the relative standard deviation being less than 2.85%, indicating that satisfactory results were obtained using the proposed Cu_2O-ErGO/GCE. Together with cost-effective, quick response, high sensitive and good selectivity, the Cu_2O-ErGO/GCE exhibits great prospects on the determination of sunset yellow in different real samples, including but not limiting to beverages, food, and nutrients.

Table 3. The results of determination of sunset yellow in soft drink and candies ($n = 3$).

Samples	Original (μmol/L)	Added (μmol/L)	RSD (%)	Found (μmol/L)	Recovery (%)	RSD (%)
Sodas	ND [1]	4	1.28	4.06	102.0	1.50
Orange juice	0.085	0.080	2.46	0.082	102.5	2.31
Candies	0.162	0.160	3.25	0.158	98.75	2.85

[1] ND: Not detected.

3. Materials and Methods

3.1. Chemical and Solution

Graphite powder, $NaNO_3$, concentrated H_2SO_4, NaOH, $KMnO_4$, H_2O_2, $CuSO_4 \cdot 5H_2O$, Na_2HPO_4, NaH_2PO_4, hydrochloric acid (HCl), polyvinylpyrrolidone (PVP), hydrate hydrazine ($N_2H_4 \cdot H_2O$), and ethyl alcohol were analytic grade and supplied by Sinopharm Chemical Reagent Co., Ltd. (Shanghai, China). Sunset yellow was purchased from Aladdin (http://www.aladdin-e.com). All of these chemicals were directly used without further purification. A series of sunset yellow standard solutions were prepared by diluting appropriately the stock sunset yellow (1×10^{-3} mol/L). These standard solutions were kept in 4 °C refrigerator when not in use. Deionized water (18.2 MΩ) was used throughout the experiments.

3.2. Synthesis of Cu_2O Nanoparticles

Cu_2O nanoparticles were synthesized by the hydrothermal method referred to our previous work [18]. Specifically, 50 mg of $CuSO_4 \cdot 5H_2O$ and 24 mg of PVP were added into 10 mL deionized water and then stirred with ultrasonication for 30 min. Afterwards, 2 mL of 0.2 mol/L NaOH solution was added and stirred for 30 min at room temperature to obtain blue $Cu(OH)_2$ precipitates. Subsequently, 6 μL of $N_2H_4 \cdot H_2O$ was added as reductant and then stirred for 20 min at room temperature to form a brick red suspension. The precipitate was separated by centrifugation at 5000 rpm, and washed repeatedly with deionized water and ethanol for three times, and vacuum-dried at 60 °C to obtain Cu_2O nanoparticles.

3.3. Preparation Cu_2O-GO Nanocomposite Dispersion

Graphene oxide (GO) is prepared from cheap graphite powder by a modified Hummers method according to our previous reports [18,19,21]. The as-prepared GO was dispersed in 100 mL of deionized water under ultrasonication for 2 h, and then centrifuged twice to obtain a golden yellow GO solution (1 mg/mL). 2 mg of Cu_2O nanoparticles were added to 5 mL of the above GO solution, and then ultrasonically dispersed for 2 h to obtain a Cu_2O-GO nanocomposite dispersion.

3.4. Preparation of Cu_2O-ErGO/GCE

The glassy carbon electrode (GCE, φ = 3 mm) was polished to a form mirror-like surface with 0.05 μm alumina slurry. Then, the electrode was rinsed by deionized water and ethanol alternately (each for 1 min), and then dried by ultrapure N_2 gas. Firstly, 5 μL of Cu_2O-GO dispersion were transferred and coated on the surface of the GCE, and then dried under an infrared lamp to obtain Cu_2O-graphene oxide nanocomposite modified glass carbon electrode (Cu_2O-GO/GCE). Then, the GO component was electrochemically reduced by potentiostatic method. Specially, the Cu_2O-GO/GCE was immersed into 0.1 mol/L PBS solution, and electrochemically reduced at −1.2 V for 120 s. For comparison, GO modified glass carbon electrode (GO/GCE), electrochemically reduced graphene oxide modified glass carbon electrode (ErGO/GCE) and Cu_2O-GO nanocomposite modified glass carbon electrode (Cu_2O-GO/GCE) were also fabricated by a similar method.

3.5. Electrochemical Measurements

All of the electrochemical measurements were carried out using a standard three-electrode assemble, comprising of bare or modified electrodes as working electrode, saturated calomel electrode (SCE) as reference electrode, and platinum wire electrode as counter electrode. 0.1 M PBS solution was used as supporting electrolytes in all electrochemical experiments. The electrochemical performances of 1×10^{-5} mol/L sunset yellow on various modified electrodes were investigated by cyclic voltammetry (CV) and second-derivative linear sweep voltammetry (SDLSV), with the potential scanning range of 0.4 ~1.2 V. Prior to all electrochemical measurements, an accumulation was performed under stirring at 500 rpm to improve the sensitivity. After 5 s rest, the CV or SDLSV were recorded at a scan rate of 100 mV/s, except where stated otherwise. All of the electrochemical tests were carried on the electrochemical workstation (CHI 760E, Shanghai Chenhua Inc., Shanghai, China).

3.6. Analysis of Real Samples

Carbonated drinks, orange juice, and candy samples were purchased from a local supermarket. The same amount of candies was taken out from five packages and carefully grounded into fine powder. Then, the candies powder (about 1.0 g) was accurately weighed and dispersed in 10 mL deionized water under sonication for 1 h. Subsequently, the mixture was centrifuged at 4000 rpm for 10 min to remove insoluble substances. The 0.5 mL as-obtained supernatant was diluted to 10 mL with 0.1 mol/L PBS solution. The liquid samples (carbonated drinks and orange juice) were added into beaker and ultrasonicated for degasification. Then, 1.0 mL of the liquid samples was diluted to 10 mL with 0.1 mol/L PBS solution. Prior to electrochemical detection, accumulation step was performed in the sample solution to enhance the electrochemical response. Then, these sample solutions were detected by second-derivative linear sweep voltammetry while using our proposed Cu_2O-ErGO/GCE. After each test, Cu_2O-ErGO/GCE was scanned by cyclic voltammetry a 0.1 mol/L blank PBS solution for serval cycles to remove any adsorbents. The modified electrode can reused only when the response peak disappeared in the blank PBS solution.

4. Conclusions

In this study, a promising electrochemical sensor based on the Cu_2O-ErGO/GCE was developed toward sensing sunset yellow. The Cu_2O-ErGO nanocomposites not only possess the advantages from individual component materials, but also exhibit obvious synergistic enhancement effects toward sunset yellow. The anodic peak current of sunset yellow on the Cu_2O-ErGO/GCE increases by about 25 times as compared to that on the bare GCE. The proposed Cu_2O-ErGO/GCE exhibits two linear regions, namely 2.0×10^{-8} mol/L–2.0×10^{-5} mol/L and 2.0×10^{-5} mol/L–1.0×10^{-4} mol/L, and the limit of detection is 6.0×10^{-9} mol/L (S/N = 3). The sensing performances in terms of linear response ranges and detection limit are comparable to, and even exceed the most reported modified electrodes, such as precious metal-based modified electrodes. Obviously, the Cu_2O-ErGO/GCE have outstanding advantages over precious metal-based modified electrodes in term of the cost. Moreover, the response current is basically not affected by potential interfering compounds, suggesting the Cu_2O-ErGO shows good selectivity. Besides, the good reproducibility was also obtained on the Cu_2O-ErGO/GCE. Finally, Cu_2O-ErGO/GCE have been successfully used for the quantitative detection of sunset yellow in real samples (i.e., carbonated beverage, orange juice, and candies) while using second-derivative linear sweep voltammetry. The satisfactory results are obtained with recovery rate is 98.75 ~102.5% and RSD is less than 2.85%. When compared with conventional analytical techniques (Table 4), the proposed method does not require expensive equipment, and time-consuming and complicated pretreatment procedures. Considering the considerable merits including low cost, rapid response, high sensitivity, as well as good selectivity and good reproducibility, the Cu_2O-ErGO/GCE will have broad application prospects in the detection of sunset yellow in diverse beverages, foods, and nutrients.

Table 4. Advantages and drawbacks of previous reported protocols and our proposed protocol.

Protocols	Advantages	Drawbacks
Liquid chromatography	Reliable; good repeatability; high sensitivity; low LOD	Limited separation ability; Time-consuming; expensive equipment
Thin layer chromatography	Low cost apparatus	Organic solvents are often used; they have intensive disagreeable smell and cancerogenic activity
Spectrophotometry	Simultaneous identification and quantification; simple technique	Low sensitivity; extraction separation is needed for detection of dyes in complex product composition
Capillary electrophoresis	High column efficiency, short analysis time and minimal amounts of samples	Limited sensitivity & selectivity; severe matrix interferences
Electrochemical analysis (This work)	Low cost, rapid response, facile operate, high sensitivity and good selectivity	Portability needs to be improved; not disposable

Author Contributions: Q.H., G.L. and D.C. conceived and designed the experiments; J.L., X.L., and Y.X. performed the experiments; G.L., P.D., and X.L. analyzed the data; Q.H., J.L. and G.L. wrote the manuscript; Q.H. and D.C. contributed reagents and materials; All authors read and approved the final manuscript.

References

1. Ni, Y.; Wang, Y.; Kokot, S. Simultaneous kinetic spectrophotometric analysis of five synthetic food colorants with the aid of chemometrics. *Talanta* **2009**, *78*, 432–441. [CrossRef] [PubMed]
2. Al-Degs, Y.S. Determination of three dyes in commercial soft drinks using HLA/GO and liquid chromatography. *Food Chem.* **2009**, *117*, 485–490. [CrossRef]
3. Mao, Y.; Fan, Q.; Li, J.; Yu, L.; Qu, L. A novel and green ctab-functionalized graphene nanosheets electrochemical sensor for Sudan I determination. *Sens. Actuators B Chem.* **2014**, *203*, 759–765. [CrossRef]
4. Chung, K.-T.; Cerniglia, C.E. Mutagenicity of azo dyes: Structure-activity relationships. *Mutat. Res.* **1992**, *277*, 201–220. [CrossRef]
5. Aguilar, F.; Charrondiere, U.R.; Dusemund, B.; Galtier, P.; Gilbert, J.; Gott, D.M.; Grilli, S.; Guertler, R.; Koenig, J.; Lambre, C.; et al. Scientific Opinion on the re-evaluation of Sunset Yellow FCF (E 110) as a food additive on request from the European Commission. *EFSA J.* **2009**, *7*, 1–44.
6. Ye, X.; Du, Y.; Lu, D.; Wang, C. Fabrication of β-cyclodextrin-coated poly (diallyldimethylammonium chloride)-functionalized graphene composite film modified glassy carbon-rotating disk electrode and its application for simultaneous electrochemical determination colorants of sunset yellow and tartrazine. *Anal. Chim. Acta* **2013**, *779*, 22–34. [PubMed]
7. Ghoreishi, S.M.; Behpour, M.; Golestaneh, M. Simultaneous determination of Sunset yellow and Tartrazine in soft drinks using gold nanoparticles carbon paste electrode. *Food Chem.* **2012**, *132*, 637–641. [CrossRef] [PubMed]
8. Minioti, K.S.; Sakellariou, C.F.; Thomaidis, N.S. Determination of 13 synthetic food colorants in water-soluble foods by reversed-phase high-performance liquid chromatography coupled with diode-array detector. *Anal. Chim. Acta* **2007**, *583*, 103–110. [CrossRef] [PubMed]
9. Soponar, F.; Cătălin Moţ, A.; Sârbu, C. Quantitative determination of some food dyes using digital processing of images obtained by thin-layer chromatography. *J. Chromatogr.* **2008**, *1188*, 295–300. [CrossRef] [PubMed]
10. De León-Rodríguez, L.M.; Basuil-Tobias, D.A. Testing the possibility of using UV–vis spectrophotometric techniques to determine non-absorbing analytes by inclusion complex competition in cyclodextrins. *Anal. Chim. Acta* **2005**, *543*, 282–290. [CrossRef]
11. Yuan, Y.; Zhao, X.; Qiao, M.; Zhu, J.; Liu, S.; Yang, J.; Hu, X. Determination of sunset yellow in soft drinks based on fluorescence quenching of carbon dots. *Spectrochim. Acta A* **2016**, *167*, 106–110. [CrossRef] [PubMed]
12. Ryvolová, M.; Táborský, P.; Vrábel, P.; Krásenský, P.; Preisler, J. Sensitive determination of erythrosine and other red food colorants using capillary electrophoresis with laser-induced fluorescence detection. *J. Chromatogr.* **2007**, *1141*, 206–211. [CrossRef] [PubMed]
13. Wang, M.; Zhao, J. Facile synthesis of Au supported on ionic liquid functionalized reduced graphene oxide for simultaneous determination of Sunset yellow and Tartrazine in drinks. *Sens. Actuators B Chem.* **2015**, *216*, 578–585. [CrossRef]
14. Rovina, K.; Siddiquee, S.; Shaarani, S.M. Highly sensitive electrochemical determination of sunset yellow in commercial food products based on CHIT/GO/MWCNTs/AuNPs/GCE. *Food Control* **2017**, *82*, 66–73. [CrossRef]
15. Wang, J.; Yang, B.; Wang, H.; Yang, P.; Du, Y. Highly sensitive electrochemical determination of Sunset Yellow based on gold nanoparticles/graphene electrode. *Anal. Chim. Acta* **2015**, *893*, 41–48. [CrossRef] [PubMed]
16. Thirumalraj, B.; Rajkumar, C.; Chen, S.M.; Veerakumar, P.; Perumal, P.; Liu, S.B. Carbon aerogel supported palladium-ruthenium nanoparticles for electrochemical sensing and catalytic reduction of food dye. *Sens. Actuators B Chem.* **2018**, *257*, 48–59. [CrossRef]
17. Wang, J.; Yang, B.; Zhang, K.; Bin, D.; Shiraishi, Y.; Yang, P.; Du, Y. Highly sensitive electrochemical determination of Sunset Yellow based on the ultrafine Au-Pd and reduced graphene oxide nanocomposites. *J. Colloid Interface Sci.* **2016**, *481*, 229–235. [CrossRef] [PubMed]
18. He, Q.; Liu, J.; Liu, X.; Li, G.; Deng, P.; Liang, J. Preparation of Cu_2O-reduced graphene nanocomposite modified electrodes towards ultrasensitive dopamine detection. *Sensors* **2018**, *18*, 199. [CrossRef] [PubMed]

19. He, Q.; Liu, J.; Liu, X.; Li, G.; Chen, D.; Deng, P.; Liang, J. Fabrication of amine-modified magnetite-electrochemically reduced graphene oxide nanocomposite modified glassy carbon electrode for sensitive dopamine determination. *Nanomaterials* **2018**, *8*, 194. [CrossRef] [PubMed]
20. He, Q.; Li, G.; Liu, X.; Liu, J.; Deng, P.; Chen, D. Morphologically tunable MnO_2 nanoparticles fabrication, modelling and their influences on electrochemical sensing performance toward dopamine. *Catalysts* **2018**, *8*, 323. [CrossRef]
21. He, Q.; Liu, J.; Liu, X.; Li, G.; Deng, P.; Liang, J.; Chen, D. Sensitive and selective detection of Tartrazine based on TiO_2-electrochemically reduced graphene oxide composite-modified electrodes. *Sensors* **2018**, *18*, 1911. [CrossRef] [PubMed]
22. Chen, K.; Wang, M.; Li, G.; He, Q.; Liu, J.; Li, F. Spherical α-MnO_2 supported on N-KB as efficient electrocatalyst for oxygen reduction in Al–Air battery. *Materials* **2018**, *11*, 601. [CrossRef] [PubMed]
23. Wang, M.; Chen, K.; Liu, J.; He, Q.; Li, G.; Li, F. Efficiently enhancing electrocatalytic activity of α-MnO_2 nanorods/N-doped ketjenblack carbon for oxygen reduction reaction and oxygen evolution reaction using facile regulated hydrothermal treatment. *Catalysts* **2018**, *8*, 138. [CrossRef]
24. Dorraji, P.S.; Jalali, F. Electrochemical fabrication of a novel ZnO/cysteic acid nanocomposite modified electrode and its application to simultaneous determination of sunset yellow and tartrazine. *Food Chem.* **2017**, *227*, 73–77. [CrossRef] [PubMed]
25. Gan, T.; Sun, J.; Meng, W.; Song, L.; Zhang, Y. Electrochemical sensor based on graphene and mesoporous TiO_2 for the simultaneous determination of trace colourants in food. *Food Chem.* **2013**, *141*, 3731–3737. [CrossRef] [PubMed]
26. Wu, X.; Zhang, X.; Zhao, C.; Qian, X. One-pot hydrothermal synthesis of ZnO/RGO/ZnO@Zn sensor for sunset yellow in soft drinks. *Talanta* **2018**, *179*, 836–844. [CrossRef] [PubMed]
27. Hara, M.; Kondo, T.; Komoda, M.; Ikeda, S.; Kondo, J.N.; Domen, K.; Hara, M.; Shinohara, K.; Tanaka, A. Cu_2O as a photocatalyst for overall water splitting under visible light irradiation. *Chem. Commun.* **1998**, *3*, 357–358. [CrossRef]
28. Shao, F.; Sun, J.; Gao, L.; Luo, J.; Liu, Y.; Yang, S. High Efficiency Semiconductor-liquid junction solar cells based on Cu/Cu_2O. *Adv. Funct. Mater.* **2012**, *22*, 3907–3913. [CrossRef]
29. Meng, H.; Yang, W.; Ding, K.; Feng, L.; Guan, Y. Cu_2O nanorods modified by reduced graphene oxide for NH3 sensing at room temperature. *J. Mater. Chem. A* **2014**, *3*, 1174–1181. [CrossRef]
30. Wang, Y.; Ji, Z.; Shen, X.; Zhu, G.; Wang, J.; Yue, X. Facile growth of Cu_2O hollow cubes on reduced graphene oxide with remarkable electrocatalytic performance for non-enzymatic glucose detection. *New J. Chem.* **2017**, *41*, 9223–9229. [CrossRef]
31. Wang, A.; Li, X.; Zhao, Y.; Wei, W.; Chen, J.; Hong, M. Preparation and characterizations of Cu_2O/reduced graphene oxide nanocomposites with high photo-catalytic performances. *Powder Technol.* **2014**, *261*, 42–48. [CrossRef]
32. Tran, P.D.; Batabyal, S.K.; Pramana, S.S.; Barber, J.; Wong, L.H.; Loo, S.C. A cuprous oxide-reduced graphene oxide (Cu_2O-rGO) composite photocatalyst for hydrogen generation: employing rGO as an electron acceptor to enhance the photocatalytic activity and stability of Cu_2O. *Nanoscale* **2012**, *4*, 3875–3878. [CrossRef] [PubMed]
33. Xie, H.; Duan, K.; Xue, M.; Du, Y.; Wang, C. Photoelectrocatalytic analysis and electrocatalytic determination of hydroquinone by using a Cu_2O-reduced graphene oxide nanocomposite modified rotating ring-disk electrode. *Analyst* **2016**, *141*, 4772–4781. [CrossRef] [PubMed]
34. Li, J.; Wang, X.; Duan, H.; Wang, Y.; Bu, Y.; Luo, C. Based on magnetic graphene oxide highly sensitive and selective imprinted sensor for determination of sunset yellow. *Talanta* **2016**, *147*, 169–176. [CrossRef] [PubMed]
35. Deng, K.; Li, C.; Li, X.; Huang, H. Simultaneous detection of sunset yellow and tartrazine using the nanohybrid of gold nanorods decorated graphene oxide. *J. Electroanal. Chem.* **2016**, *780*, 296–302. [CrossRef]
36. Jampasa, S.; Siangproh, W.; Duangmal, K.; Chailapakul, O. Electrochemically reduced graphene oxide-modified screen-printed carbon electrodes for a simple and highly sensitive electrochemical detection of synthetic colorants in beverages. *Talanta* **2016**, *160*, 113–124. [CrossRef] [PubMed]
37. Tkalya, E.E.; Ghislandi, M.; With, G.D.; Koning, C.E. The use of surfactants for dispersing carbon nanotubes and graphene to make conductive nanocomposites. *Curr. Opin. Colloid Interface Sci.* **2012**, *17*, 225–232. [CrossRef]

38. Songyang, Y.; Yang, X.; Xie, S.; Hao, H.; Song, J. Highly-sensitive and rapid determination of sunset yellow using functionalized montmorillonite-modified electrode. *Food Chem.* **2015**, *173*, 640–644. [CrossRef] [PubMed]
39. Qiu, X.; Lu, L.; Leng, J.; Yu, Y.; Wang, W.; Jiang, M.; Bai, L. An enhanced electrochemical platform based on graphene oxide and multi-walled carbon nanotubes nanocomposite for sensitive determination of Sunset Yellow and Tartrazine. *Food Chem.* **2016**, *190*, 889–895. [CrossRef] [PubMed]
40. Laviron, E. General expression of the linear potential sweep voltammogram in the case of diffusionless electrochemical systems. *J. Electroanal. Chem.* **1979**, *101*, 19–28. [CrossRef]
41. Vladislavić, N.; Buzuk, M.; Rončević, I.Š.; Brinić, S. Electroanalytical Methods for Sunset Yellow Determination—A Review. *Int. J. Electrochem. Sci.* **2018**, *13*, 7008–7019. [CrossRef]
42. Chao, M.; Ma, X. Convenient electrochemical determination of sunset yellow and tartrazine in food samples using a poly(L-phenylalanine)-modified glassy carbon electrode. *Food Anal. Method* **2015**, *8*, 130–138. [CrossRef]
43. Arvand, M.; Zamani, M.; Sayyar Ardaki, M. Rapid electrochemical synthesis of molecularly imprinted polymers on functionalized multi-walled carbon nanotubes for selective recognition of sunset yellow in food samples. *Sens. Actuators B Chem.* **2017**, *243*, 927–939. [CrossRef]
44. Yu, L.; Shi, M.; Yue, X.; Qu, L. A novel and sensitive hexadecyltrimethyl ammonium bromide functionalized graphene supported platinum nanoparticles composite modified glassy carbon electrode for determination of sunset yellow in soft drinks. *Sens. Actuators B Chem.* **2015**, *209*, 1–8. [CrossRef]
45. Qin, C.; Guo, W.; Liu, Y.; Liu, Z.; Qiu, J.; Peng, J. A novel electrochemical sensor based on graphene oxide decorated with silver nanoparticles–molecular imprinted polymers for determination of sunset yellow in soft drinks. *Food Anal. Method* **2017**, *10*, 1–9. [CrossRef]
46. Sierra-Rosales, P.; Toledo-Neira, C.; Squella, J.A. Electrochemical determination of food colorants in soft drinks using MWCNT-modified GCEs. *Sens. Actuators B Chem.* **2017**, *240*, 1257–1264. [CrossRef]
47. Yu, L.; Zheng, H.; Shi, M.; Jing, S.; Qu, L. A novel electrochemical sensor based on poly (diallyldimethylammonium chloride)-dispersed graphene supported palladium nanoparticles for simultaneous determination of sunset yellow and tartrazine in soft drinks. *Food Anal. Method* **2016**, *10*, 1–10. [CrossRef]
48. Arvand, M.; Erfanifar, Z.; Ardaki, M.S. A new core@shell silica-coated magnetic molecular imprinted nanoparticles for selective detection of sunset yellow in food samples. *Food Anal. Method* **2017**, *10*, 1–14. [CrossRef]

14

Use of ^{1}H NMR to Detect the Percentage of Pure Fruit Juices in Blends

Lucia Marchetti [1,2], Federica Pellati [1,*], Stefania Benvenuti [1] and Davide Bertelli [1]

[1] Department of Life Sciences, University of Modena and Reggio Emilia, Via G. Campi 103, 41125 Modena, Italy

[2] Doctorate School in Clinical and Experimental Medicine (CEM), University of Modena and Reggio Emilia, 41125 Modena, Italy

* Correspondence: federica.pellati@unimore.it

Abstract: The consumption of high-nutritional-value juice blends is increasing worldwide and, considering the large market volume, fraud and adulteration represent an ongoing problem. Therefore, advanced anti-fraud tools are needed. This study aims to verify the potential of ^{1}H NMR combined with partial least squares regression (PLS) to determine the relative percentage of pure fruit juices in commercial blends. Apple, orange, pineapple, and pomegranate juices were selected to set up an experimental plan and then mixed in different proportions according to a central composite design (CCD). NOESY (nuclear Overhauser enhancement spectroscopy) experiments that suppress the water signal were used. Considering the high complexity of the spectra, it was necessary to pretreat and then analyze by chemometric tools the large amount of information contained in the raw data. PLS analysis was performed using venetian-blind internal cross-validation, and the model was established using different chemometric indicators (RMSEC, RMSECV, RMSEP, R^2_{CAL}, R^2_{CV}, R^2_{PRED}). PLS produced the best model, using five factors explaining 94.51 and 88.62% of the total variance in X and Y, respectively. The present work shows the feasibility and advantages of using ^{1}H NMR spectral data in combination with multivariate analysis to develop and optimize calibration models potentially useful for detecting fruit juice adulteration.

Keywords: fruit juice; blends; adulteration; ^{1}H NMR; PLS; chemometrics

---

1. Introduction

Fruit juice consumption is increasing worldwide since fruit-based products are promoted as healthy foods and, in addition, producers try to be more and more innovative by developing the market segment of blends and mixed juices of high nutritional value. According to the European Fruit Juice Association (AIJN), consumption of 38.5 million L of fruit juice was registered worldwide in 2015 and, in the same year, the consumption estimated per head of population was about 19 L in the European Union, 26 L in North America, 6 L in South America and 2 L in Pacific Asia [1]. Considering the large market volume, economic frauds and adulteration in this sector represent an ongoing problem and have often been reported; hence, there is a need for advanced and suitable anti-fraud tools [2,3]. Tackling the total cost of analysis to assess fruit juice authenticity can be very expensive; mostly the aim is to detect water dilution, the addition of inexpensive juice blends to higher-value fruit juice, or the addition of pure beet sugar. Currently, the most common techniques applied to reveal fraud in this field involve whole-food profiling or the search for a number of compounds (targeted analysis). The main disadvantages of these analyses are the high cost and time-consuming nature, the fact that adulterants at concentrations lower than 10% are difficult to detect, and the fact that, in most cases, only one type of adulteration can be unmasked at a time [4].

For this reason, advanced analytical methods are highly recommended to prevent fraudulent practices and to protect the rights of producers as well as those of consumers, with particular attention to safety issues. Since many factors contribute to the variation in juice composition, e.g., fruit geographical origin and climate, maturity degree, technological processes, and storage conditions, a comprehensive method based on the fingerprinting approach is recommended, in which a large number of unknown metabolites is included in the analysis and contributes to the results. For this purpose, the possibility of using nuclear magnetic resonance (NMR) spectroscopy for monitoring fruit juice and other fruit-derived products is well known and has been demonstrated to be an efficient tool in beverage authentication, since the spectral data cover a wide range of compounds [5]. NMR has many advantages over the most common separation methods based on GC and HPLC, which may be cheaper but they seem to be more appropriate for target analysis [6]. Indeed, NMR can detect many different compounds in one sample run, it is non-destructive, stable over time, and it requires only a limited sample preparation [6]. Various applications of NMR spectroscopy are now available for beverage quality control, as reported for wines, spirits, and juices [7,8]. ^{1}H NMR spectroscopy has shown great potential for determining the country of origin of green tea samples [9]. Moreover, the same technique has been demonstrated to be very accurate in the determination of the origin of fruit, and it can be used to examine the source of the raw material used in the preparation of juices [10]. NMR spectroscopy has also been recently applied to alcoholic beverages for authentication purposes, taking into account their high prices and the high risk of fraud by adulteration or deliberate mislabeling. The potential of NMR spectroscopy has also been evaluated to verify the composition of beer and to correlate it to the brewing site and to the date of production, as well as for quality control [11]. In particular, the use of NMR spectroscopy has become an indispensable tool for authenticity studies on fruit juice and has now reached the commercial level. The NMR technique is able to evaluate simultaneously, from a single dataset, a multitude of parameters related to the quality and authenticity of juice, providing targeted and non-targeted multi-marker analysis. In addition, the spectral results can be compared with databases of reference juice [10,12,13]. However, to the best of our knowledge, NMR application for the qualitative and quantitative determination of the composition in juice blends has not been previously described in the literature.

The application of ^{1}H NMR to complex matrices, as in the case of fruit juice, which normally contains a large amount of natural or added sugars and organic acids, and lower but significant amounts of other organic substances (phenolics, terpenes, amino acids, etc.), may result in complex spectra, with a lot of crowded signals. This condition often makes it difficult to define the assignments of substances and to correctly proceed with the integration without the use of deconvolution. Deconvolution is an algorithm-based process that allows us to resolve or decompose a set of overlapping peaks into their separate additive components. In these cases, a non-targeted approach based on a sample fingerprint is advantageous. In particular, the use of the entire ^{1}H NMR spectrum profile without any kind of targeted measurement was evaluated in this work. Fingerprinting techniques require the use of adequate statistical methods to extract information from raw data. Chemometrics can reveal latent correlations in the data and can be useful for both qualitative and quantitative purposes.

Partial least squares (PLS) modeling is a powerful multivariate statistical tool that has often been applied to spectral analysis. PLS is related to other multivariate calibration methods, such as classical least squares (CLS), inverse least squares (ILS), and principal component regression (PCR) methods [14]. The main scope of PLS is to eliminate multicollinearity in the set of explanatory variables X of a regression model, reducing the dimension of the set in such a way that the resulting subset of descriptive variables is optimal for predicting the dependent variable Y. The values of Y typically represent the analyte concentrations (or any sample properties) [14]. For the quantitative approach, PLS is the most frequently used multivariate statistical method. Hence, this study aims to verify the potential of proton NMR (^{1}H NMR) combined with PLS to construct models for the determination of the relative percentage of pure fruit juices in a blend.

2. Results and Discussion

The use of ^{1}H NMR coupled with PLS was proposed here to develop and optimize multivariate calibration models to determine the relative concentration of four pure juices in blends. The PLS model was established by using different chemometric indicators (root mean square error of calibration (RMSEC); root mean square error of cross-validation (RMSECV); root mean square error of prediction (RMSEP); coefficient of determination for calibration (R^2_{CAL}), cross-validation (R^2_{CV}) and prediction (R^2_{PRED}).

Figure 1 shows the NMR spectra of the four pure juices considered, and Figure 2 shows the typical spectrum of a mixture containing equal percentages of apple, orange, pineapple, and pomegranate juice. As expected, the spectra present wide regions with overlapping phenomena, which makes it difficult to proceed with the peak integration. The typical ^{1}H NMR spectrum of a juice shows three defined regions. In the first, ranging from 0.5 to 3.0 ppm, protons of organic acids (citric and malic) and amino acids (alanine, valine, and proline) are present. The second region (3.0–6.0 ppm) is typical for carbohydrates, with sucrose, α-glucose, β-glucose, and fructose being the most abundant [7]. The last region, ranging from 6.0 to 8.5 ppm, shows phenolic metabolites and aromatic protons. Moreover, each juice presents the peak assigned to the methyl group of ethanol at 1.17 ppm [6,15]. Spectra of pure juice have similar shapes in the aliphatic region and limited quantitative differences, except for the two peaks at 5.40 and 4.22 ppm related to sucrose, which is absent in pomegranate. The region ranging from 6 to 10 ppm is meanwhile more typical; here, aromatic and phenolic compounds are normally present. In this case, the signals are lower in intensity with respect to the aliphatics. As regards the mixtures, the enlargement of the aromatic region in Figure 2 appears even more complex, due to the fact that the weak aromatic signals of each single juice are more diluted in the blend and the number of signals is increased, owing to the simultaneous presence of many compounds. All these considerations led us to choose an untargeted approach combined with the consolidated PLS chemometric method [16].

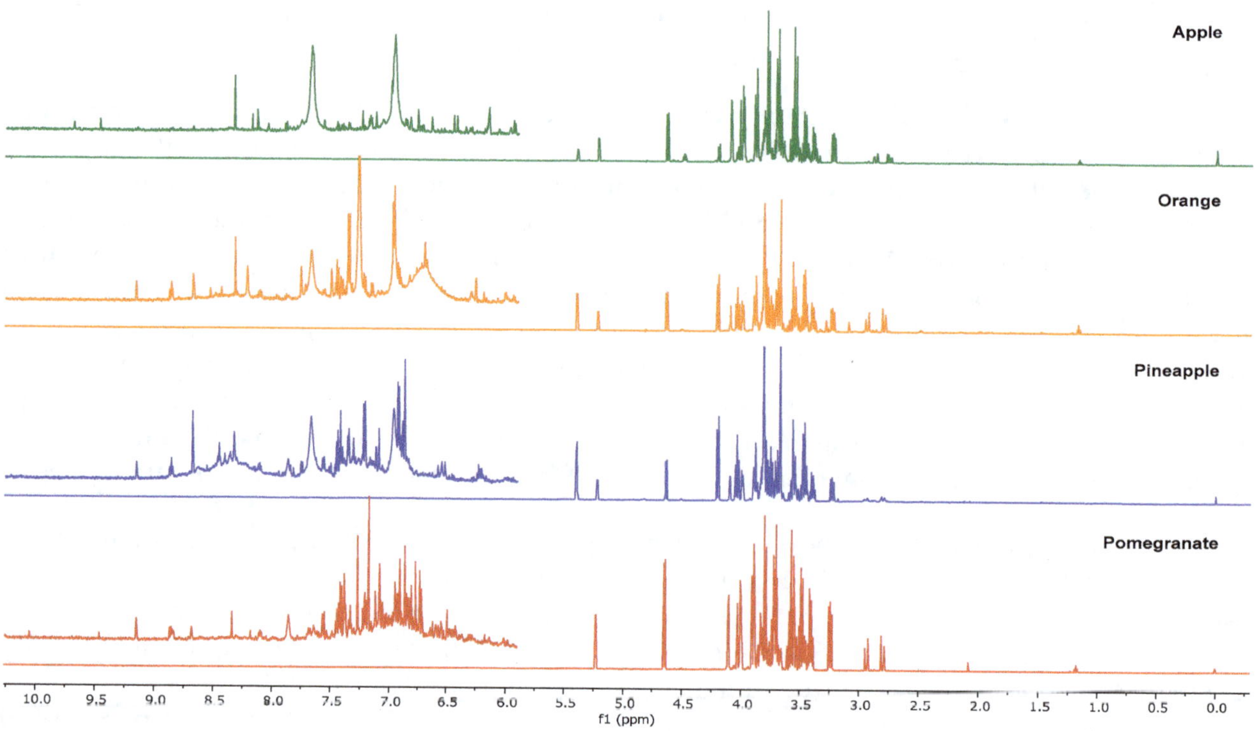

Figure 1. 1D NOESY spectra of four pure juices, with the 6–10 ppm regions enlarged.

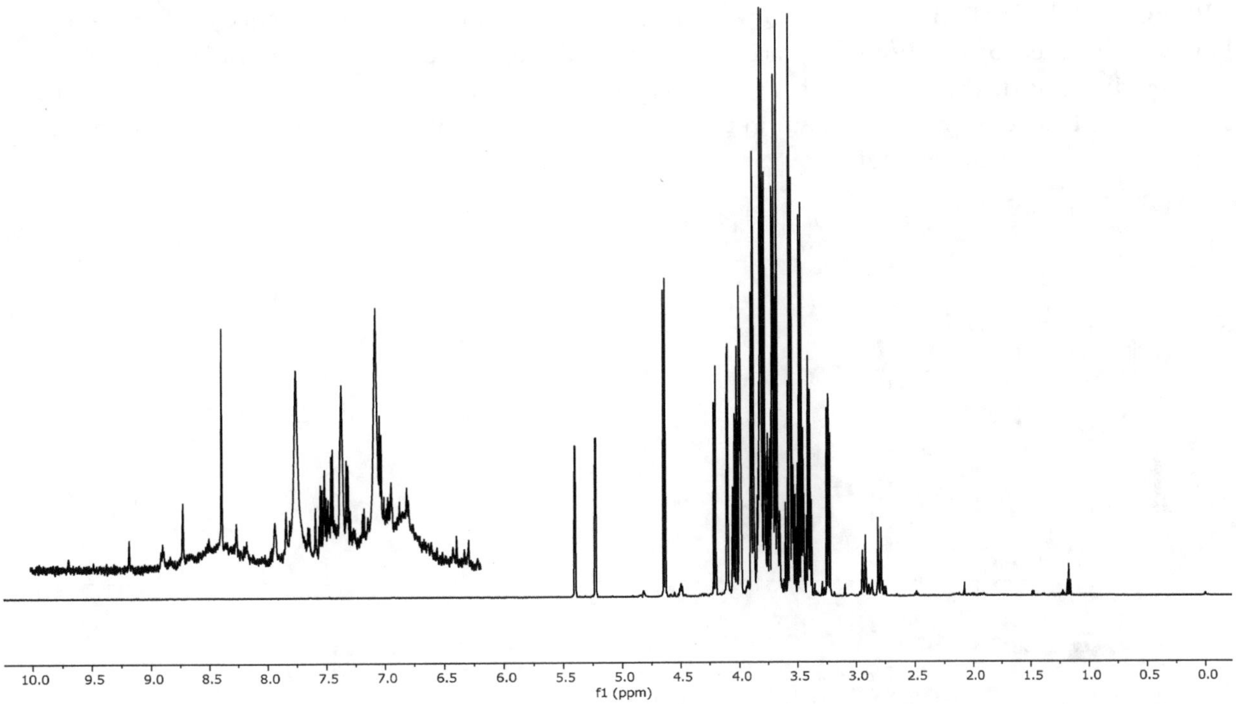

Figure 2. Typical 1D NOESY spectrum of mixture containing equal percentages of pure apple, orange, pineapple, and pomegranate juice.

The number of latent variables to be included in the PLS model was selected in such a way that the RMSEC, RMSECV (obtained from calibration and internal cross-validation, respectively) and RESEP (obtained from external validation test set) were reduced to the lowest values, ensuring at the same time the highest possible predictive capacity. Other statistical parameters to be considered are R^2 and the amount of explained variance. The best model was built by using five factors, explaining 94.51% and 88.62% of total variance in X and Y, respectively. All the calculated RMSEs (except RMSECV for pineapple and pomegranate) are lower than 10 and each R^2 value is acceptable, especially for prediction values. All the chemometric indicators are reported in Table 1.

Table 1. PLS model results of four fruit juice samples.

Model	**Number of Factors: 5 Variance: Y = 88.62%; X = 94.51%**			
	RMSEC [a]	**RMSECV** [b]	**RMSEP** [c]	$\mathbf{R^2}$ [d]
Apple	6.869	8.732	2.324	$R^2_{CAL} = 0.912$ $R^2_{CV} = 0.899$ $R^2_{PRED} = 0.987$
Orange	6.333	9.435	4.435	$R^2_{CAL} = 0.914$ $R^2_{CV} = 0.882$ $R^2_{PRED} = 0.950$
Pineapple	8.634	12.631	5.438	$R^2_{CAL} = 0.885$ $R^2_{CV} = 0.821$ $R^2_{PRED} = 0.946$
Pomegranate	7.182	10.511	7.092	$R^2_{CAL} = 0.950$ $R^2_{CV} = 0.860$ $R^2_{PRED} = 0.929$

[a] RMSEC: root mean square error of calibration; [b] RMSECV: root mean square error of cross-validation; [c] RMSEP: root mean square error of prediction. [d] R^2: coefficient of determination for calibration (CAL), cross-validation (CV) and prediction (P).

In Figure 3 the Hotelling's T^2 vs. Q residuals graphic is shown. The presence of potential spectral outliers was verified by applying a 95% confidence interval, any outlier is present. Four samples with high Q residuals and three with high T^2 scores are present, but none of them was excluded from the model. All the test set samples (shown in red) are in the 95% confidence area, demonstrating the high predictive capacity of the model.

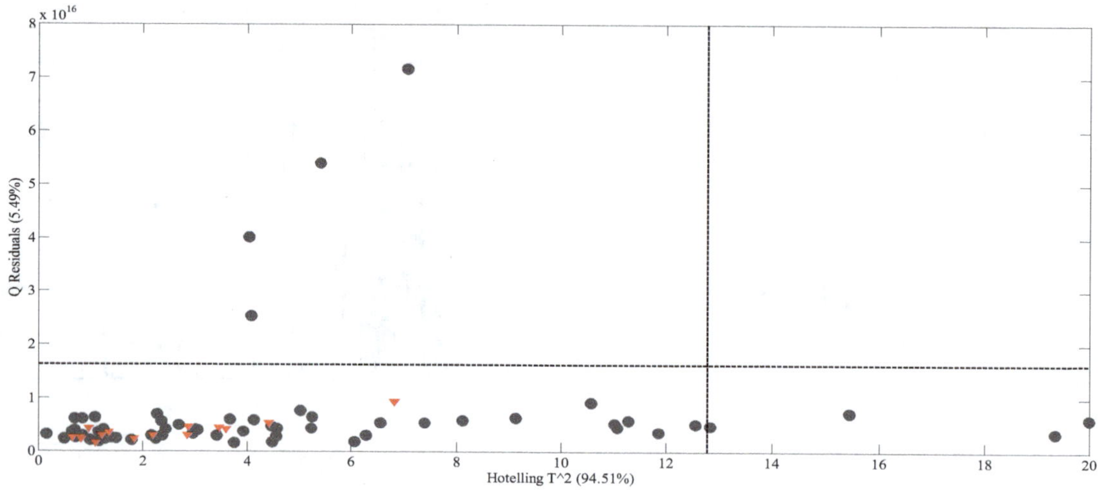

Figure 3. Q residuals versus Hotelling's T^2 plot for the PLS model of 60 fruit juice samples (•) and test set samples (▾).

In Figures 4 and 5 the loadings for the five extracted latent variables and the VIP (variable importance in the projection) scores, respectively, are shown for each juice. A VIP score is a measure of a variable importance in the model. It summarizes the contribution of a variable to the model. The VIP score is calculated as a weighted sum of the squared correlations between the PLS components and the original variables. As regards loadings, highly correlated variables have similar weights in the loading vectors. Thus, a different pattern of variables is significant for each extracted factor. In this case, for all the latent variables any significance arises for the aromatic signals, even though this spectral region (6–8 ppm) seems to be more characteristic for each juice in comparison with the aliphatic areas (3–4 ppm), which appear quite similar (Figure 1). Considering VIP scores, apple, pineapple, and pomegranate share the same pattern of variables, while in the case of orange juice the most important signals are those located at low frequencies, corresponding to acid compounds. In any case, and also for orange juice, the significance of the aromatic and phenolic region is limited. Some different pre-treatment procedures were attempted to increase the significance of the aromatic spectral region, but no substantial improvement was achieved.

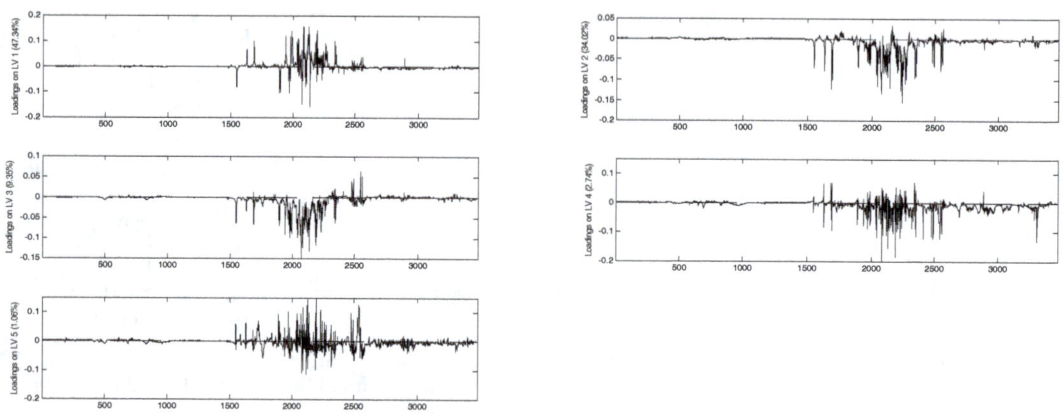

Figure 4. Loadings for the five extracted latent variables.

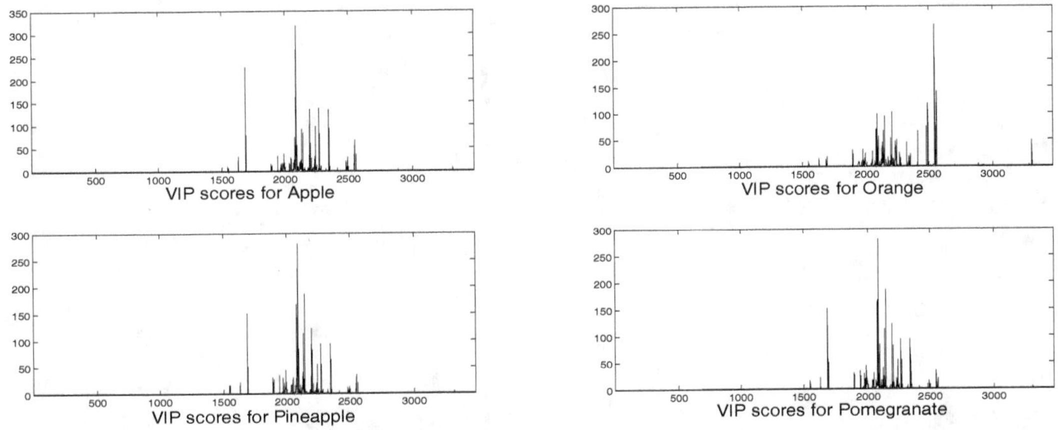

Figure 5. VIP scores obtained from the selected model for the four juices.

The regression vectors for each juice are shown in Figure 6. The four plots are different from each other, confirming the same tendency of VIP scores, thus in the definition of regression models different variables play a significant role for each juice.

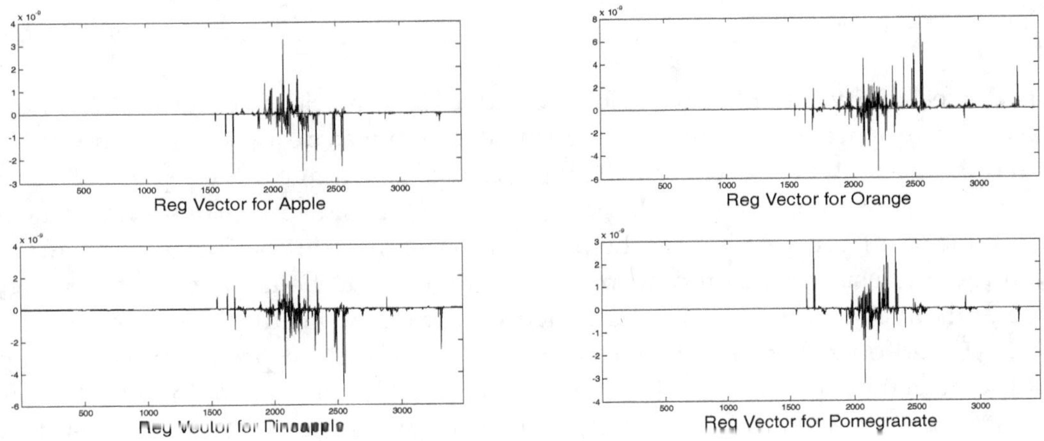

Figure 6. Regression vectors for the four juices considered in the model.

Figure 7 shows the scatter plot of measured values vs. predicted values in mixture samples for calibration and test sets (regression data are reported in Table 1). The regression models are similar in quality and predictive capacity. The best-performing model is apple, while the worst is pomegranate (Figure 7: top left and bottom right, respectively).

The method discussed here is suitable and sufficiently effective for the quantification of the considered mixtures. To validate this approach, the next step will be the evaluation of the predictive capacity of mixtures prepared with juices of other brands and also of mixtures in which one or more components are different from those covered by the model. Moreover, this method showed its effectiveness in a wide range of component percentages (6.25–100%), thus proving useful for both high and low concentrations. As a consequence, it should also be able to detect small differences from the composition declared on the labels of the commercial products.

The results presented herein show that the NMR analysis coupled with chemometrics provides adequate results in a comparable time with respect to other analytical approaches. The main advantages of the method proposed include the reduced sample preparation and lack of extraction and purification steps. In addition, this technique is effective at reducing the amounts of reagents and solvents, making it more competitive and environmentally sustainable than the most common separative techniques.

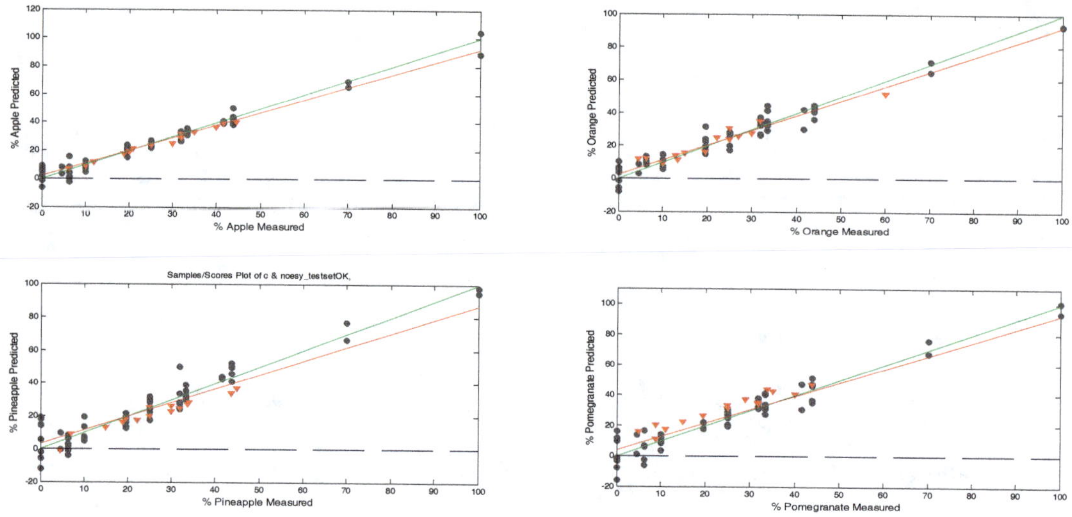

Figure 7. Correlation between the prediction and measured values of juice samples, showing calibration (•) and test set (▼) samples.

3. Materials and Methods

3.1. Samples

To verify the feasibility of ^{1}H NMR combined with chemometrics to determine the percentage of pure fruit juices in a mixture, the following experimental plan was designed. To minimize the system complexity in this step of the research, four fruit juices (apple, orange, pineapple, and pomegranate) from a single brand were purchased from a local marketplace. These samples represent some of the most consumed fruit juices, often combined for the preparation of more attractive mixtures for consumers.

To obtain the highest significant distribution of mixtures to be analyzed, they were prepared in accordance with an experimental plan, obtained by means of a central composite design (CCD), with $\alpha = 1.4826$, based on a 2^4 full factorial design, plus eight axial points, plus two replicates in the center of the domain and four pure juice samples [17]. The ratios provided by the model were transformed into a percentage composition to get the final experimental plan and then were used as dependent variables for the subsequent PLS analysis (Table 2).

Table 2. Percentage composition of juice samples.

Sample	% Apple	% Orange	% Pineapple	% Pomegranate
1	100	0	0	0
2	0	100	0	0
3	0	0	100	0
4	0	0	0	100
5	43.75	6.25	43.75	6.25
6	25	25	25	25
7	6.25	6.25	43.75	43.75
8	43.75	6.25	6.25	43.75
9	25	25	25	25
10	19.5	19.5	19.5	41.5
11	10	10	10	70
12	33.33	33.33	0	33.33
13	31.82	4.54	31.82	31.82
14	33.33	33.33	33.33	0
15	6.25	43.75	43.75	6.25
16	0	33.33	33.33	33.33
17	10	70	10	10
18	19.51	41.46	19.51	19.51

Table 2. *Cont.*

Sample	% Apple	% Orange	% Pineapple	% Pomegranate
19	19.51	19.51	41.46	19.51
20	10	10	70	10
21	25	25	25	25
22	31.82	31.82	4.54	31.82
23	6.25	43.75	6.25	43.75
24	4.54	31.82	31.82	31.82
25	33.33	0	33.33	33.33
26	43.75	43.75	6.25	6.25
27	31.82	31.82	31.82	4.54
28	41.46	19.51	19.51	19.51
29	25	25	25	25
30	70	10	10	10

Analyses were performed in duplicate, making sure to repeat the analysis on different days. In this way, the 60 experiments were conducted in a randomized order. The intra- and inter-day variability, previously evaluated on one sample, were analyzed 10 times on the same and on different days, giving 1.4 and 4.0 CV%, respectively. In addition, an external validation test set was constructed including five randomly selected CCD replicates (newly prepared and analyzed) and 10 extra mixtures outside the CCD, with random percentages of the four juices (Table 3).

Table 3. Percentage composition of test set samples.

Sample	% Apple	% Orange	% Pineapple	% Pomegranate
TS1	20	15	30	35
TS2	44.44	22.22	22.22	11.11
TS3	40	13	7	40
TS4	18.92	13.51	33.78	33.78
TS5	20	60	15	5
TS6	30	30	25	15
TS7	12	60	19	9
TS8	35	10	30	25
TS9	21	25	45	9
TS10	10	27	34	29
7 *	6.25	6.25	43.75	43.75
9 *	25	25	25	25
13 *	31.82	4.54	31.82	31.82
22 *	31.82	31.82	4.54	31.82
28 *	41.46	19.51	19.51	19.51

* CCD samples randomly selected.

For the preparation of samples, each juice was first centrifuged at 5000 rpm for 5 min to remove all the material in suspension. Subsequently, for NMR analysis, different juices were mixed together in the ratios indicated by the CCD.

For organic acids, amino acids, and each compound whose ionization changes depending on pH, the chemical shift varies accordingly. These spectral variations may affect or invalidate the results of the multivariate analysis, whose basic principle is that the corresponding signals in different samples must have the same chemical shift. To minimize shifts, the pH of the samples with a buffered solution was standardized. To 9 mL of each mixture, 1 mL of 1 M KH_2PO_4 was added and the pH was measured; eventually, it was adjusted to 3.1–3.2 by using small amounts of 1 M HCl. An aliquot of 630 μL of this buffered mixture were then added to 70 μL of 13 mM 3-(trimethylsilyl)propionic-2,2,3,3-d_4 acid sodium salt (TSP) and 0.1% sodium azide in 99.9% deuterium oxide (D_2O). TSP was employed for internal referencing of ^1H chemical shifts, while sodium azide as the preserving agent. The final solution

(700 μL) was transferred into a WILMAD® NMR tube, 5 mm, Ultra-Imperial grade, L 7 in., 528-PP purchased from Sigma-Aldrich (Milan, Italy). All chemicals and solvents were of analytical grade and they were purchased from Sigma-Aldrich.

3.2. *NMR Spectroscopy*

One-dimensional ^{1}H NMR spectra of juice mixtures were acquired with a Bruker FT-NMR Avance III HD 600 MHz spectrometer (Ettlingen, Germany), and all the NMR experiments were performed at 300 °K. After 10 min of thermal equilibration inside the probe, the solvent (D_2O) was locked to assure the maximum sensitivity, the probe was manually tuned and matched, and the 90° pulse was calibrated; subsequently, the power level for pre-saturation was calculated, and finally the receiver gain was automatically set. 1D NOESY experiments were acquired by using the Bruker sequence "noesygppr1d" in order to suppress the water signal. The pre-saturation frequency was adjusted for each sample before acquisition. The acquisition parameters were as follows: time domain (number of data points), 64 K; dummy scans, 0; number of scans, 32; acquisition time, 3.90 s; delay time, 8 s; spectral width, 14 ppm (8403.4 Hz), fid resolution, 0.2564 Hz; digitization mode, baseopt, total acquisition time, 6 min 22 s.

3.3. *Spectra Pretreatment*

The application of the ^{1}H NMR technique to juice samples generates complex and crowded spectra, which need to be previously pre-treated and then analyzed by chemometric tools to handle the high amount of information contained in the raw data. Assuming that peak intensities are directly proportional to the concentration of compounds, each spectral point was used as an absolute intensity value, without performing any type of signal integration. This approach allows for overcoming issues related to the understanding and integration of overlapped signals. First, ^{1}H NMR spectra were phased and calibrated using the TSP signal for chemical shifts referencing, and the baseline was adjusted. All spectra processing was performed using TopSpin 3.5 software package (Bruker Biospin GmbH Rheinstetten). Each spectrum generates a 64 K data point file. In order to perform the spectra alignment, the files from each sample were exported and collected in a dataset consisting in 64 K spectral variables for 75 samples (60 calibration samples plus 15 test set samples). The alignment was performed by using Icoshift 1.0 toolbox for MATLAB® (Mathworks Inc., Natick, MA, USA) to reduce the lack of homogeneity in chemical shifts that principally occurs for pH-dependent signals [18]. Afterwards, in order to reduce the number of data points and manipulate datasets more easily, non-significant spectral regions were deleted to obtain a new dataset with 34,740 data points and, finally, the resolution was reduced by selecting one out of every 10 points, thus obtaining 3474 data points for each spectrum.

3.4. *Statistical Analysis*

With a view to performing the statistical analysis, the dataset used for the alignment and resolution reduction (75 samples) was again split into two datasets, i.e., one containing the 60 samples for calibration and the second containing the 15 samples for the external validation test set. All multivariate data analyses and calculations were performed by using the software PLS_Toolbox 5.2.2 (Eigenvector Research Inc., Manson, WA, USA) for MATLAB®. The PLS method is based on the SIMPLS algorithm. All data were mean-centered and scaled by applying the Pareto scaling method, which is useful when spectral noise is expected to be proportional to the standard deviation square root of variables. Pareto scaling can reduce the relative importance of large values, keeping the data structure partially intact [19].

Partial least squares regression (PLS) is an extension of the multiple linear regression model that does not impose the restrictions employed by discriminant analysis, principal components regression, or canonical correlation. In PLS, prediction functions are represented by factors extracted from the Y'XX'Y matrix, and it is probably the least restrictive of the multiple linear regression extensions. PLS analysis was performed using a venetian-blind internal cross-validation method with the number of data splitting set to 7. The number of factors (latent variables) in the models and the model performance were assessed by using the root mean square error of calibration (RMSEC) and the root mean square error of cross-validation (REMSCV). The presence of potential spectral outliers was verified by applying a 95% confidence interval to the Q residuals and Hotelling's T^2 scores (Figure 3).

The final regression model was assessed by the coefficients of determination for calibration, cross-validation and test sets (R^2_{CAL}, R^2_{CV}), the root mean square error of calibration and the root mean square error of cross-validation (RMSEC, RMSECV). The prediction ability of the model was evaluated by external validation set (test set), through the coefficient of determination for the test set (R^2_{PRED}) and the root mean square error of prediction (RMSEP).

4. Conclusions

The present work proposes the use of ^{1}H NMR applied to the development and optimization of multivariate calibration models to determine the relative concentration of four juices in blends. The PLS model was established by using different chemometric indicators (RMSEC, RMSECV, RMSEP, R^2_{CAL}, R^2_{CV} and R^2_{PRED}). The results demonstrate the feasibility and several advantages of using ^{1}H NMR spectral data in combination with multivariate data analysis to build a model capable of detecting adulterations in fruit juice.

Author Contributions: Conceptualization, D.B.; methodology, D.B. and L.M.; software, D.B. and L.M.; validation, D.B. and L.M.; formal analysis, L.M.; investigation, L.M.; resources, D.B.; data curation, F.P.; writing—original draft preparation, L.M.; writing—review and editing, D.B. and F.P.; visualization, L.M.; supervision, S.B.; project administration, D.B.

Acknowledgments: The authors want to express their thanks to the C.I.G.S. staff (Centro Interdipartimentale Grandi Strumenti, Modena, Italy) for assistance during the experimental work and to Fondazione Cassa di Risparmio di Modena for the purchase of a Bruker FT-NMR Avance III HD 600 MHz spectrometer.

References

1. Market Report 2016. Available online: https://aijn.eu/en/publications/market-reports-1/market-report-2016 (accessed on 9 May 2019).
2. Bat, K.B.; Eler, K.; Mazej, D.; Vodopivec, B.M.; Mulič, I.; Kump, P.; Ogrinc, N. Isotopic and elemental characterisation of Slovenian apple juice according to geographical origin: Preliminary results. *Food Chem.* **2016**, *203*, 86–94.
3. Kamiloglu, S. Authenticity and traceability in beverages. *Food Chem.* **2019**, *277*, 12–24. [CrossRef] [PubMed]
4. GAO/RCED-96-18-Fruit Juice Adulteration: Detection Is Difficult, and Enhanced Efforts Would Be Costly -Content Details-GAOREPORTS-RCED-96-18. Available online: https://www.govinfo.gov/app/details/GAOREPORTS-RCED-96-18 (accessed on 9 May 2019).
5. Cuny, M.; Vigneau, E.; Le Gall, G.; Colquhoun, I.; Lees, M.; Rutledge, D.N. Fruit juice authentication by 1H NMR spectroscopy in combination with different chemometrics tools. *Anal. Bioanal. Chem.* **2008**, *390*, 419–427. [CrossRef] [PubMed]
6. Salazar, M.O.; Pisano, P.L.; Sierra, M.G.; Furlán, R.L. NMR and multivariate data analysis to assess traceability of argentine citrus. *Microchem. J.* **2018**, *141*, 264–270. [CrossRef]
7. Cusano, E.; Simonato, B.; Consonni, R. Fermentation process of apple juice investigated by NMR spectroscopy. *LWT* **2018**, *96*, 147–151. [CrossRef]
8. Jamin, E.; Thomas, F. SNIF-NMR Applications in an Economic Context: Fraud Detection in Food Products. In *Modern Magnetic Resonance*; Springer Science and Business Media LLC: Berlin/Heidelberg, Germany, 2018; pp. 1405–1416.
9. Le Gall, G.; Colquhoun, I.J.; Defernez, M. Metabolite Profiling Using 1H NMR Spectroscopy for Quality Assessment of Green Tea, Camellia sinensis (L.). *J. Agric. Food Chem.* **2004**, *52*, 692–700. [CrossRef] [PubMed]

10. Spraul, M.; Schütz, B.; Rinke, P.; Koswig, S.; Humpfer, E.; Schäfer, H.; Mörtter, M.; Fang, F.; Marx, U.C.; Minoja, A. NMR-Based Multi Parametric Quality Control of Fruit Juices: SGF Profiling. *Nutrients* **2009**, *1*, 148–155. [CrossRef] [PubMed]
11. Almeida, C.; Duarte, I.F.; Barros, A.; Rodrigues, J.E.; Spraul, M.; Gil, A.M. Composition of Beer by1H NMR Spectroscopy: Effects of Brewing Site and Date of Production. *J. Agric. Food Chem.* **2006**, *54*, 700–706. [CrossRef] [PubMed]
12. Minoja, A.P.; Napoli, C. NMR screening in the quality control of food and nutraceuticals. *Food Res. Int.* **2014**, *63*, 126–131. [CrossRef]
13. Mixture Analysis by NMR as Applied to Fruit Juice Quality Control-Spraul-2009-Magnetic Resonance in Chemistry -Wiley Online Library. Available online: https://onlinelibrary.wiley.com/doi/abs/10.1002/mrc.2528 (accessed on 9 May 2019).
14. Haaland, D.M.; Thomas, E.V. Partial least-squares methods for spectral analyses. 1. Relation to other quantitative calibration methods and the extraction of qualitative information. *Anal. Chem.* **1988**, *60*, 1193–1202. [CrossRef]
15. Villa-Ruano, N.; Pérez-Hernández, N.; Zepeda-Vallejo, L.G.; Quiroz-Acosta, T.; Mendieta-Moctezuma, A.; Montoya-García, C.O.; García-Nava, M.L.; Martínez, E.B. 1 H-NMR Based Metabolomics Profiling of Citrus Juices Produced in Veracruz, México. *Chem. Biodivers.* **2019**, *16*, e1800479. [CrossRef] [PubMed]
16. Ebrahimi, P.; Viereck, N.; Bro, R.; Engelsen, S.B. Chemometric Analysis of NMR Spectra. In *Modern Magnetic Resonance;* Springer Science and Business Media LLC: Berlin/Heidelberg, Germany, 2018; pp. 1649–1668.
17. Lundstedt, T.; Seifert, E.; Abramo, L.; Thelin, B.; Nyström, Å.; Pettersen, J.; Bergman, R. Experimental design and optimization. *Chemom. Intell. Lab. Syst.* **1998**, *42*, 3–40. [CrossRef]
18. Savorani, F.; Tomasi, G.; Engelsen, S.B. icoshift: A versatile tool for the rapid alignment of 1D NMR spectra. *J. Magn. Reson.* **2010**, *202*, 190–202. [CrossRef] [PubMed]
19. Winning, H.; Roldán-Marín, E.; Dragsted, L.O.; Viereck, N.; Poulsen, M.; Sanchez-Moreno, C.; Cano, M.P.; Engelsen, S.B. An exploratory NMR nutri-metabonomic investigation reveals dimethyl sulfone as a dietary biomarker for onion intake. *Analyst* **2009**, *134*, 2344–2351. [CrossRef] [PubMed]

15

Essential Oils from *Humulus Lupulus* $scCO_2$ Extract by Hydrodistillation and Microwave-Assisted Hydrodistillation

Katarzyna Tyśkiewicz [1,*], Roman Gieysztor [1], Marcin Konkol [1], Jan Szałas [2] and Edward Rój [1]

1 Supercritical Extraction Department, New Chemical Syntheses Institute, Al. Tysiąclecia Państwa Polskiego 13A, 24-110 Puławy, Poland; roman.gieysztor@gmail.com (R.G.); marcin.konkol@ins.pulawy.pl (M.K.); edward.roj@ins.pulawy.pl (E.R.)

2 Import Export J.A. Szałas Company, Garbarska 125A, 26-600 Radom, Poland; jan.szalas@chmiel.com.pl

* Correspondence: katarzyna.tyskiewicz@ins.pulawy.pl

Abstract: Two different extraction methods were used for a comparative study of essential oils obtained from the *Humulus lupulus* $scCO_2$ (sc-supercritical) extract: microwave-assisted hydrodistillation (MAHD) and conventional hydrodistillation (HD). As a result, the best conditions for the maximum essential oil production were determined for the MAHD method at 335 W microwave power for 30 min at water to raw material ratio of 8:3. The obtained essential oils were enriched in β-myrcene in the amount of 74.13%–89.32% (wt%). Moreover, the first application for determination of the above-mentioned volatile compounds by supercritical fluid chromatography (SFC) with photo-diode array detection (PDA) is presented, which in comparison with gas chromatography coupled with mass spectrometry (GC-MS/MS) resulted in similar values for β-myrcene and α-humulene in obtained samples within less than 1 min.

Keywords: essential oil; extraction techniques; hops extracts; hydrodistillation; Marynka strain; microwave-assisted hydrodistillation

---

1. Introduction

Hops (*Humulus lupulus*) belong to a perennial plant, which is mainly used in the brewing industry due to its valuable properties [1,2]. These properties are attributed to alpha- and beta-acids (responsible for bitterness), essential oils, and polyphenols, which constitute 24, 1, and 5 wt%, respectively [3,4]. The above-mentioned groups of compounds are considered as secondary metabolites. The phenolic compounds present in hops have been investigated in terms of their anticarcinogenic, antioxidant, and anti-inflammatory properties [5,6]. Secondary metabolites' composition affects the assessment of the diversity of a given species. However, according to Biendl et al. [3] the exact functions of these metabolites are yet unknown. It is believed that they protect the plant against pests and pathogens. Due to their favorable physiological and pharmacological properties, secondary metabolites are used in the pharmaceutical industry as medicines and dietary supplements. The antibacterial properties were also evaluated for hops extract obtained by the supercritical fluid extraction [7].

The main methods used to obtain essential oils are, among other, hydrodistillation, steam distillation, maceration, absorption, and supercritical extraction. These methods are referred to as conventional methods [8]. In order to shorten the extraction time and improve the extraction efficiency, another form of hydrodistillation has gained a lot of interest. Microwave-assisted hydrodistillation (MAHD), which is an advanced distillation method using microwave heating, has an advantage over conventional hydrodistillation as was indicated by a number of researchers [8–11]. It is a technique widely used to extract essential oils from medicinal plants and herbs due to its economic and environmental advantages ('green' technology) [12]. It is used in the extraction of various chemical compounds from the group of aromatic compounds, phenolic compounds, pesticides and, especially, essential oils. It is proven that the 30-minute duration of the microwave hydrodistillation process corresponds to a duration of conventional hydrodistillation that is eight times longer [9]. Hydrodistillation and microwave-assisted hydrodistillation have been compared in terms of extraction time, extraction efficiency, chemical composition, and quality of the essential oil as well as energy consumption. Microwave action is based mainly on the weakening of cell membranes and facilitating the extraction of biologically active compounds [13].

Literature data confirm the superiority of the microwave assisted hydrodistillation over conventional hydrodistillation. In previous studies, microwave-assisted hydrodistillation was carried out on such materials as ginger [12], lemongrass [12], rosemary [14,15], cinnamon [8,16], mint [10], thyme [9], sandalwood [11], and orange [17]. However, there is no information available on the use of microwave hydrodistillation in order to obtain essential oils from hop extracts, and also to compare this method with hydrodistillation with the example of the mentioned materials. The mixture of hop extract and water required the addition of residues as a scattering element obtained after the $scCO_2$ extraction of *Humulus lupulus*.

β-myrcene and α-humulene, which belong to a group of terpenes, are characterized by their high oxidative sensitivity. During the brewing processes, thermal oxidation products of β-myrcene and α-humulene may affect both the quality of beer and its organoleptic properties [18]. Therefore, their use as beer additives is limited. However, if not in brewing, the oil rich in β-myrcene and α-humulene might be used in medicine [19]. For instance, β-myrcene is responsible for the anti-inflammatory effects in human osteoarthritis treatment [20]. The activity of α-humulene is described as similar to that of dexamethasone (anti-inflammatory, antiallergic, immunosuppressive) [21]. Figure 1 presents the chemical structures of β-myrcene and α-humulene.

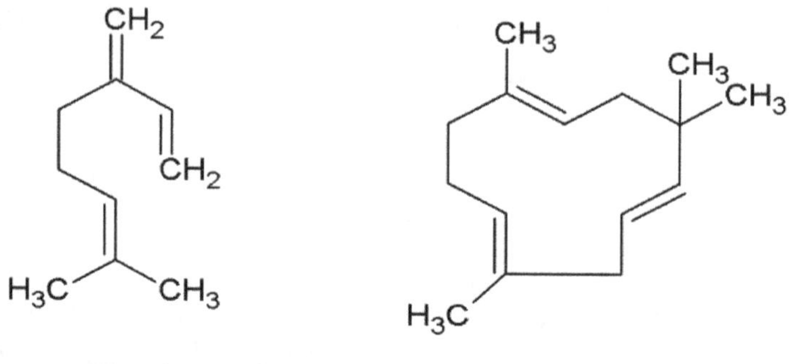

Figure 1. The chemical structures of β-myrcene and α-humulene.

2. Materials and Methods

2.1. Plant Material

Humulus lupulus $scCO_2$ extracts were used as experimental raw materials in this study. The extracts were obtained with the use of supercritical fluid extraction with carbon dioxide in a supercritical state from the Polish hops variety, Marynka. Hop cones from the 2017 harvest were used to produce the hop extract, "Marynka". The geographic origin of crops was the Lublin province, Poland. Cones with a moisture content of 10.6% were milled on a Retsch SM100 mill using a sieve with a mesh size of 1.5 mm. The average particle size was 0.8 mm. The produced material was subjected to the extraction with supercritical carbon dioxide using a pilot plant. A total of 4 kg of ground hop cones were loaded into a 20 dm^3 extractor. The extraction was carried out with the following parameters: Pressure −250 bar, extraction temperature −50 °C, and CO_2 consumption −50 kg/kg dm. The extraction yield was 11.4% by weight.

2.2. *Sample Preparation*

The samples for microwave-assisted hydrodistillation (MAHD) and hydrodistillation (HD) were prepared in the following way. The extract (ca. 150 g) was weighed separately in a 1 L flask and approx. 400 g of distilled water was added and the resulting emulsion was stirred for half an hour. Water to extract ratio was 8:3. For the experiments, 50 g of residues (residues obtained after the extraction of hops with $scCO_2$) were also added to the mixture. Such preparation of the material assures an optimal increase of the material contact with the solvent and enables satisfactory efficiency. Table 1 summarizes the exact compositions of the mixtures for microwave-assisted extraction and hydrodistillation for Marynka extract. For the sake of clarity and discussion, the oils were named according to the scheme: the name of the extract + power at which the extraction was carried out or the name of hydrodistillation apparatus (e.g., Marynka_275 W).

Table 1. Preparation of samples for Marynka extract.

Sample Name	Charge Mass (g)	Extract Mass (g)	Water Mass (g)	Residue Mass (g) *
Marynka_275W	603.34	150.25	402.51	50.58
Marynka _295W	605.99	150.99	404.99	50.01
Marynka_335W	602.71	150.47	402.16	50.08
Marynka_395W	608.65	156.00	401.94	50.71
Marynka_Deryng	602.61	151.83	400.78	50.00

* Residue mass is the mass of the residue obtained after the extraction of *Humulus lupulus* with $scCO_2$

2.3. *Microwave-Assisted Hydrodistillation (MAHD)*

The MAHD extractions of the essential oils from Marynka hops varieties were performed at atmospheric pressure using a NEOS-GR (Milestone Technologies) system, equipped with a glycol cooled condenser, lab-grade glycol chiller, collected specimens, as well as a video system for the visual control of the process. The extraction was continued at 100 °C until no more essential oil was obtained. The essential oils were dried over anhydrous sodium sulfate, and filtered and stored in dark vials at 4 °C. During the process, two parameters were tested, namely, microwave power as well as the influence of the extraction time (HD and MAHD) on the oils composition and obtained yields.

The efficiency of the processes was calculated from the amount of the extracts loaded into the extraction process (m_e) to the amount of obtained essential oils (m_{oil}) [4]. The yields (Y) of extractions were calculated according to the following equation (Equation (1)):

$$Y\ (\%) = \frac{m_e}{m_{oil}} \times 100\% \qquad (1)$$

The total energy used for the extraction of essential oils by means of MAHD method was calculated according to the following equation (Equation (2)):

$$EC = \frac{P \times t}{1000} \qquad (2)$$

in which, EC (kWh) is the energy consumed, P (W) is the power consumed during the extraction, and t (h) is the extraction time.

2.4. *Hydrodistillation (HD)*

The mixture was subjected to hydrodistillation at 150 °C until no more essential oils were obtained, which was approx. 276 min for Marynka using a Deryng apparatus. The heating was performed using a heater equipped with a magnetic stirrer (Heidolph, Schwabach, Germany) with 825 W and working temperature up to 300 °C. The extracted essential oils were dried over anhydrous sodium sulfate, and filtered, weighed and stored in dark sealed vials at 4 °C.

3. Analytical Procedure

3.1. *Supercritical Fluid Chromatography (SFC)*

The separation of β-myrcene and α-humulene was conducted on a Waters Acquity Ultra-Performance Convergence Chromatography (UPC^2) system (Waters, Milford, MA, USA) equipped with a PDA detector. For the data analysis, Empower 3 software was used. The column was Acquity UPC^2 HSS C18 SB (100 mm × 3.0 mm; 1.8 μm), packed with a chemically modified, highly strengthened silica with C18 groups. Carbon dioxide (99.998% purity), which was used as the mobile phase in SFC, was purchased from Air Liquide. The separation was performed with an isocratic elution with pure CO_2 and a flow rate of 2.3 mL/min as well as a column temperature of 35 °C and ABPR (automated back pressure regulator) set to 12.6 MPa. The compensation wavelength was set at 500—600 nm throughout the analyses. The wavelength for the compounds was set to 220 nm. The authentic analytical standards of β-myrcene and α-humulene were purchased from Sigma-Aldrich and were dissolved in hexane. Figure 2 presents the chromatogram of β-myrcene and α-humulene by UPC^2 method.

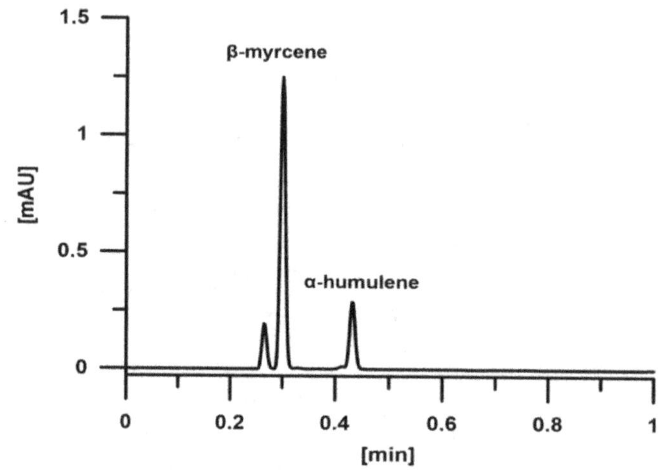

Figure 2. The chromatogram of β-myrcene and α-humulene by UPC^2 (ultra-performance convergence chromatography) on HSS C18 SB (100 mm × 3.0 mm; 1.8 μm) column; 35 °C, ABPR (automated back pressure regulator): 12.6 MPa; pure CO_2; isocratic elution.

3.1.1. Quantification

In order to quantify β-myrcene and α-humulene in hops oils, an external calibration method was applied. Individual stock solutions of authentic standards of β-myrcene and α-humulene were prepared in hexane. The appropriate amounts of β-myrcene and α-humulene were weighed and hexane was increased to the volume of 1 mL. The standard curves were obtained from five concentrations in the amount of 1000, 500, 250, 125, and 62.5 μg for β-myrcene and 500, 250, 125, 62.5, and 31.125 μg for α-humulene, adjusted to the real content of β-myrcene and α-humulene in the analyzed samples.

3.1.2. Partial Method Validation

Limits of detection (LOD), limits of quantification (LOQ), as well as repeatability and intermediate precision, were determined for the mixture of β-myrcene and α-humulene in order to evaluate the correctness of the provided method. Limits of detection (LOD) were 0.78 μg/mL for β-myrcene and 0.89 μg/mL for α-humulene, whereas limits of quantification were, respectively, 2.23 μg/mL and 1.95 μg/mL. The coefficients of determination (R^2) for calibration curves were 0.998 for both compounds. The repeatability was measured by 3 replicate analyses of a mixture of β-myrcene and α-humulene at three different concentrations at the same day, whereas the intermediate precision on three consecutive days. The RSD values, which are calculated according to the equation: RSD% = (standard deviation of the peak area/mean) × 100, show that both repeatability (RSD% < 3) and intermediate precision (RSD% < 6) are acceptable (Table 2).

Table 2. Calibration data as well as repeatability and intermediate precision for analyzing β-myrcene and α-humulene by UPC^2 system.

Compound	Linearity Range (μg/mL)	R2	LOD (μg/mL)	LOQ (μg/mL)	Concentration (μg/mL)	RSD % (n = 3)	
						Repeatability	Intermediate Precision
β-myrcene	62.5–1000	0.998	0.89	2.23	75	2.16	3.87
					100	1.02	2.91
					200	0.72	2.36
α-humulene	31.125–500	0.998	0.78	1.95	75	2.00	3.12
					100	1.27	2.74
					200	0.50	2.03

3.2. Gas Chromatography (GC-MS/MS)

The qualitative analyses of obtained hops oils were performed with a GC-MS instrument (Agilent 7890) equipped with mass spectrometry and NIST 2011 MS spectra library. As for the acquired results, Masshunter software (ver. C.01.03) was used. The following analyses parameters were used for *Humulus lupulus* essential oil samples: Capillary GC column DB-EUPAH (60 m × 250 μm; 0.25 μm) with helium as the carrier gas, and flow rate of 2 mL/min in a split injection mode (40:1). Oven temperature was set initially at 50 °C (held for 1 min) and increased to 260 °C at the rate of 6 °C/min (held for 6 min) and then finally reached 300 °C at the rate of 3 °C/min and held for 20 min. The system was operating in the EI mode (electron energy 70 eV) with the temperature of the source ion set to 230 °C and the scanning range from 40 to 650 amu.

3.2.1. Quantification

Concerning the quantitative analysis, internal standard calibration method was used with 4-androstene-3, and 17-dione as an internal standard. About 5 mg of the standard was weighed into a 5 mL volumetric flask. Then 5 mL of methanol/chloroform (9:1, v/v) mixture was added and 10 μL of such prepared standard was added to each oil sample in order to quantify β-myrcene and α-humulene. The standard curves were obtained from five concentration points: 200, 100, 50, 25, and 10 μg for β-myrcene and 50, 25, 10, 5, and 1 μg for α-humulene, adjusted to the real content of β-myrcene and α-humulene in the analyzed samples. The quantification was achieved in the MRM mode with different GC conditions as were used in the qualitative analyses. The quantitative method was shortened to 38 min with an oven temperature of 80 °C at the starting point, held for 1 min, and then ramped up to 300 °C with the rate of 10 °C/min and held for 15 min with nitrogen as a collision gas at the flow rate of 1.5 mL/min. The fragment ions for quantification and identification of each analyte were 136→121 (6) and 136→107 (6) for β-myrcene, 204→189 (10) and 204→161 (10) for α-humulene, as well as 286→148 (5), 124→109 (10) and 124→96 (10) for internal standard. The collision energies are provided in the brackets.

3.2.2. Partial Method Validation

Limits of detection (LOD) were 1.8 ng/mL and 2.5 ng/mL, whereas limits of quantification were 4.6 ng/mL and 8.0 ng/mL, respectively, for β-myrcene and α-humulene. The coefficients of determination (R^2) for the calibration curves were 0.999 for both compounds. The mixture of β-myrcene and α-humulene at three different concentrations was injected on the same day as well as on three consecutive days to evaluate the repeatability and intermediate precision of the developed method. The RSD values indicate that both repeatability (RSD% < 3) and intermediate precision (RSD% < 6) are acceptable (Table 3).

Table 3. Calibration data as well as repeatability and intermediate precision for analyzing β-myrcene and α-humulene by GC-MS/MS system.

Compound	Linearity Range [μg/mL]	R2	LOD [ng/mL]	LOQ [ng/mL]	Concentration [μg/mL]	RSD % (n = 3)	
						Repeatability	Intermediate Precision
β-myrcene	10–200	0.999	1.8	4.6	50	2.36	3.25
					100	1.54	2.84
					200	0.98	1.99
α-humulene	1–50	0.999	2.5	8.0	50	2.47	3.17
					100	1.69	2.99
					200	1.12	2.39

4. Results and Discussion

4.1. *Material for Extraction*

During our previous studies different forms of hops were tested before the final material was chosen for the extraction. One of the materials was post-extraction residue (residue after hops extraction using $scCO_2$) as a scattering element of the extract mixture with water. However, hop extract is characterized by its poor solubility in water [10,16]. The microwave hydrodistillation process was very rapid with the accompanying overheating of the raw material in an aqueous environment. During the optimization of the extraction process, attempts were made to replace the residues with an emulsifying substance. For this purpose, four different emulsifiers were used, including polysorbate 80 (E433), emulsifying biobase, soya lecithin, and cetyl alcohol. None of the used emulsifiers brought the expected results, meaning no adequate degree of mixing and homogenization was achieved. The local overheating of the material was also observed. Our studies were also performed on other forms of materials, such as raw hops and hops granules. In any case, no satisfactory effect and no amount of oils were obtained. It might have been caused by the low concentration of desirable compounds in the analyzed material.

4.2. *Extraction Results*

The microwave-assisted hydrodistillation (MAHD) with various microwave power and hydrodistillation (HD) were performed on extracts of Marynka hops variety. In case of the Marynka extract, the yield obtained by MAHD with the microwave power of 295 W was similar to the yield obtained by means of the HD method (1.80% for MAHD vs. 1.90% for HD). The extraction time of MAHD was significantly shorter than that of the conventional HD (40 vs. 276 min). The maximum yield of the essential oil obtained from the Marynka extract was using MAHD with microwave power of 335 W (3.03%). Table 4 summarizes the results of MAHD and HD methods for the Marynka extract.

Table 4. The results for Marynka extract essential oils microwave-assisted extraction.

Sample Name	Raw Material Mass (g)	Total Oil Amount (mg)	Yield (wt%)	Extraction Time (min)	Energy Consumption (kWh)
Marynka_275W	603.34	2248.14	1.50	46	0.215
Marynka_295W	605.99	2720.97	1.80	40	0.200
Marynka_335W	602.71	5676.44	3.77	30	0.170
Marynka_395W	608.65	4723.37	3.03	20	0.134
Marynka_Deryng	602.61	2807.31	1.90	276	1.897

According to Routray and Orsat [22], the higher the extraction temperature, the better the yield can be obtained. However, when the optimal temperature is achieved, the extraction efficiency decreases. During the studies, a relationship between the extraction power and the extraction time was observed. With increasing microwave power, the extraction time was shortened. When the extraction was performed at 395 W (3.03%) the extraction yield was relatively lower in a comparison with the process at 335 W (3.77%). Figure 3 presents the influence of the extraction power on the extraction time and yield.

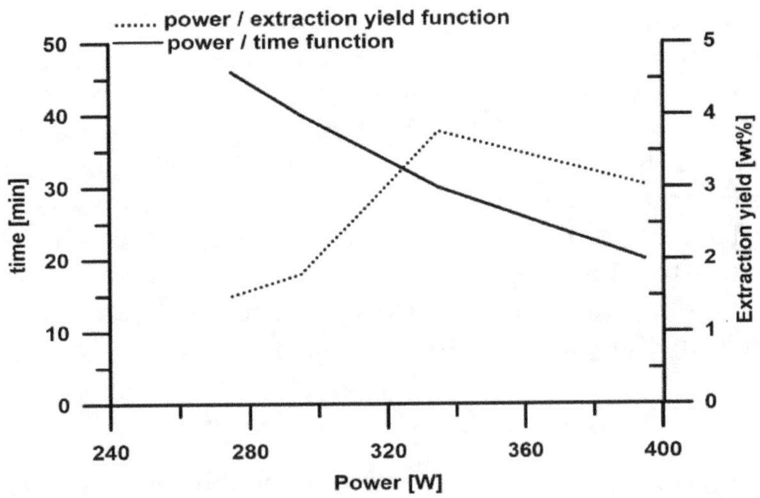

Figure 3. The influence of microwave power on the extraction time (———) and the extraction yield (…… …. ….).

The function of the extraction time from the microwave power for the microwave-assisted hydrodistillation of the Marynka extract was determined. From the graph, the curve equation for the experimental data was read, from which the values for the extraction time with the applied microwave power values were determined. The obtained data showed that with the increase of the microwave power by 5 W, the extraction time is shorter by 1.07 min (Figure 4).

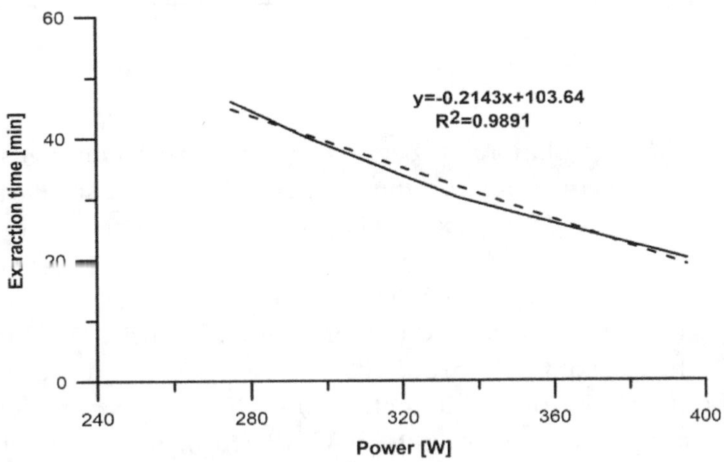

Figure 4. The function of the extraction time from the microwave power for the microwave-assisted hydrodistillation.

4.3. β-Myrcene and α-Humulezne Determination

Ligor et al. [23] obtained the essential oils from hops with a use of different techniques, such as supercritical fluid extraction, steam distillation, accelerated solvent extraction (ASE), and solid phase microextraction (SPME). The values for β-myrcene and α-humulene in obtained essential oils ranged from 15.7% to 21.1% and from 11.1% to 33.4%, respectively. Our study involves the use of microwave-assisted hydrodistillation, which is considered as a safe method [9]. The method resulted in obtaining the essential oil from *Humulus lupulus* scCO_2 extract at optimized parameters of 30 min and 335 W, which corresponds to one third of the maximal working power of the system.

The first application of the supercritical fluid chromatography with a photo-diode array detector to determine volatile compounds such as β-myrcene and α-humulene is provided in this study. This method was compared with a traditional method involving gas chromatography equipped with mass spectrometer (GC-MS/MS). As was suggested by Flament et al. [24], the use of scCO_2 for the separation of volatile compounds is inappropriate. However, in this study the obtained results for β-myrcene and α-humulene were comparable. The SFC method appeared to be significantly shorter than GC-MS/MS. Moreover, the use of supercritical carbon dioxide is understandable here due to the non-polar nature of volatile compounds being a subject of analysis. According to SFC quantitative analysis of β-myrcene and α-humulene, confirmed by GC-MS/MS analysis, the obtained oils were enriched in β-myrcene in the amount of approx. 74.13%–89.32%, which was several times higher in a comparison with the values obtained by Ligor et al. [23]. In the case of α-humulene, the concentration ranged from 7.36% to 10.14% in the essential oils obtained by microwave-assisted hydrodistillation, whereas only 1.59% in the essential oil obtained by hydrodistillation. Table 5 summarizes the concentrations of β-myrcene and α-humulene in obtained essential oils analyzed by UPC^2 and GC-MS/MS systems.

Table 5. The results of quantitative analyses of β-myrcene and α-humulene in essential oils obtained by microwave-assisted hydrodistillation (MAHD) and hydrodistillation (HD).

Sample Name	β-Myrcene		α-Humulene	
	UPC2 (wt%)	GC-MS/MS (wt%)	UPC2 (wt%)	GC-MS/MS (wt%)
Marynka-275W	89.32	88.89	10.14	10.56
Marynka-295W	74.13	75.02	8.23	8.01
Marynka-335W	77.24	77.36	9.33	9.47
Marynka-395W	75.12	74.66	7.36	7.96
Marynka-Deryng	84.73	84.99	1.59	1.87

5. Conclusions

Microwave-assisted hydrodistillation has been proven to offer advantages over hydrodistillation in determination of essential oils in *Humulus lupulus* $scCO_2$ extract. A similar extraction yield of the essential oil from *Humulus lupulus* $scCO_2$ extract is achieved in shorter extraction time (30 min for MAHD vs. 276 min for HD) and at a lower energy consumption (0.17 kWh/g of essential oil for MAHD vs. 1.89 kWh/g of essential oil for HD). Increasing the extraction time increases efficiency of the process as well as the amount of obtained compounds. Moreover, a significantly higher temperature and longer extraction time in the case of the hydrodistillation may have an influence on the compound's thermal degradation. Both chromatographic methods, such as supercritical fluid chromatography (SFC) and gas chromatography-mass spectrometer (GC-MS/MS), are comparable with the advantage of the SFC method in terms of shortened analysis time.

Author Contributions: Conceptualization, R.G., J.S. and E.R.; Methodology, K.T. and E.R.; Software, R.G. and J.S.; Validation, K.T. and M.K.; Formal Analysis, K.T., M.K. and R.G.; Investigation, K.T., R.G. and J.S; Resources, J.D. and E.R.; Data Curation, R.G.; Writing-Original Draft Preparation, K.T.; Writing, Reviewing and Editing, M.K.; Visualization, M.K.; Supervision, M.K. and E.R; Project Administration, E.R.; Funding Acquisition, E.R.

References

1. Cermak, P.; Paleckova, V.; Houska, M.; Strohalm, J.; Novotna, P.; Mikyska, A.; Jurkova, M.; Sikorova, M. Inhibitory effects of fresh hops on *Helicobacter pylori* strains. *Czech J. Food Sci.* **2015**, *33*, 302 307. [CrossRef]
2. Skorek, U.; Hubicki, Z.; Rój, E. Intensification of the use of hop extract for beer production. *Chemik Sci.-Tech.-Market.* **2011**, *65*, 160–163.
3. Biendl, M.; Engelhard, B.; Forster, A.; Gahr, A.; Lutz, A.; Mitter, W.; Schmidt, R.; Schönberger, C. *Hops: Their Cultivation, Composition and Usage*; Fachverlag Hans Carl: Nuremberg, Germany, 2014.
4. Biendl, M.; Pinzl, C. Hops and health. *German Hop Museum Wolnzach* **2013**, 1–123. [CrossRef]

5. Luzak, B.; Kassassir, H.; Rój, E.; Stańczyk, L.; Watala, C.; Golanski, J. Xanthohumol from hop cones (*Humulus lupulus* L.) prevents ADP-induced platelet reactivity. *Arch. Physiol. Biochem.* **2017**, *123*, 54–60. [CrossRef] [PubMed]
6. Yamaguchi, N.; Yamaguchi, K.; Ono, M. In vitro evaluation of antibacterial anticollagenase and antioxidant activities of hop components (*Humulus lupulus*) addressing acne vulgaris. *Phytomedicine* **2009**, *16*, 369–376. [CrossRef] [PubMed]
7. Rój, E.; Tadić, V.M.; Mišić, D.; Žižović, I.; Arsić, I.; Dobrzyńska-Inger, A.; Kostrzewa, D. Supercritical carbon dioxide hops extract with antimicrobial properties. *Open Chem.* **2017**, *13*, 1157–1171. [CrossRef]
8. Jeyaratnam, N.; Abdurahman, H.N.; Ramesh, K.; Azhari, H.N.; Yuvaraj, A.R.; Akindoyo, J.O. Essential oil from *Cinnamomum cassia* bark through hydrodistillation and advanced microwave assisted hydrodistillation. *Ind. Crop. Prod.* **2016**, *92*, 57–66. [CrossRef]
9. Golmakani, M.T.; Rezaei, K. Comparison of microwave-assisted hydrodistillation with the traditional hydrodistillation method in the extraction of essential oils from *Thymus vulgaris* L. *Food Chem.* **2008**, *109*, 925–930. [CrossRef] [PubMed]
10. Gavahian, M.; Farahnaky, A.; Farhoosh, R.; Javidnia, K.; Shahidi, F. Extraction of essential oils from *Mentha piperita* using advanced techniques: Microwave versus ohmic assisted hydrodistillation. *Food Bioprod. Process.* **2015**, *94*, 50–58. [CrossRef]
11. Kusuma, H.S.; Mahfud, M. Kinetic studies on extraction of essential oil from sandalwood (*Santalum album*) by microwave air-hydrodistillation method. *Alex. Eng. J.* **2017**, *57*, 1163–1172. [CrossRef]
12. Abdurahman, H.N.; Ranitha, M.; Azhari, H.N. Extraction and characterization of essential oil from Ginger (*Zingiber Officinale* Roscoe) and Lemongrass (*Cymbopogon citratus*) by microwave-assisted hydrodistillation. *Int. J. Chem. Environ.* **2013**, *4*, 221–226.
13. Farzaneh, V.; Carvalho, I.S. Modelling of microwave assisted extraction (MAE) of anthocyanins (TMA). *J. Appl. Res. Med. Aromat. Plants* **2017**, *6*, 92–100. [CrossRef]
14. Karakaya, S.; El, S.N.; Karagozlu, N.; Sahin, S.; Sumnu, G.; Bayramoglu, B. Microwave-assisted hydrodistillation of essential oil from rosemary. *J. Food. Sci. Technol.* **2014**, *51*, 1056–1065. [CrossRef] [PubMed]
15. Fargat, A.; Benmoussa, H.; Bachoual, R.; Nasfi, Z.; Elfalleh, W.; Romdhane, M.; Bouajila, J. Efficiency of the optimized microwave assisted extractions on the yield, chemical composition and biological activities of *Tunisian Rosmarinus officinalis* L. essential oil. *Food Bioprod. Process.* **2017**, *105*, 224–233.
16. Wei, L.; Zhang, Y.; Jiang, B. Comparison of microwave-assisted hydrodistillation with the traditional hydrodistillation method in the extraction of essential oils from dwarfed *Cinnamomum camphora* var. *Linaolifera Fujita leaves and twigs. Adv. J. Food Sci. Technol.* **2013**, *5*, 1436–1442.
17. Franco-Vega, A.; Ramirez-Corona, N.; Palou, E.; Lopez-Malo, A. Estimation of mass transfer coefficients of the extraction process of essential oil from orange peel using microwave assisted extraction. *J. Food Eng.* **2016**, *170*, 136–143. [CrossRef]
18. Yang, X.; Lederer, C.; McDaniel, M.; Deinzer, M. Hydrolysis products of caryophyllene oxide in hops and beer. *J. Agric. Food Chem.* **1993**, *41*, 2082–2085. [CrossRef]
19. Hartsel, J.A.; Eades, J.; Hickory, B.; Makriyannis, A. Cannabis sativa and hemp. In *Nutraceuticals*; Elsevier: Amsterdam, The Netherlands, 2016.
20. Rufino, A.T.; Ribeiro, M.; Sousa, C.; Judas, F.; Salgueiro, L.; Cavaleiro, C.; Mendes, A.F. Evaluation of the anti-inflammatory, anti-catabolic and pro-anabolic effects of E.-caryophyllene, myrcene and limonene in a cell model of osteoarthritis. *Eur. J. Pharmacol.* **2015**, *750*, 141–150. [CrossRef] [PubMed]
21. Fernandes, E.S.; Passos, G.F.; Medeiros, R.; da Cunha, F.M.; Ferreira, J.; Campos, M.M.; Pianowski, L.F.; Calixto, J.B. Anti-inflammatory effects of compounds alpha-humulene and (−)-trans-caryophyllene isolated from the essential oil of Cordia verbenacea. *Eur. J. Pharmacol.* **2007**, *569*, 228–236. [CrossRef] [PubMed]
22. Routray, W.; Orsat, V. Microwave-assisted extraction of flavonoids: A review. *Food Bioprocess Technol.* **2011**, *5*, 409–424. [CrossRef]
23. Ligor, M.; Stankevičius, M.; Wenda-Piesik, A.; Obelevičius, K.; Ragažinskienė, O.; Stanius, Ž.; Maruška, A.; Buszewski, B. Comparative gas chromatographic-mass spectrometric evaluation of hop (*Humulus lupulus* L.) essential oils and extracts obtained using different sample preparation methods. *Food Anal. Method* **2014**, *7*, 1433–1442. [CrossRef]

24. Flament, I.; Chevalier, C.; Keller, U. Extraction and chromatography of food constituents with supercritical CO_2. In *Flavour Science and Technology*; Martens, M., Dalen, G.A., Russwurm, H., Chichester, H.R., Jr., Eds.; Wiley: New York, NY, USA, 1987; pp. 151–163.

PERMISSIONS

All chapters in this book were first published by MDPI; hereby published with permission under the Creative Commons Attribution License or equivalent. Every chapter published in this book has been scrutinized by our experts. Their significance has been extensively debated. The topics covered herein carry significant findings which will fuel the growth of the discipline. They may even be implemented as practical applications or may be referred to as a beginning point for another development.

The contributors of this book come from diverse backgrounds, making this book a truly international effort. This book will bring forth new frontiers with its revolutionizing research information and detailed analysis of the nascent developments around the world.

We would like to thank all the contributing authors for lending their expertise to make the book truly unique. They have played a crucial role in the development of this book. Without their invaluable contributions this book wouldn't have been possible. They have made vital efforts to compile up to date information on the varied aspects of this subject to make this book a valuable addition to the collection of many professionals and students.

This book was conceptualized with the vision of imparting up-to-date information and advanced data in this field. To ensure the same, a matchless editorial board was set up. Every individual on the board went through rigorous rounds of assessment to prove their worth. After which they invested a large part of their time researching and compiling the most relevant data for our readers.

The editorial board has been involved in producing this book since its inception. They have spent rigorous hours researching and exploring the diverse topics which have resulted in the successful publishing of this book. They have passed on their knowledge of decades through this book. To expedite this challenging task, the publisher supported the team at every step. A small team of assistant editors was also appointed to further simplify the editing procedure and attain best results for the readers.

Apart from the editorial board, the designing team has also invested a significant amount of their time in understanding the subject and creating the most relevant covers. They scrutinized every image to scout for the most suitable representation of the subject and create an appropriate cover for the book.

The publishing team has been an ardent support to the editorial, designing and production team. Their endless efforts to recruit the best for this project, has resulted in the accomplishment of this book. They are a veteran in the field of academics and their pool of knowledge is as vast as their experience in printing. Their expertise and guidance has proved useful at every step. Their uncompromising quality standards have made this book an exceptional effort. Their encouragement from time to time has been an inspiration for everyone.

The publisher and the editorial board hope that this book will prove to be a valuable piece of knowledge for researchers, students, practitioners and scholars across the globe.

LIST OF CONTRIBUTORS

Dario Donno, Maria Gabriella Mellano, Marta De Biaggi, Isidoro Riondato, Giovanni Gamba and Gabriele Loris Beccaro
Dipartimento di Scienze Agrarie, Forestali e Alimentari, Università degli Studi di Torino, 10095 Grugliasco, Italy

Saandia Hassani
École National de Cuisine et d'Application-Codcom, 167 Moroni, Comoros

Cristina Giacoma
Dipartimento di Scienze della Vita e Biologia dei Sistemi, Università degli Studi di Torino, 10123 Torino, Italy

Bo Wang
College of Veterinary Medicine, Yangzhou University, Yangzhou 225009, China
Joint International Research Laboratory of Agriculture & Agri-Product Safety, Yangzhou University, Yangzhou 225009, China

Xing Xie
Key Laboratory of Veterinary Biological Engineering and Technology, Ministry of Agriculture, Institute of Veterinary Medicine, Jiangsu Academy of Agricultural Sciences, Nanjing 210014, China

Xia Zhao, Kaizhou Xie, Zhixiang Diao, Genxi Zhang, Tao Zhang and Guojun Dai
Joint International Research Laboratory of Agriculture & Agri-Product Safety, Yangzhou University, Yangzhou 225009, China
College of Animal Science and Technology, Yangzhou University, Yangzhou 225009, China

Danilo Giusepponi, Fabiola Paoletti, Carolina Barola, Simone Moretti, Giorgio Saluti and Roberta Galarini
Istituto Zooprofilattico Sperimentale dell'Umbria e delle Marche "Togo Rosati", 06126 Perugia, Italy

Federica Ianni and Roccaldo Sardella
Department of Pharmaceutical Sciences, University of Perugia, 06123 Perugia, Italy

Maykel Hernández-Mesa
Department of Analytical Chemistry, Faculty of Sciences, University of Granada, Campus Fuentenueva s/n, E-18071 Granada, Spain
INRA, UR1268 Biopolymers Interactions Assemblies, F-44316 Nantes, France
Laboratoire d'Etude des Résidus et Contaminants dans les Aliments (LABERCA), Oniris, INRA UMR 1329, Route de Gachet-CS 50707, F-44307 Nantes CEDEX 3, France

Ana M. García-Campaña
Department of Analytical Chemistry, Faculty of Sciences, University of Granada, Campus Fuentenueva s/n, E-18071 Granada, Spain

David Ropartz and Hélène Rogniaux
INRA, UR1268 Biopolymers Interactions Assemblies, F-44316 Nantes, France

Gaud Dervilly-Pinel and Bruno Le Bizec
Laboratoire d'Etude des Résidus et Contaminants dans les Aliments (LABERCA), Oniris, INRA UMR 1329, Route de Gachet-CS 50707, F-44307 Nantes CEDEX 3, France

Luying Shan, Dazhou Wang, Yinjiao Li and Shi Zheng
Yunnan Institute of Food Safety, Kunming University of Science and Technology, Yunnan 650500, China

Wentao Xu
Beijing Laboratory of Food Quality and Safety, College of Food Science and Nutritional Engineering, China Agricultural University, Beijing 100083, China

Ying Shang
Yunnan Institute of Food Safety, Kunming University of Science and Technology, Yunnan 650500, China
Beijing Laboratory of Food Quality and Safety, College of Food Science and Nutritional Engineering, China Agricultural University, Beijing 100083, China

Yingbin Shen and Junning Fu
Department of Food Science and Engineering, Jinan University, Guangzhou 510632, China

Liyou Zheng, Jun Jin and Xiaojing Li
State Key Laboratory of Food Science and Technology, Synergetic Innovation Center of Food Safety and Nutrition, School of Food Science and Technology, Jiangnan University, 1800 Lihu Road,Wuxi 214122, China

Mingzhong Wang
Shenzhen Kivita Innovative Drug Discovery Institute, Shenzhen 518110, China

Yifu Guan
Research Institute for Marine Drugs, Guangxi University of Chinese Medicine, Nanning 530200, China

Xun Song
School of Pharmaceutical Science, Shenzhen University, Shenzhen 518060, China

Zhou Xu, Shiling Feng, Ming Yuan, Yan Huang, Jinqiu Liao, Lijun Zhou, Hongyu Yang and Chunbang Ding
College of Life Sciences, Sichuan Agricultural University, Yaan 625014, China

Liang Fu
Dazhou Institute of Agricultural Sciences, Dazhou 635000, China

Laura Chronopoulou, Chiara Dal Bosco, Fabrizio Di Caprio, Letizia Prosini and Francesca Pagnanelli
Chemistry Department, University of Rome La Sapienza, 00185 Rome, Italy

Alessandra Gentili and Cleofe Palocci
Chemistry Department, University of Rome La Sapienza, 00185 Rome, Italy
CIABC, University of Rome La Sapienza, 00185 Rome, Italy

Ana V. González de Peredo, Mercedes Vázquez-Espinosa, Estrella Espada-Bellido, Marta Ferreiro-González, Miguel Palma and Gerardo F. Barbero
Department of Analytical Chemistry, Faculty of Sciences, University of Cadiz, Agrifood Campus of International Excellence (ceiA3), IVAGRO, 11510 Puerto Real, Cadiz, Spain

Antonio Amores-Arrocha and Ana Jiménez-Cantizano
Department of Chemical Engineering and Food Technology, Faculty of Sciences, University of Cadiz, Agrifood Campus of International Excellence (ceiA3), IVAGRO, 11510 Puerto Real, Cadiz, Spain

Raluca Popescu, Roxana Elena Ionete, Oana Romina Botoran, Diana Costinel, Felicia Bucura and Elisabeta Irina Geana
National Research and Development Institute for Cryogenics and Isotopic Technologies-ICSI Rm. Valcea, ICSI Analytics Group, 4 Uzinei Street, RO-240050 Râmnicu Vâlcea, Romania

Yazan Falah Jadee 'Alabedallat
University of Craiova, Faculty of Horticulture, Department of Horticulture and Food Science, 13 Al. I. Cuza Street, RO-200585 Craiova, Romania

Mihai Botu
University of Craiova, Faculty of Horticulture, Department of Horticulture and Food Science, 13 Al. I. Cuza Street, RO-200585 Craiova, Romania
University of Craiova, Fruit Growing Research Station (SCDP) Vâlcea, 464 Calea lui Traian Street, RO-240273 Râmnicu Vâlcea, Romania

Reem Khan
BAE: Biocapteurs-Analyses-Environnement, Universite de Perpignan Via Domitia, 52 Avenue Paul Alduy, 66860 Perpignan CEDEX, France
Department of chemistry, COMSATS University Islamabad, Abbottabad Campus, Abbottabad 22060, Pakistan
Interdisciplinary Research Centre in Biomedical Materials (IRCBM), COMSATS University Islamabad, Lahore Campus, Lahore 54000, Pakistan

Gaelle Catanante and Jean Louis Marty
BAE: Biocapteurs-Analyses-Environnement, Universite de Perpignan Via Domitia, 52 Avenue Paul Alduy, 66860 Perpignan CEDEX, France

Tauqir A. Sherazi
Department of chemistry, COMSATS University Islamabad, Abbottabad Campus, Abbottabad 22060, Pakistan

Akhtar Hayat
Interdisciplinary Research Centre in Biomedical Materials (IRCBM), COMSATS University Islamabad, Lahore Campus, Lahore 54000, Pakistan

Sondes Ben Aissa
BAE: Biocapteurs-Analyses-Environnement, Universite de Perpignan Via Domitia, 52 Avenue Paul Alduy, 66860 Perpignan CEDEX, France
Université de Tunis El Manar, Faculté des Sciences de Tunis, Laboratoire de Chimie Analytique et Electrochimie (LR99ES15), Sensors and Biosensors Group, Campus Universitaire de Tunis El Manar, Tunis 2092, Tunisia

Quoc Toan Tran, Minh Quan Pham and Quoc Long Pham
Institute of Natural Products Chemistry, Vietnam Academy of Science and Technology, Hanoi 122100, Vietnam
Graduate University of Science and Technology, Vietnam Academy of Science and Technology, Hanoi 122100, Vietnam

Tien Lam Do
Institute of Natural Products Chemistry, Vietnam Academy of Science and Technology, Hanoi 122100, Vietnam

Thi Thanh Tra Le
Graduate University of Science and Technology, Vietnam Academy of Science and Technology, Hanoi 122100, Vietnam
Department of Chemical Engineering, Faculty of Environment, Thuy loi University, Hanoi 122100, Vietnam

Manh Hung Vu
Institute of Marine Environment and Resources, Vietnam Academy of Science and Technology, Hanoi 122100, Vietnam

Duy Chinh Nguyen and Le Minh Bui
NTT Hi-Tech Institute, Nguyen Tat Thanh University, Ho Chi Minh City 700000, Vietnam

Long Giang Bach
NTT Hi-Tech Institute, Nguyen Tat Thanh University, Ho Chi Minh City 700000, Vietnam
Center of Excellence for Biochemistry and Natural Products, Nguyen Tat Thanh University, Ho Chi Minh City 700000, Vietnam

Quanguo He, Jun Liu and Guangli Li
School of Materials Science and Energy Engineering, Foshan University, Foshan 528000, China
College of Life Sciences and Chemistry, Hunan University of Technology, Zhuzhou 412007, China

Dongchu Chen
School of Materials Science and Energy Engineering, Foshan University, Foshan 528000, China

Xiaopeng Liu
College of Life Sciences and Chemistry, Hunan University of Technology, Zhuzhou 412007, China

Yonghui Xia
Zhuzhou Institute for Food and Drug Control, Zhuzhou 412000, China

Peihong Deng
Department of Chemistry and Material Science, Hengyang Normal University, Hengyang 421008, China

Federica Pellati, Stefania Benvenuti and Davide Bertelli
Department of Life Sciences, University of Modena and Reggio Emilia, Via G. Campi 103, 41125 Modena, Italy

Lucia Marchetti
Department of Life Sciences, University of Modena and Reggio Emilia, Via G. Campi 103, 41125 Modena, Italy
Doctorate School in Clinical and Experimental Medicine (CEM), University of Modena and Reggio Emilia, 41125 Modena, Italy

Katarzyna Tyśkiewicz, Roman Gieysztor, Marcin Konkol and Edward Rój
Supercritical Extraction Department, New Chemical Syntheses Institute, Al. Tysiąclecia Państwa Polskiego 13A, 24-110 Puławy, Poland

Jan Szałas
Import Export J.A. Szałas Company, Garbarska 125A, 26-600 Radom, Poland

Index

A

Aminobenzoic Acid, 175-176, 180-181

Antibiotics, 20-21, 33-34, 36, 38, 45-47, 49, 51-54

Aptasensor, 33, 173-183

B

Bioactive Compounds, 2, 16-17, 55, 137, 139, 146, 149, 152, 155-158, 171

Biogenic Amines, 64, 73, 78, 161

Biomass, 125-128, 134-135, 137-138, 199

C

Carotenoids, 2, 96, 125-128, 131-132, 135-137

Chemical Composition, 1, 17, 62, 108, 111, 122-124, 134, 141, 156, 161, 166, 170-171, 228, 235

Chemometrics, 12, 15, 19, 62, 65-67, 73, 79, 161, 171, 213, 216-217, 221-222, 225

Chia Seed Oil, 95-96, 98, 100-104, 106-109

Chloroform, 169, 196, 231

Clam Species, 186-189

Cultivars, 92, 157, 160-168, 170-171

Cytotoxicity, 102, 106-107

D

Diazonium Salt, 173, 175, 181-182

Dietary Supplements, 186, 227

Dilution, 31, 208, 216

E

Electrochemical Response, 179, 181, 203, 205-207

Electrode Surface, 175-178, 182, 204

Electrostatic Repulsion, 175, 177

Essential Oil, 17, 156, 227-229, 231-235

Ethanol, 102, 111-112, 117, 119-120, 125-127, 134-136, 156, 211, 218

Extraction Techniques, 127, 140, 227

F

Fat-soluble Vitamins, 125, 127-128, 135

Fatty Acids, 66, 95-102, 104, 107-109, 111, 113, 116, 118, 123, 157, 160-161, 163, 165, 169-171, 185-190, 196-200

Fatty Acids Composition, 118, 123, 160, 163, 170

Ferulic Acid, 5, 7-8, 15, 63

Florfenicol Amine, 20-21, 33-35, 39, 44, 49

Food Adulteration, 55, 66, 74, 83-84

Food Authentication, 55, 61, 66, 73-74, 78, 92

Food Composition, 55-56, 61-64, 71, 73, 84

Food Process Control, 55, 64-65, 73

Food Production, 1, 16

Food Quality, 55, 64, 75, 83, 160, 183, 202

Food Safety, 55-56, 60-61, 67-69, 71, 75, 83-84, 95, 202

Food Security, 1-3, 16

Fruit Jams, 1-3, 7, 9, 11, 15-17

G

Genotype, 2, 10-11, 16, 108, 160-161

H

Hard Clams, 185-188, 190, 197

Harvest Discrimination, 161, 166

Hydrodistillation, 227-229, 232-235

I

Immobilization, 176-177, 179, 181-182

Impedimetric Aptasensor, 173-175, 178-179, 182

Inflammation, 96, 137, 161, 174, 186

Ion Mobility, 55-57, 59-60, 62-63, 65, 69-70, 75-82

Ion Source, 118, 120, 197

Ionization, 18, 34, 50, 53-54, 61, 63-64, 66, 69, 77-78, 105, 135, 197, 223

Isotopic Composition, 160-161, 164-165, 170

Isotopic Fingerprint, 164-165, 170

L

Lactarius Deliciosus, 111-112, 117, 122

Linear Equations, 26, 206-208

Linear Sweep Voltammetry, 175, 201, 203, 207-208, 211

Lipid, 8, 61, 64-65, 71, 76, 95-96, 101-102, 106-107, 109, 116, 137, 140, 171-172, 185-188, 196-200

Lipid Classes, 61, 185-188, 196-197, 199-200

Liquid Chromatography, 1-2, 4, 18-22, 24, 33-36, 51-56, 63, 76-78, 80, 84, 108-109, 119, 122, 124, 135-137, 152, 158, 183, 197-198, 202, 213

M

Mass Spectrometry, 4, 18-19, 21, 34-36, 38, 51-56, 68, 71, 75-82, 84, 106, 109, 118, 120, 135-137, 152, 174, 186, 197-198, 200, 227, 231

Meretrix Lyrata, 185-187, 197, 199

Microalgae, 125-127, 131-132, 134, 136-138

Modified Electrode, 176, 201-203, 214-215

Molecular Species, 185-187, 189-190, 194-195, 197-200

Mycotoxin, 68, 79-80, 173-174, 182-183

Myrtle, 139-141, 147-158

N

Nutraceutical Traits, 1, 3, 11, 16

Nutritional Traits, 1, 9, 11-12, 15-16

Nutritional Value, 4, 55, 61, 96, 111-112, 118, 122-123, 160-161, 170-171, 186, 199, 216

P

Phenolic Acids, 1-2, 7, 11, 63-64, 77

Phospholipids, 185-186, 188-190, 196-197, 200

Pol Gene, 83-84, 86-88, 90-91

Polyphenolic Compounds, 3, 5, 7, 9, 11

Polyunsaturated Fatty Acid, 95, 97, 99, 198

Poultry Eggs, 20-22, 24-28, 32

Poverty Reduction, 1, 12, 16

Q

Quercetin, 1, 5, 8, 15, 140

R

Reduced Graphene Oxide, 201-205, 211, 213-214

Rural Development, 1, 3, 16

S

Scenedesmus, 125-127, 132, 134, 137

Sterols, 96, 99-100, 104-105, 108, 137, 161, 187-188, 200

Supercritical Fluid Extraction, 125, 127-128, 135, 137, 157, 227-228, 233

Sweep Rate, 201, 203, 206-207

Synergistic Enhancement, 201, 204

T

Thiamphenicol, 20-21, 33-35, 41, 44, 49

Tocopherols, 96, 98-99, 103, 105, 107-108, 110

Tricholoma Matsutake, 63, 77, 83, 85-86, 91-92, 94, 112, 122

U

Ultrasound Amplitude, 139, 145, 154-155

Ultrasound-assisted Extraction, 139-140, 151, 157-159

V

Vegetable Oils, 67, 95, 98, 101-102, 108-109, 123, 161, 171-172

W

Wavelengths, 13, 24, 31

www.ingramcontent.com/pod-product-compliance
Lightning Source LLC
LaVergne TN
LVHW081020210926
848485LV00004B/913